REMOTE SENSING OF THE TERRESTRIAL WATER CYCLE

Remote Sensing of the Terrestrial Water Cycle

Editor:

Maria Tattaris, Ph.D.

Delve Publishing
www.delvepublishing.com

Remote Sensing of the Terrestrial Water Cycle
Editor: **Maria Tattaris, Ph.D.**

708 3rd Avenue, 6th Floor
New York, NY 10017
United States of America

www.delvepublishing.com

ISBN: 978-1-68095-832-4

Library of Congress Control Number: 2016951574

About the Editor

Maria Tattaris, Ph.D.

Maria Tattaris has a background in Mathematics and a PhD in remote sensing within the area of atmospheric sciences, awarded in 2012. Her interests include field spectroscopy, image processing, spectral analysis, field data collection, and investigating new data analysis methods for remote sensing techniques, all within the areas of environmental, plant, atmospheric and related sciences.

Contents

List of Contributors

Komeil Rokni
Department of Geoinformation, Faculty of Geoinformation and Real Estate, Universiti Teknologi Malaysia, 81310 UTM Johor Bahru, Johor, Malaysia

Anuar Ahmad
Department of Geoinformation, Faculty of Geoinformation and Real Estate, Universiti Teknologi Malaysia, 81310 UTM Johor Bahru, Johor, Malaysia

Ali Selamat
UTM-IRDA Digital Media, K-Economy Research Alliance UTM and Faculty of Computing, Universiti Teknologi Malaysia, 81310 UTM Johor Bahru, Johor, Malaysia

Sharifeh Hazini
Institute of Geospatial Science and Technology (INSTeG), Faculty of Geoinformation and Real Estate, Universiti Teknologi Malaysia, 81310 UTM Johor Bahru, Johor, Malaysia

Frederick Bloetscher
Florida Atlantic University, Civil Engineering Department, Boca Raton, FL, USA

Fernando J. Pleitez
Florida Atlantic University, Civil Engineering Department, Boca Raton, FL, USA

Thomas Romah
Florida Atlantic University, Civil Engineering Department, Boca Raton, FL, USA

Ahmed Albasri
Florida Atlantic University, Civil Engineering Department, Boca Raton, FL, USA

Chase Dickinson
Florida Atlantic University, Civil Engineering Department, Boca Raton, FL, USA

Husayn El Sharif
Florida Atlantic University, Civil Engineering Department, Boca Raton, FL, USA

Kevin Matthews
Florida Atlantic University, Civil Engineering Department, Boca Raton, FL, USA

Trung-Duong Nguyen
Florida Atlantic University, Civil Engineering Department, Boca Raton, FL, USA

Lucrece Riche
Florida Atlantic University, Civil Engineering Department, Boca Raton, FL, USA

Frank Youngman
Florida Atlantic University, Civil Engineering Department, Boca Raton, FL, USA

Thomas P. Carsey
Atlantic Oceanographic and Meteorological Laboratory, Ocean Chemistry Division, Miami, FL, USA

Jack Stamates
Atlantic Oceanographic and Meteorological Laboratory, Ocean Chemistry Division, Miami, FL, USA

John R. Proni
Applied Research Center (ARC), Florida International University, Miami, FL, USA

James H. Churnside
Earth System Research Laboratory, National Oceanic and Atmospheric Administration, 325 Broadway, Boulder, CO 80305, USA

Brandi J. McCarty
Cooperative Institute for Research in Environmental Sciences, University of Colorado, 325 Broadway, Boulder, CO 80305, USA

Xiaomei Lu
Langley Research Center, National Aeronautics and Space Administration, Hampton, VA 23666, USA

Palanisamy Shanmugam
Department of Ocean Engineering, Indian Institute of Technology Madras, Chennai, India

Muniyandi Tholkapiyan
Department of Ocean Engineering, Indian Institute of Technology Madras, Chennai, India

Ji JIAN
College of Marine Science, University of South Florida, 140 Seventh Avenue, South, St. Petersburg, FL 33701, USA
International Institute for Earth System Science, Nanjing University, Nanjing, China

Peifen PAN
Institute of RS and GIS, Chengdu University of Technology, Chengdu, China

Yuanyuan CHEN
College of Marine Science, University of South Florida, 140 Seventh Avenue, South, St. Petersburg, FL 33701, USA

Wunian YANG
Institute of RS and GIS, Chengdu University of Technology, Chengdu, China

Jun Zhao
College of Marine Science, University of South Florida, 140 Seventh Avenue, South, St. Petersburg, FL 33701, USA

Chuanmin Hu
College of Marine Science, University of South Florida, 140 Seventh Avenue, South, St. Petersburg, FL 33701, USA

Brian Lapointe
Harbor Branch Oceanographic Institute, Florida Atlantic University, 5600 US 1 North, Fort Pierce, FL 34946, USA

Nelson Melo
Cooperative Institute for Marine and Atmospheric Studies, University of Miami, 4600 Rickenbacker Causeway, Miami, FL 33149, USA
NOAA Atlantic Oceanographic & Meterological Laboratory, 4301 Rickenbacker Causeway, Miami, FL 33149, USA

Elizabeth M. Johns
NOAA Atlantic Oceanographic & Meterological Laboratory, 4301 Rickenbacker Causeway, Miami, FL 33149, USA

Ryan H. Smith
NOAA Atlantic Oceanographic & Meterological Laboratory, 4301 Rickenbacker Causeway, Miami, FL 33149, USA

Scarla Weeks
Biophysical Oceanography Group, School of Geography, Planning and Environmental Management, University of Queensland, Brisbane, QLD 4072, Australia

P. Jeremy Werdell
Ocean Biology Processing Group, Goddard Space Flight Center, NASA Greenbelt, MD 20771, USA

Britta Schaffelke
Australian Institute of Marine Science, PMB 3, Townville, QLD 4810, Australia

Marites Canto
Biophysical Oceanography Group, School of Geography, Planning and Environmental Management, University of Queensland, Brisbane, QLD 4072, Australia

Zhongping Lee
Department of Environmental, Earth and Ocean Sciences, University of Massachusetts, Boston, MA 02125, USA

John G. Wilding
Ocean Biology Processing Group, Goddard Space Flight Center, NASA Greenbelt, MD 20771, USA

Gene C. Feldman
Ocean Biology Processing Group, Goddard Space Flight Center, NASA Greenbelt, MD 20771, USA

Archana Sarkar
National Institute of Hydrology, Roorkee, India

R. D. Garg
Indian Institute of Technology, Roorkee, India

Nayan Sharma
Indian Institute of Technology, Roorkee, India

Rong-Rong Li
Naval Research Laboratory, Code 7231, Washington, DC 20375, USA

Mark David Lewis
Naval Research Laboratory, Code 7331, Stennis Space Center, MS 39529, USA

Richard W. Gould Jr.
Naval Research Laboratory, Code 7331, Stennis Space Center, MS 39529, USA

Adam Lawson
Naval Research Laboratory, Code 7331, Stennis Space Center, MS 39529, USA

Ruhul Amin
Naval Research Laboratory, Code 7331, Stennis Space Center, MS 39529, USA

Sonia C. Gallegos
Naval Research Laboratory, Code 7331, Stennis Space Center, MS 39529, USA

Sherwin Ladner
Naval Research Laboratory, Code 7331, Stennis Space Center, MS 39529, USA

Ana Cleide Bezerra Amorim
Universidade Federal do Rio Grande do Norte (UFRN), Programa de Pós-Graduação em Ciências Climáticas (PPGCC), Natal, Brasil

Rosane Rodrigues Chaves
Universidade Federal do Rio Grande do Norte (UFRN), Programa de Pós-Graduação em Ciências Climáticas (PPGCC), Natal, Brasil

Cláudio Moisés Santos e Silva
Universidade Federal do Rio Grande do Norte (UFRN), Programa de Pós-Graduação em Ciências Climáticas (PPGCC), Natal, Brasil

Jian Peng
Max Planck Institute for Meteorology, 20146 Hamburg, Germany

Jonathan Niesel
Max Planck Institute for Meteorology, 20146 Hamburg, Germany

Alexander Loew
Max Planck Institute for Meteorology, 20146 Hamburg, Germany
Department of Geography, University of Munich (LMU), 80333 Munich, Germany

Shiqiang Zhang
College of Urban and Environmental Sciences, Northwest University, Xi'an 710027, China

Jie Wang
Yunnan Institute of Water Resources and Hydropower Research, Kunming 650228, China

Vagner G. Ferreira
School of Earth Sciences and Engineering, Hohai University, Nanjing 210098, China

Zheng Gong
State Key Laboratory of Hydrology-Water Resources and Hydraulic Engineering, Hohai University, Nanjing 210098, China

Xiufeng He
School of Earth Sciences and Engineering, Hohai University, Nanjing 210098, China

Yonglei Zhang
School of Earth Sciences and Engineering, Hohai University, Nanjing 210098, China

Samuel A. Andam-Akorful
School of Earth Sciences and Engineering, Hohai University, Nanjing 210098, China

Mariana A. Soppa
Alfred Wegener Institute, Bussestraße 24, D-27570 Bremerhaven, Germany

Christoph Völker
Alfred Wegener Institute, Am Handelshafen 12, D-27570 Bremerhaven, Germany

Astrid Bracher
Alfred Wegener Institute, Bussestraße 24, D-27570 Bremerhaven, Germany
Institute of Environmental Physics, University of Bremen, D-28334 Bremen, Germany

Ayodeji Richard Balogun
Department of Meteorology, Federal University of Technology, Akure, Nigeria

Zechariah Debo Adeyewa
Department of Meteorology, Federal University of Technology, Akure, Nigeria
Vice Chancelor'S Office, Redeemer's University (RUN), Lagos, Nigeria

Yin Zhang
Key Laboratory of Data Mining & Information Sharing of Ministry of Education, Fuzhou University, Fuzhou, China

Xiaoqin Wang
Key Laboratory of Data Mining & Information Sharing of Ministry of Education, Fuzhou University, Fuzhou, China

Yunzhi Chen
Key Laboratory of Data Mining & Information Sharing of Ministry of Education, Fuzhou University, Fuzhou, China

Preface

The terrestrial water cycle incorporates the transportation of water (solid/liquid/water vapour) within all land and atmospheric processes on the Earth. This includes precipitation, surface run-off, river flow, surface-water and groundwater storage and fluxes, and evapotranspiration, to name just a few. The study of this subject is key to many research areas, for example impacts on climate change, agriculture, ecology and biogeochemistry, as well as to practical human needs, such as irrigation and hydroelectricity. The processes involved in the terrestrial water cycle are highly complicated, as well as extremely variable. Standard in-situ measurements are often unable to capture this variability, as they are usually point measurements. Remote sensing offers the potential for large scale, high-resolution, multi-temporal, multi-sensor measurements, with the possibility of large amounts of data stretching over space and time. When combined with rapidly developing computational and modelling resources, remote sensing is an invaluable tool. This book presents numerous examples of how remote sensing can be applied in a way to aid in the study of some areas of the terrestrial water cycle. Specific topics covered include soil moisture, water vapour content, storm structure, sea surface temperature, precipitation, ocean colour, river dynamics, water clarity, ocean outfall plumes, diatom ocean phenology, river discharge, black water events and ocean signals. Remote sensing data types and techniques that are described include the derivation of spectral indices from multispectral satellite imagery, the use of data derived from satellite imagery as inputs for atmospheric correction, radar and lidar data, microwave and scatterometer data, spatial analysis of ocean properties and feature extraction and change detection.

Editor

Maria Tattaris, Ph.D.

INTRODUCTION

Terrestrial Water Cycle is the Latin word terra, meaning "earth," "land," or "ground," gives us the root terr. Words from the Latin terra have something to do with the earth or land. A territory is an area of land ruled by a particular government. A terrace is a piece of land with a flat top. Anything terrestrial lives on land and not in water.

Territorial waters or a territorial sea as defined by the 1982 United Nations Convention on the Law of the Sea, is a belt of coastal waters extending at most 12 nautical miles (22.2 km; 13.8 mi) from the baseline (usually the mean low-water mark) of a coastal state. The territorial sea is regarded as the sovereign territory of the state, although foreign ships (both military and civilian) are allowed innocent passage through it, or transit passage for straits; this sovereignty. Adjustment of these boundaries is called, in international law, maritime delimitation. The term "territorial waters" is also sometimes used informally to refer to any area of water over which a state has jurisdiction, including internal waters, the contiguous zone, the exclusive economic zone and potentially the continental shelf.

Territorial waters

Territorial waters, in international law, that area of the sea immediately adjacent to the shores of a state and subject to the territorial jurisdiction of that state. Territorial waters are thus to be distinguished on the one hand from the high seas, which are common to all countries, and on the other from internal or inland waters, such as lakes wholly surrounded by the national territory or certain bays or estuaries.

Historically, the concept of territorial waters originated in the controversy over the status of the sea in the formative period of modern international law in the 17th century. Although the doctrine that the sea by its nature must be free to all was eventually upheld, most commentators did recognize that, as a practical matter, a coastal state needed to exercise some jurisdiction in the waters adjacent to its shores. Two different concepts developed—that the area of jurisdiction should be limited to cannon-shot range, and that the area should be a much greater belt of uniform width adjacent to the coast—and in the late 18th century these concepts coalesced in a compromise view that proposed a fixed limit of 3 nautical miles (1 marine league, or 3.45 statute miles [5.5 km]). In 1793 the United States adopted three miles for neutrality purposes, but although many other maritime states during the 19th century came to recognize the same limit, it never won such universal acceptance as to become an undisputed rule of international law.

In the course of this historical development, it became settled that the belt of territorial waters, together with the seabed and subsoil beneath it and the airspace above, is under the sovereignty of the coastal state. This sovereignty is qualified only by a right of innocent passage—that is, peaceful transit not prejudicial to the good order or security of the coastal state—for merchant vessels of other nations. The right of innocent passage does not apply to submerged submarines or to aircraft, nor does it include a right to fish.

On the width of the belt there has developed no universal agreement except that every state is entitled to a minimum of three nautical miles. Claims in excess of 12 nautical miles (22 km) commonly meet widespread opposition from other states, though in the 1960s and '70s a trend to a 12-nautical-mile limit was evident; among about 40 states taking this view were China, India, Mexico, Pakistan, Egypt, and the Soviet Union.

Distinct from territorial waters proper are zones in the adjacent high seas in which coastal states claim no territorial rights but assert limited jurisdiction for one or more special purposes. These contiguous zones of 6 to 12 nautical miles (11 to 22 km) beyond territorial waters are most commonly claimed for the enforcement of customs and sanitary regulations, but in some instances they may be established for fishery protection or for security. Also distinct from territorial waters are the claims made after 1945 by many states to the continental shelf off their shores, in or on which potentially valuable resources might exist. Such claims met with little objection from other states when confined to the shelf itself, without affecting the status as high seas of the waters above, but actions by some states, such as Chile, Ecuador, and Peru, that asserted jurisdiction over waters as well as shelf for as much as 200 nautical miles (370 km) offshore evoked wide protest as amounting to unacceptable extensions of territorial waters.

A United Nations Conference on the Law of the Sea that was convened at Geneva in 1958 and attended by 86 nations developed a convention affirming the commonly accepted principles of the legal nature of the territorial sea and the right of innocent passage. This convention took effect in 1964 and by 1970 had been ratified by almost 40 states. A more comprehensive Law of the Sea treaty was signed by 117 nations in 1982.

What is Surface Water?

Surface water is any water that collects on the surface of the earth. This includes oceans, seas, lakes, rivers, or wetlands. Fresh surface water is maintained by rainfall or other precipitation, and it's lost through seepage through the ground, evaporation, or use by plants and animals. Man-made bodies of water are not considered surface water since they generally rest on artificial surfaces, not the ground itself.

Properties of Surface Water

There are many important properties of surface water: temperature, saltiness (called salinity), turbidity, and levels of dissolved nutrients, such as oxygen and carbon dioxide. These factors all affect climate and the biodiversity in and around a body of water.

Temperature, scientifically, is the average kinetic energy of the molecules in a substance. The temperature of surface water is warmest at the top, and it gets cooler as you go deeper. The deep oceans, for example, are extremely cold, dark places. Surface water temperature varies much more by season as well. So, when we look at temperature, we tend to look at averages.

While were talking about temperature, an interesting fact is that sea water doesn't freeze at 0 degrees Celsius; the salt in the water allows it to get colder before it does. Also, warmer water tends to have less oxygen, which can make it harder for some species of marine life to survive.

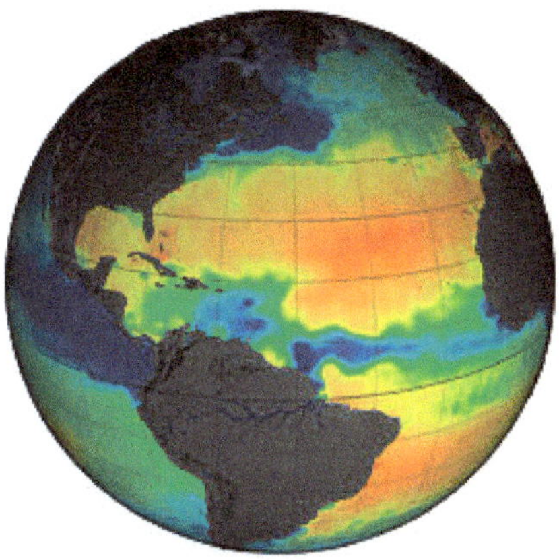

Representation of Sea Surface Salinity from Aquarius Spacecraft.

Sea Surface Salinity from Aquarius spacecraft

Salinity is the saltiness of water. It measures the amount of dissolved sodium, potassium, and other salts in the water. Higher salinity leads to denser water, which has an impact on water currents around the world. Areas with a lot of evaporation have higher salinity and denser water because when water evaporates, it leaves the salts behind.

Dissolved nutrients are the amounts of various chemicals that are dissolved into water. This includes nitrogen, phosphorous, oxygen, carbon dioxide, and many more, and they each have different effects. The presence of nitrogen and phosphorus increases the success of plankton, which is important food for ocean ecosystems. Oxygen is also really important because fish and other sea life need it to survive.

Water is continually moving around, though, and above the Earth. It moves as water vapor, liquid water, and ice. It is constantly changing its form. Water on Earth is known by different terms, depending on where it is and where it came from.

- Meteoric water - is water in circulation
- Connate water - “fossil” water, often saline.
- Juvenile water - water that comes from the interior of the earth.
- Surface water - water in rivers, lakes, oceans and so on.
- Subsurface water - Groundwater, connate water, soil, capillary water
- Groundwater - exists in the zone of saturation, and may be fresh or saline.

The movement of water is referred to as the global water cycle (hydrologic cycle). Precipitation, evaporation/transpiration, and runoff (surface runoff and subsurface infiltration) are the primary phases in the hydrologic cycle. The global water budget is based on the recycling (movement, storage, and transfer) of the Earth’s water supply.

The direct process by which water changes from a liquid state to a vapor state is called evaporation. In transpiration, water passes from liquid to vapor through plant metabolism. Plants are classified as hydrophytes, phreatophytes, mesophytes, or xerophytes. Hydrophytes take their nutrients directly from the water. Mesophytes are plants that grow under well-balanced moisture supplies. Xerophytes are plants that are adapted to dry conditions. Phreatophytes are long rooted plants that absorb water from the water table or directly above it. Golden tamarisk and mesquite are phreatophytes.

How Much Water is There In and On the Earth?

The volume of the Earth’s water supply is about 326 million cubic miles. Each cubic mile is greater than 1 trillion gallons. Although water is abundant on a global scale, more than 99% is unavailable for our use. A mere 0.3% is usable by humans, with an even smaller amount accessible! The oceans, ice caps, and glaciers contain most of the Earth’s water supplies. Ocean water is too saline to be economically useful, while glaciers and icecaps are “inconveniently located.” Click here to see a chart with these data.

Surface water supplies, primarily river runoff, are about 300 cubic miles. That means we have about 1/10,000th of 1% to use! Conservation is important!

Surface runoff plays an important role in the recycling process. Not only does it replenish lakes, streams, and groundwater; it also creates the landscape by eroding topography and transporting the material elsewhere.

A stream typically transports three types of sediment- dissolved load, suspended load, and bed load. Chemical weathering of rocks produces ions in solution (examples- Ca^{2+}, Mg^{+}, and HCO_3^{+}). Hence, a dissolved load. High concentrations of Ca_2^{+} and Mg+ are also known by another name - hard water. Some of you may be very familiar with hard water! Take a look at some water chemistry. Suspended sediment makes water look cloudy or opaque. The greater the suspended load, the muddier the water. Bed load (silt- to boulder-sized, but mostly sand and gravel) settles on the bottom of the channel. Bed load sediment moves by bouncing or rolling along the bottom. The distance that bed load travels depends on the velocity of the water.

Factors Affecting Surface Runoff

Several factors can affect surface runoff. The extent of runoff is a function (*f*) of geology, slope, climate, precipitation, saturation, soil type, vegetation, and time. Geology includes rock and soil types and characteristics, as well as degree of weathering. Porous material (sand, gravel, and soluble rock) absorbs water far more readily than does fine-grained, dense clay or unfeatured rock. Well-drained material (porous) has a lower runoff potential therefore has a lower drainage density. Poorly-drained material (non-porous) has a higher runoff potential, resulting in greater drainage density. Drainage density is a measure of the length of channel per unit area. Many channels per unit area means that more water is moving off of the surface, rather than soaking into the soil.

Which Type of Drainage Basin Has the Greatest Effect on Surface Runoff?

Long, narrow drainage basins generally display the most dramatic effects of surface runoff. They have straight stream channels and short tributaries. Storm waters reach the main channels far more rapidly in long narrow basins than in other types of basins. Flash floods are common in long, narrow drainage basins, resulting in greater erosion potential. Topography (relief) and slope (gradient) are additional factors affecting water velocity, infiltration rate, and overland flow rate. Water velocity, infiltration rate and overland flow rate affect surface and subsurface runoff rates.

Runoff Paths

There are three runoff paths that water follows to reach a stream channel- through flow, overland flow, and groundwater flow. Through flow is a shallow subsurface flow that occurs above the groundwater table. A major requirement for through flow is a good infiltration capacity. Through flow commonly occurs in humid climates containing thick soil layers and good vegetation cover. In such locations, saturated soil conditions result in surface runoff (overland flow). Overland flow occurs when precipitation exceeds infiltration rates. Overland flow is common in semi-arid regions, sparsely vegetated and/or disturbed areas, and locations containing dense, clay-rich layers. Climate is also important. Precipitation (type, duration, and intensity) is the key climatic factor. Infrequent torrential downpours easily erode sediment-laden topography, while soft drizzly rain infiltrates the soil. Vegetation aids in slope stability. Removal of vegetation by fire, clear-cutting (logging), or animal grazing often results in soil erosion. The eroded material is washed into streams, adding to the sediment load.

Surface Water /Groundwater Interaction

Surface streams have an effect on the groundwater table. Influent streams recharge groundwater supplies. Influent streams, located above the groundwater table, flow in direct response to precipitation. Water percolates down through the vadose zone to the water table, forming a recharge mound. Effluent streams are discharge zones for groundwater. Effluent streams are generally perennial (flow year round). Groundwater seeps into stream channels, maintaining water flow during dry seasons. Groundwater supplies 30% of the water present in our streams. Recall that effluent streams act as discharge zones for groundwater during dry seasons. This phenomenon is known as base flow. Groundwater overdraft reduces the base flow, which results in the reduction of water supplied to our streams.

1
CHAPTER

WATER FEATURE EXTRACTION AND CHANGE DETECTION USING MULTITEMPORAL LANDSAT IMAGERY

Komeil Rokni, Anuar Ahmad, Ali Selamat and Sharifeh Hazini

Keywords: NDWI, Landsat, Surface water, Change detection

INTRODUCTION

The monitoring of change using remote sensing technology is widely used for various applications, such as land use/cover change [1,2], disaster monitoring [3,4], forest and vegetation change [5,6], urban sprawl [7,8], and hydrology [9,10]. Surface water is an irreplaceable strategic resource for human survival and social development [11]. It is essential for humans, food crops, and ecosystems [12]. Reliable information about the spatial distribution of open surface water is critically important in various scientific disciplines, such as the assessment of present and future water resources, climate models, agriculture suitability, river dynamics, wetland inventory, watershed analysis, surface water survey and management, flood mapping, and environment monitoring [13–16]. Spaceborne satellites at varying spatial, spectral, and temporal resolutions provide an enormous amount of data that has extensively been used for detecting and extracting surface water and its changes in recent decades [17–23].

In recent decades, several image processing techniques have been introduced for the extraction of water features from satellite data. Single-band methods utilize a selected threshold value to extract water features. In general, for these types of approaches,

Citation: Rokni, K, Ahmad, A, Selamat, A, Hazini, S. Water Feature Extraction and Change Detection Using Multitemporal Landsat Imagery. Remote Sens. 2014, 6, 4173-4189, doi:10.3390/rs6054173.

errors are common because of mixing of water pixels with those of different cover types [15]. Classification techniques adopted to extract surface water are usually more accurate compared to single-band methods [15]. Multi-band methods combine different reflective bands for improved surface water extraction [15]. For instance, the Normalized Difference Water Index (NDWI) was developed for the extraction of water features from Landsat imagery [24]. Since water features extracted using the NDWI include false positives from built-up land, a modified NDWI (MNDWI) was developed in which the middle infrared (MIR) band was replaced with the near infrared (NIR) band [18]. The MNDWI extracts surface water while suppressing errors from built-up land, as well as vegetation and soil [18]. Surface water change detection is usually conducted by extracting water features individually from multi-date satellite images, before making comparisons to detect changes [15,19,20,25–28].

In this study, surface area changes of Lake Urmia, Iran in the period 2000–2013 were investigated. Lake Urmia, with an area varying between 5200–6000 km^2 in the 20th century, is the 20th largest lake and the second largest hyper saline lake (before September 2010) in the world. It is also the largest inland body of salt water in the Middle East [29]. The lake is the habitat for a unique bisexual Artemia (a species of brine shrimp), and becomes a host for more than 20,000 pairs of Flamingo and approximately 200–500 pairs of White Pelican every winter [30,31]. Lake Urmia forms a rare and important ecologic, economic and geo-touristic zone and was recognized as a Biosphere Reserve by the United Nations Educational, Scientific and Cultural Organization (UNESCO) in 1975. In addition, the lake helps moderate the temperature and humidity of the region, providing a suitable place for agricultural activities [31]. The lake is currently in a critical situation as a consequence of decreasing surface water area and increasing salinity. To facilitate expanding agricultural activities, 35 dams have been constructed on 21 rivers flowing to the lake [29]. These dams lead to the exploitation of ground water and increased water demand for industrial and domestic uses, the major factors contributing to the shrinkage of the lake surface area, which has been intensified by a long period of drought [31]. It is vital to monitor the decline of the surface area of the lake, as regular monitoring can provide the basis for understanding human influences on the lake in order to improve management.

This study aimed to model the spatiotemporal changes of Lake Urmia in the period 2010–2013 using multi-temporal Landsat ETM+, TM, and OLI data. Different surface water extraction techniques were initially examined and the most suitable technique was used to detect and map the spatiotemporal changes of the lake. In addition, a novel surface water change detection approach, based on Principal Components of the multi-temporal Normalized Difference Water Index (NDWI-PCs), was developed and evaluated.

MATERIALS AND METHODS

In order to achieve the aims of the study, the following tasks were performed: study area definition, data collection, image pre-processing, comparison of different satellite-derived indices for surface water detection, extraction of the lake surface area in each image, developing the NDWI-PCs approach, and change detection and modelling.

Figure 1 shows the methods adopted in this study to detect the lake surface area changes.

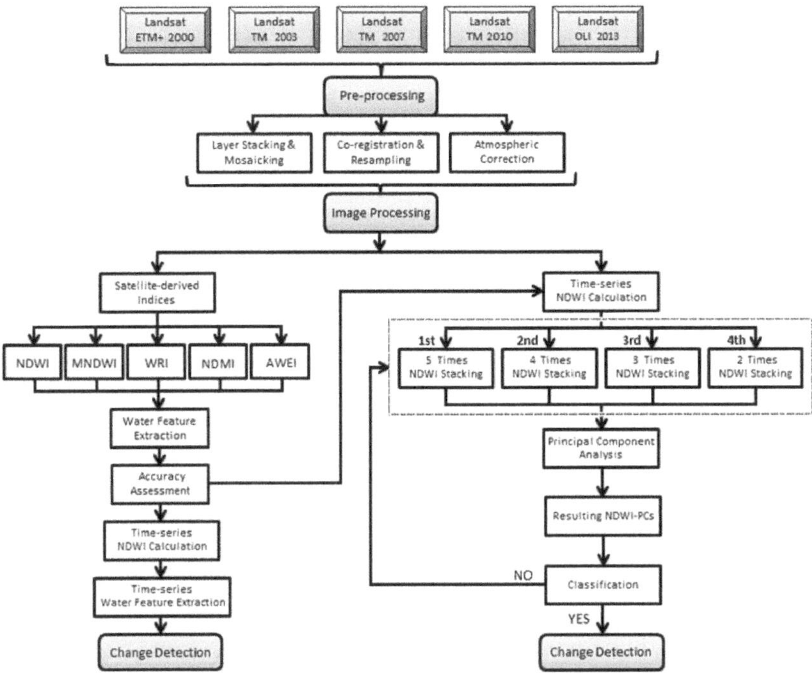

Figure 1. Flowchart showing the overall methods adopted in the study.

Study Area

Lake Urmia (Urumiyeh in Persian), bound by 37°03′N to 38°17′N and 44°59′E to 45°59′E, has a maximum depth of 16 m and is approximately 140 km long and 40–55 km wide [29]. The lake is a shallow and saline lake that is located in the northwest of Iran, with a total catchment area of approximately 51,876 km^2, about 3.2% of the size of Iran, and it represents about 7% of the country's surface water [30]. The maximum surface area of the lake has been estimated to be around 6100 km^2, however since 1995, it has been constantly declining and reached 2366 km^2 in August of 2011 [32]. Around 60 rivers (permanent and episodic) are found in the catchment area of the lake, all of which flow through agricultural, industrial and urban areas normally without waste water treatment [30,33]. The climate of the lake basin is characterized by cold winters and relatively temperate summers, being influenced by the mountains surrounding the lake [33].

Data Set

Three Landsat-7 ETM+ scenes acquired in August 2000, twoLandsat-5 TM scenes acquired in September 2003, three Landsat-5 TM scenes acquired in September 2007, three Landsat-5 TM scenes acquired in July 2010, and one Landsat-8 OLI image acquired in August 2013 were obtained from the US Geological Survey (USGS) Global Visualization Viewer. All the images are obtained for the same season (summer). The obtained Landsat data (Level 1 Terrain Corrected (L1T) product) were pre-georeferenced to the WGS-84 UTM zone 38 North projection. The remaining necessary corrections were performed in this study. Table 1 presents the specifications of the Landsat TM, ETM+ and OLI images.

Table 1. Specifications of Landsat TM, ETM+ and OLI data

Satellite	Sensor	Path/Row	Year	Resolution (m)	Wavelength (μm)
Land-sat-5	TM	169/33 169/34 168/34	2003 2007 2010	30	Band 1: 0.45–0.52
					Band 2: 0.52–0.60
					Band 3: 0.63–0.69
					Band 4: 0.76–0.90
					Band 5: 1.55–1.75
					Band 7: 2.08–2.35
Land-sat-7	ETM+	169/33 169/34 168/34	2000	30	Band 1: 0.45–0.515
					Band 2: 0.525–0.605
					Band 3: 0.63–0.69
					Band 4: 0.75–0.90
					Band 5: 1.55–1.75
					Band 7: 2.09–2.35
Land-sat-8	OLI	169/33	2013	30	Band 1: 0.435–0.451
					Band 2: 0.452–0.512
					Band 3: 0.533–0.590
					Band 4: 0.636–0.673
					Band 5: 0.851–0.879
					Band 6: 1.566–1.651
					Band 7: 2.107–2.294
					Band 9: 1.363–1.384

Image Pre-Processing

To prepare the input satellite images for further processing, the following pre-processing steps were performed: radiometric calibration, atmospheric correction, mosaicking, co-registration, and resampling. Radiometric calibration and atmospheric correction were conducted according to [34]. In doing so, the obtained images were converted to at-satellite radiance using the Landsat calibration tool in ENVI 4.8. After conversion to at-satellite radiance, each image was converted to at-satellite reflectance. The required information, including the data acquisition date and sun elevation, was obtained from the Landsat header files. Subsequently, the Dark Object Subtraction (DOS) method [35] was employed to convert at-satellite reflectance to surface reflectance for full absolute correction.

The corrected images for each year were mosaicked to generate new images, covering the entire study area, using the mosaicking tool for georeferenced images. For co-registration of the multi-temporal images, one of the images was used as the reference to register the other images. The input images were co-registered with a Root Mean Square Error (RMSE) of less than 0.5 pixels, using the manual image to image co-registration method. Around 25 control points were selected for co-registration of each image with the reference image. Finally, the dataset was resampled to the same size of the study area using the Nearest Neighbour method.

Comparison of Different Feature Extraction Techniques for Surface Water Extraction

In order to detect the surface area changes of Lake Urmia in the period of 2000–2013, the water surface of the lake in each temporal image was extracted individually. In doing so, the performances of different satellite-derived indices, including the Normalized Difference Water Index (NDWI) [24], the Normalized Difference Moisture Index (NDMI) [36], the Modified Normalized Difference Water Index (MNDWI) [18], the Water Ratio Index (WRI) [37], the Normalized Difference Vegetation Index (NDVI) [38], and the Automated Water Extraction Index (AWEI) [39] were examined for the extraction of surface water from Landsat data (Table 2).

In this respect, the NDWI, NDMI, MNDWI, WRI, NDVI, and AWEI indices were calculated from the Landsat ETM+ 2000 and the TM 2010 images (as a sample) to evaluate their performances for the extraction of surface water. A land-water threshold was manually applied to classify the images into two classes, land and water. Suitable land-water thresholds for each index were determined through trial and error and comparison to reference maps generated using visual interpretation. For visual interpretation of water bodies, the near-infrared (NIR) band is usually preferred, because NIR is strongly absorbed by water and is strongly reflected by the terrestrial vegetation and dry soil [16]. Thus, band 4 of Landsat data was selected due to its higher ability to discriminate water and dry/land areas. The reference maps were generated utilizing careful on-screen digitizing of the lake surface area in multi-temporal Landsat ETM+ 2000 and TM 2010 images (band 4) using visual interpretation.

Table 2. Satellite-derived indices used for water features extraction (in Landsat imagery: Green = Band 2, Red = Band 3, NIR (near-infrared) = Band 4, MIR (middle-infrared) = Band 5, SWIR (shortwave-infrared) = Band 7)

Index	Equation	Remark	Reference
Normalized Difference Water Index	NDWI = (Green − NIR)/ (Green + NIR)	Water has positive value	[24]

Normalized Difference Moisture Index	NDMI = (NIR − MIR)/(NIR + MIR)	Water has positive value	[36]
Modified Normalized Difference Water Index	MNDWI = (Green − MIR)/(Green + MIR)	Water has positive value	[18]
Water Ratio Index	WRI = (Green + Red)/(NIR + MIR)	Value of water body is greater than 1	[37]
Normalized Difference Vegetation Index	NDVI = (NIR − Red)/(NIR + Red)	Water has negative value	[38]
Automated Water Extraction Index	AWEI = 4 × (Green-MIR) − (0.25 × NIR + 2.75 × SWIR)	Water has positive value	[39]

Surface Water Change Detection Using the Most Suitable Water Feature Extraction Technique

The feature extraction technique, identified as the best option for water features extraction, was employed to model the spatiotemporal changes of Lake Urmia in the period 2000–2013 using the multi-temporal Landsat ETM+ 2000, TM 2003, TM 2007, TM 2010, and OLI 2013 images. To achieve this, the selected index was calculated, analysed and classified independently (using image-specific thresholds) to extract the lake surface area in each year. Finally, the generated maps were overlaid to produce the lake surface area changes map in the period 2000–2013.

Developing a New Approach for Surface Water Change Detection (NDWI-PCs)

As manual thresholding of several images is time consuming, a novel surface water change detection approach was developed and tested for the detection of the lake surface area changes between five, four, three, and two different times, respectively. The proposed approach is based on Principal Components of multi-temporal NDWI (NDWI-PCs). A change detection method, based on Principal Components of multi-temporal NDMI (also known as $NDWI_{Gao}$ [40]), was suggested by [41] for forest fire detection (NDMI-PCs). In the present study, the suitability of the NDMI-PCs was tested for detection of Lake Urmia surface area changes, but this method was found inappropriate for surface water change detection. Further investigation showed that the NDMI used in this method is not suitable for the extraction of surface water from Landsat data (see Table 3). Therefore, the performance of different satellite-derived indices including NDWI, MNDWI, WRI, NDVI, and AWEI were examined for the extraction of surface water from Landsat data. The accuracy assessment analyses indicated superiority and higher performance of the NDWI compared with other indices used for this purpose. Accordingly, the NDMI-PCs method was reformulated for surface water change detection through substituting the NDMI with the NDWI.

To perform the new NDWI-PCs approach, the NDWI was calculated from the multi-temporal Landsat data. Subsequently, the obtained NDWI images were stacked into one composite file. The Principal Component Analysis (PCA) technique was then performed to transform the composite image into a new PCA space. Finally, the resulting Principal Components of multi-temporal NDWI (NDWI-PCs) were classified and analysed for detection of the lake surface area changes. The thresholding technique, based on the trial and error method, was manually applied to classify the resulting PCs to detect the changes. The proposed approach was evaluated for detection of the lake water surface changes between five, four, three, and two different times, respectively.

In order to evaluate the effectiveness of the proposed approach for surface water change detection, different accuracy assessment analyses were performed. The performance of different methods to detect the lake surface area changes between 2000 and 2010 was first evaluated through the calculation of the absolute error, as the difference between the changed areas detected using the applied method and the reference. In this respect, the lake surface area changes between 2000 and 2010 were calculated using each method and the absolute error was calculated based on the reference change result. In addition, the overall accuracy, kappa coefficient, and producer's and user's accuracies (for change pixels) were calculated to support the accuracy assessment analysis.

RESULTS

Once the required satellite data was obtained, the image pre-processing techniques were performed to prepare the images for further processing. The dataset after pre-processing are shown in Figure 2. One image for the year 2003 was unavailable in USGS to cover the entire lake. However, the missed area is very small compared with the total surface area of the lake and we tried to reduce the error by adding the missed area to the calculated area of the lake through a simple interpolation.

Landsat ETM + 2003 Landsat TM 2003 Landsat TM 2007

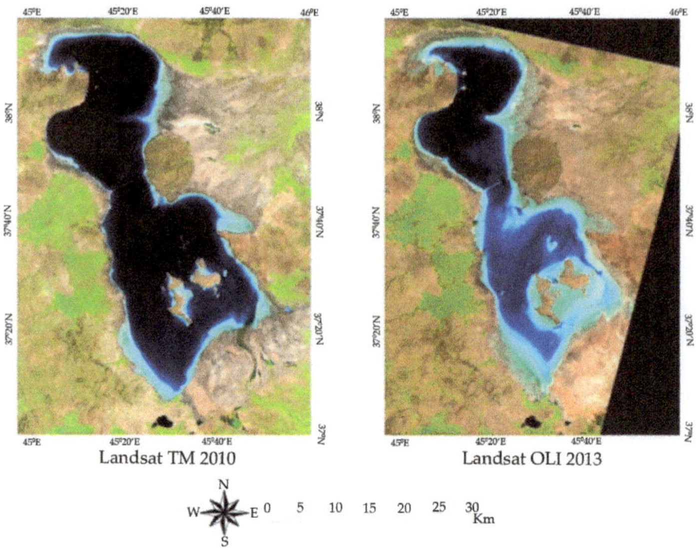

Figure 2. The dataset after pre-processing (RGB-543 false colour composite).

Surface Water Change Detection Using the Most Suitable Technique

Different satellite-derived indices, including NDWI, NDMI, MNDWI, WRI, NDVI, and AWEI were examined for the extraction of surface water from the Landsat ETM+ 2000 and TM 2010 data. The absolute error, overall accuracy and kappa coefficient were calculated to assess the accuracy of the results. The lake surface areas for the years 2000 and 2010 estimated using the selected image-specific thresholds, are presented in Table 3. The accuracy assessment analyses are shown in Table 4.

Table 3. Performance evaluation of the satellite-derived indexes used for surface water extraction

Index	Land-Water Threshold	Lake Area in 2000 (km^2)	Land-Water Threshold	Lake Area in 2010 (km^2)
Reference		4699		3328
NDWI	0.234	4711	0.153	3331
NDMI	-	-	-	-
MNDWI	0.450	4808	0.270	3607
WRI	2.330	4751	1.790	3442
NDVI	−0.210	4742	−0.290	3358
AWEI	0.000	4782	−0.150	3499

The results show that the NDMI was incapable of extracting the water surface of Lake Urmia, while the NDWI and NDVI provided the highest accuracy results. It seems like the errors in the results are mainly omission/commission of water pixels around the edges of the lake. The results further indicate the superiority and higher performance of the NDWI as compared with other indices for the extraction of surface water from Landsat data. Accordingly, the NDWI was used to model the spatiotemporal changes of Lake Urmia in the period 2000–2013. For this purpose, the NDWI was calculated from the multi-temporal Landsat 2000, 2003, 2007, 2010, and 2013 images. The lake surface area for each year was extracted through the classification of the NDWI images using image-specific thresholds. Finally, the multi-temporal lake surface area maps were overlaid to produce the lake surface area change map for the period 2000–2013 (Figure 3).

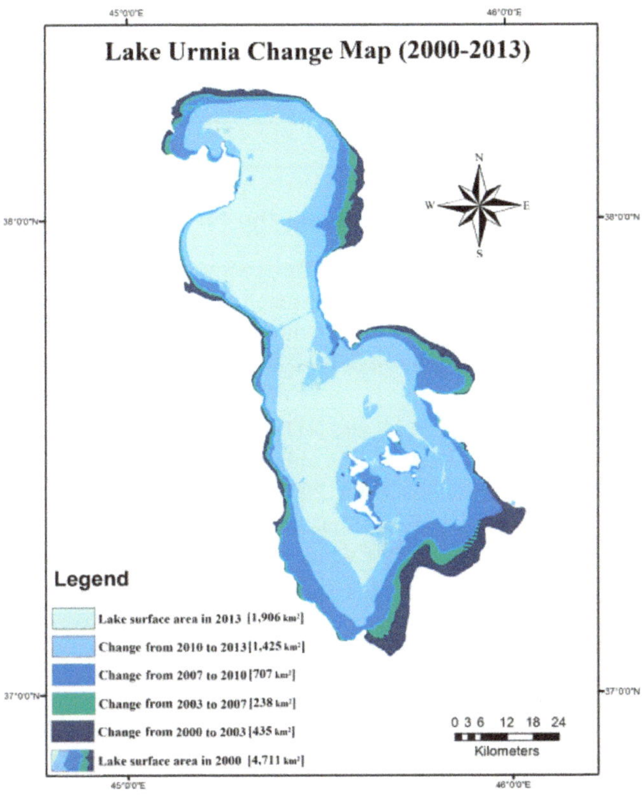

Figure 3. Lake Urmia surface area changes map in the period 2000–2013.

The statistical results, summarized in Table 5, reveal that the lake surface area was about 4711 km^2 in 2000, 4276 km^2 in 2003, 4038 km^2 in 2007, 3331 km^2 in 2010, and 1906 km^2 in 2013. The results further show that the lake surface area changes were about 435 km^2 between 2000 and 2003, 238 km^2 between 2003 and 2007, 707 km^2 between 2007 and 2010, and 1425 km^2 between 2010 and 2013, while the total surface area changes of the lake between 2000 and 2013 were about 2805 km^2. The results indicate an intense decreasing trend in the Lake Urmia surface area in the period 2000–2013. The most intense changes occurred between 2010 and 2013; within this period the lake lost about one third of its surface area compared to the year 2000 and half of its surface area compared to the year 2010. The greatest changes are observed around the southern and eastern parts of the lake.

Table 5. Statistics of the lake surface area changes

<table>
<tr><th>Year</th><th>Lake Surface Area (km^2)</th><th colspan="5">Lake Surface Area Change (km^2)</th></tr>
<tr><td>2000</td><td>4711</td><td rowspan="2">−435</td><td colspan="3"></td><td rowspan="5">−2805</td></tr>
<tr><td>2003</td><td>4276</td><td rowspan="2">−238</td><td colspan="2"></td></tr>
<tr><td>2007</td><td>4038</td><td></td><td rowspan="2">−707</td><td></td></tr>
<tr><td>2010</td><td>3331</td><td colspan="2"></td><td rowspan="2">−1425</td></tr>
<tr><td>2013</td><td>1906</td><td colspan="3"></td></tr>
</table>

Surface Water Change Detection Using the New NDWI-PCs Approach

As the NDWI was assessed as the best water extraction technique, it was used to formulate the new principal components approach (NDWI-PCs). Table 6 outlines the applicability of the NDWI-PCs for simultaneously detecting the surface water changes of multi-temporal images. This approach was initially examined to detect the surface water changes between five different times, simultaneously. The results show that the NDWI-PC2, NDWI-PC3, NDWI-PC4, and NDWI-PC5 were incapable of detecting the changes. Nevertheless, the NDWI-PC1 was able to detect the changes from 2000–2003 and 2003–2007, while the changes from 2007–2010 and 2010–2013 were partially mixed (Figure 4a). The results indicate that the NDWI-PCs approach was inefficient for simultaneously detecting the surface water changes between five different times. In the next experiment, the NDWI-PCs approach was examined for the time-series 2000, 2003, 2007, and 2010 (four different times). The results show that only the NDWI-PC1 was able to detect the surface water changes. Although the changes from 2000–2003 and 2003–2007 could be identified using NDWI-PC1, the changes from 2007–2010 were not detectable using this component (Figure 4b).

In the next experiment, the NDWI-PCs approach was examined for simultaneously detecting the surface water changes for the time-series 2000, 2010 and 2013, when most changes occurred. Contrary to the previous experiments, interpretation and classification of the resulting PCs show that the NDWI-PC2 successfully detected the changes from 2000–2010 and 2010–2013, simultaneously (Figure 4c). The results reveal that the lake surface area has decreased about 1349 km^2 between 2000 and 2010, and 1447 km^2 between 2010 and 2013. Finally, the proposed NDWI-PCs approach was investigated for detection of the lake surface area changes between 2000 and 2010. The results indicate high performance of the NDWI-PCs approach in detecting the surface water changes between two different times, especially using the NDWI-PC2, which yielded the highest accuracy output (Figure 4d). The shrinkage of the lake surface area was about 1360 km^2 in this period.

Table 6. Application of NDWI-PCs for multi-temporal change detection

Multi-Temporal Change Detection	NDWI-PCs	Can Detect the Changes	Cannot Detect the Changes	Remark
Change detection between 5 different times	NDWI-PC1	×		Only PC1 partially detected the changes
	NDWI-PC2		×	
	NDWI-PC3		×	
	NDWI-PC4		×	
	NDWI-PC5		×	
Change detection between 4 different times	NDWI-PC1	×		Only PC1 partially detected the changes
	NDWI-PC2		×	
	NDWI-PC3		×	
	NDWI-PC4		×	

Change detection between 3 different times	NDWI-PC1		×	Only PC2 well detected the changes
	NDWI-PC2	×		
	NDWI-PC3		×	
Change detection between 2 different times	NDWI-PC1	×		PC1 and PC2 well detected the changes
	NDWI-PC2	×		

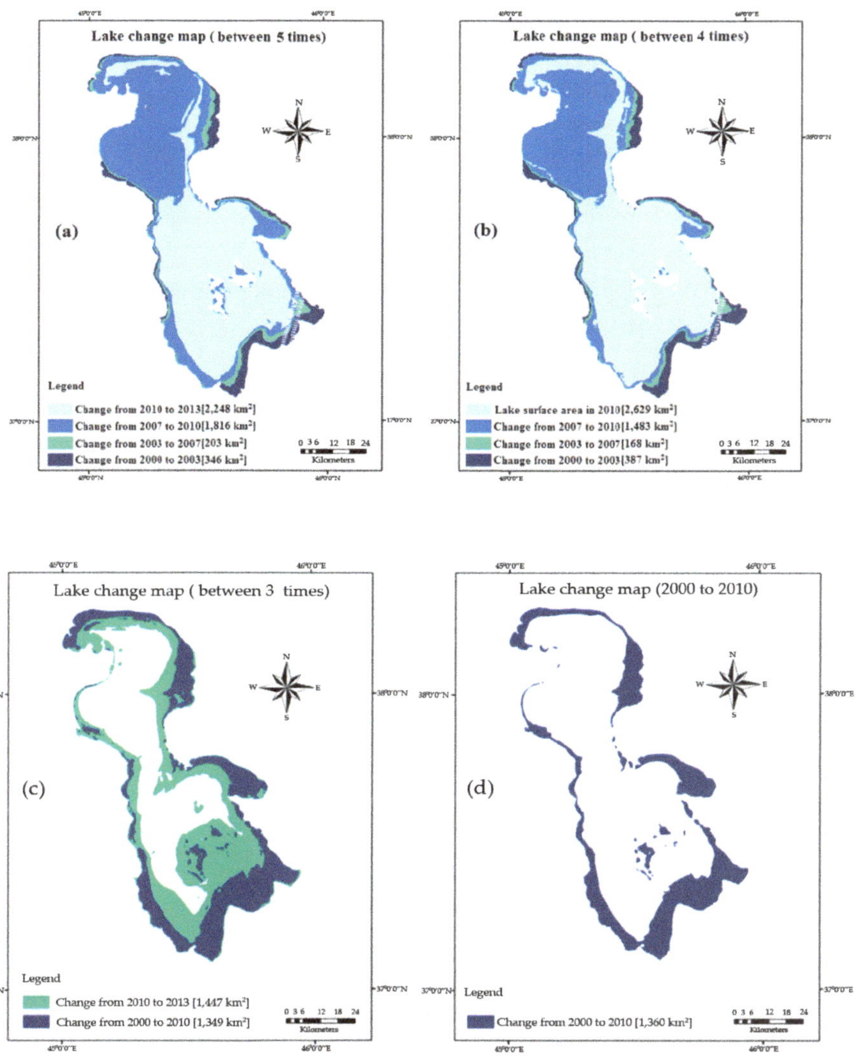

Figure 4. Lake Urmia surface area change maps, between (**a**) five, (**b**) four, (**c**) three, and (**d**) two different times, generated using the NDWI-PCs approach.

ACCURACIES, ERRORS, AND UNCERTAINTIES

A comparative analysis was performed to assess the performances of different methods adopted in this study to detect the water surface changes of Lake Urmia between 2000 and 2010. The statistical results, as shown in Table 7, indicate that the lake surface area between 2000 and 2010 decreased by approximately 1360 km^2 based on NDWI-PCs, while it decreased by about 1380 km^2 based on multi-temporal NDWI, 1384 km^2 based on multi-temporal NDVI, 1309 km^2 based on multi-temporal WRI, 1283 km^2 based on multi-temporal AWEI, and 1201 km^2 based on multi-temporal MNDWI. Based on the result of visual interpretation as the reference, the lake surface area shrunk by 1371 km^2 in this period, experiencing an approximate 29% shrinkage in 2010 compared to 2000. The results are suggestive of a high association between the changed areas detected using the NDWI-PCs, multi-temporal NDWI, and multi-temporal NDVI. Based on the reference change result, the NDWI-PCs approach underestimated the lake surface area change about 11 km^2, while the change was overestimated about 9 km^2 and 13 km^2 using the NDWI and NDVI methods, respectively.

The WRI, AWEI and MNDWI methods highly underestimated the lake surface area change between 2000 and 2010.

Table 7. Statistics of Lake Urmia surface area change between 2000 and 2010, and accuracy assessment analyses

Method	Changed Area (km^2)	Absolute Error (km^2)	Overall Accuracy (%)	Kappa Coefficient	Producer's Accuracy (%)	User's Accuracy (%)
Reference	1371	0	100	1	100	100
Proposed NDWI-PCs	1360	11	99.86	0.91	92.22	90.67
Multi-temporal NDWI	1380	9	99.88	0.91	92.54	91.06
Multi-temporal NDVI	1384	13	99.71	0.87	93.17	87.33
Multi-temporal WRI	1309	62	97.30	0.86	87.49	90.09
Multi-temporal AWEI	1283	88	96.37	0.89	85.13	88.38
Multi-temporal MNDWI	1201	170	94.81	0.86	37.31	82.45

Accuracy assessment analyses (Table 7) show superiority of the NDWI-PCs, Multi-temporal NDWI and Multi-temporal NDVI methods for surface water change detection compared with other methods. The NDWI-PCs achieved an absolute error of 11 km^2, an overall accuracy of 99.86%, a kappa coefficient of 0.91, a producer's accuracy of 92.22, and a user's accuracy of 90.67. The NDWI achieved an absolute error of 9 km^2, an overall accuracy of 99.88%, a kappa coefficient of 0.91, a producer's accuracy of 92.54, and a user's accuracy of 91.06. NDVI achieved an absolute error of 13 km^2, an overall accuracy of 99.71%, a kappa coefficient of 0.87, a producer's accuracy of 93.17, and a user's accuracy of 84.33. Only NDWI and NDWI-PCs methods produced accuracies of higher than 90% for surface water change detection. The accuracies of WRI, AWEI and MNDWI did not reach those of the NDWI-PCs, NDWI and NDVI in detecting the water surface changes in Lake Urmia. Overall, NDWI performed slightly better than the NDWI-PCs in detecting the surface water changes. Nevertheless, the NDWI-PCs have advantage over NDWI in that it detects the surface water changes of two and three different times simultaneously by applying a single threshold to the selected PC. In contrast, NDWI requires applying a threshold for each time to extract surface water, then subtracting the statistics to estimate the changes.

DISCUSSIONS

NDMI was developed for detection of vegetation water, and thus would not be efficient for extraction of water features. NDVI was developed mainly for separating green vegetation from other surfaces. However, it did perform well for surface water detection. The water features enhanced using NDWI are often mixed with built-up land and the area of extracted water can be overestimated. MNDWI was developed to modify NDWI in detecting water features for water regions with backgrounds dominated by built-up land areas. AWEI was formulated to effectively eliminate dark built surfaces in areas with urban background. In Lake Urmia, where there are no urban areas, the NDWI performed significantly better compared to MNDWI and AWEI for the extraction of open surface water.

In this study, NDWI performed significantly better compared with other indices for mapping lake water surface using Landsat data. PCA is a useful technique for the production of uncorrelated output bands, segregation of noise components, and reduction of dimensionality of data set. The present study integrated the advantages of both NDWI and PCA techniques to develop a new approach, named NDWI-PCs. The proposed approach has the advantage of simultaneously detecting the surface water changes of multi-temporal images (between two and three different times) through the application of a single threshold to the selected PC. The NDWI-PCs could be useful in detecting the decreasing or increasing changes in any open surface water type around the world, as the behaviour of water is approximately similar in different regions and different satellite images with similar band wavelengths.

As shown in the study, Lake Urmia lost more than half of its surface area in the period 2000–2013, with half of this decline occurring between 2010 and 2013. If this trend continues, it is very likely that the lake will lose all of its surface area in the near future. This is very critical as the lake provides many benefits for the society and the people living in its surroundings. Therefore, appropriate measures need to be taken by policy makers to prevent further decline of the lake surface area and to restore the lake to its original condition. It is evident that constructing dams on the rivers flowing to the lake, excessive ground water exploitation, devoting water sources to agricultural, industrial and domestic uses, and long periods of drought have all reduced the surface area of Lake Urmia. Furthermore, the changes to water supply and extraction from the catchment, for example changes in rainfall and agricultural land use, should be investigated.

CONCLUSIONS

This study aimed to model the spatiotemporal changes of Lake Urmia over the period of 2000–2013. Through a comparative analysis, NDWI was selected and employed for this purpose. The results showed an intense decreasing trend in the lake surface area over the period 2000–2013, particularly between 2010 and 2013, when the lake lost approximately 1425 km^2 (one third) of its surface area compared to the year 2000. The total change of the lake over the time period was about 2805 km^2. If such a decreasing trend in Lake Urmia continues, it is very likely that the lake will lose its entire water surface in the near future. Furthermore, a new approach was introduced for surface water change detection which is based on Principal Components of multi-temporal NDWI (NDWI-PCs). The study demonstrated high performance of the proposed approach in simultaneously detecting the surface water changes between two and three image dates. In detecting the lake water surface change between 2000 and 2010, the NDWI-PCs approach achieved an absolute error of 11 km^2, an overall accuracy of 99.86%, a kappa coefficient of 0.91, a producer's accuracy of 92.22, and a user's accuracy of 90.67. This approach has the advantage over other methods in that it detects the surface water changes of two and three different times simultaneously through applying a single threshold to the selected PC. In conclusion, the proposed NDWI-PCs approach has been proven to be effective in detecting the water surface changes in Lake Urmia, Iran. Accordingly, the method may prove useful in studying other surface waters around the world, as well as flood monitoring.

ACKNOWLEDGMENTS

The Universiti Teknologi Malaysia (UTM) under Research University Funding Scheme (Q.J130000.3009.00M19 & Q.J130000.2527.03H68) and Ministry of Education Malaysia are hereby acknowledged for providing the facilities utilized during this research work and for supporting the related research. The authors are also grateful to U.S. Geological Survey (USGS) for providing the multi-temporal Landsat data.

REFERENCES

1. Salmon, B.P.; Kleynhans, W.; van Den Bergh, F.; Olivier, J.C.; Grobler, T.L.; Wessels, K.J. Land cover change detection using the internal covariance matrix of the extended Kalman filter over multiple spectral bands. *IEEE J. Sel. Topics Appl. Earth Observations Remote Sens* 2013, *6*, 1079–1085.
2. Demir, B.; Bovolo, F.; Bruzzone, L. Updating land-cover maps by classification of image time series: A novel change-detection-driven transfer learning approach. *IEEE Trans. Geosci. Remote Sens* 2013, *51*, 300–312.
3. Volpi, M.; Petropoulos, G.P.; Kanevski, M. Flooding extent cartography with Landsat TM imagery and regularized Kernel Fisher's discriminant analysis. *Comput. Geosci* 2013, *57*, 24–31.
4. Brisco, B.; Schmitt, A.; Murnaghan, K.; Kaya, S.; Roth, A. Sar polarimetric change detection for flooded vegetation. *Int. J. Digit. Earth* 2013, *6*, 103–114.

5. Kaliraj, S.; Muthu Meenakshi, S.; Malar, V.K. Application of remote sensing in detection of forest cover changes using geo-statistical change detection matrices—A case study of devanampatti reserve forest, tamilnadu, India. *Nat. Environ. Polluti. Technol* 2012, *11*, 261–269.

6. Markogianni, V.; Dimitriou, E.; Kalivas, D.P. Land-use and vegetation change detection in plastira artificial lake catchment (Greece) by using remote-sensing and GIS techniques. In Int. J. Remote Sens; 2013; 34, pp. 1265–1281.

7. Bagan, H.; Yamagata, Y. Landsat analysis of urban growth: How Tokyo became the world's largest megacity during the last 40 years. *Remote Sens. Environ* 2012, *127*, 210–222.

8. Raja, R.A.A.; Anand, V.; Kumar, A.S.; Maithani, S.; Kumar, V.A. Wavelet based post classification change detection technique for urban growth monitoring. *J. Indian Soc. Remote Sens* 2013, *41*, 35–43.

9. Dronova, I.; Gong, P.; Wang, L. Object-based analysis and change detection of major wetland cover types and their classification uncertainty during the low water period at Poyang Lake, China. *Remote Sens. Environ* 2011, *115*, 3220–3236.

10. Zhu, X.; Cao, J.; Dai, Y. A. Decision Tree Model For Meteorological Disasters Grade Evaluation of Flood. Proceedings of 4th International Joint Conference on Computational Sciences and Optimization 2011, Kunming and Lijiang, Yunnan, China, 15–19 April 2011; Institute of Electrical and Electronics Engineers: New York NY, USA, 2011; pp. 916–919.

11. Ridd, M.K.; Liu, J. A comparison of four algorithms for change detection in an urban environment. *Remote Sens. Environ* 1998, *63*, 95–100.

12. Lu, S.; Wu, B.; Yan, N.; Wang, H. Water body mapping method with HJ-1A/B satellite imagery. *Int. J. Appl. Earth Obs. Geoinf* 2011, *13*, 428–434.

13. Desmet, P.J.J.; Govers, G. A. GIS procedure for automatically calculating the USLE LS factor on topographically complex landscape units. *J. Soil Water Conserv* 1996, *51*, 427–433.

14. Zhou, W.; Wu, B. Assessment of soil erosion and sediment delivery ratio using remote sensing and GIS: A case study of upstream chaobaihe river catchment, north China. *Int. J. Sediment Res* 2008, *23*, 167–173.

15. Du, Z.; Linghu, B.; Ling, F.; Li, W.; Tian, W.; Wang, H.; Gui, Y.; Sun, B.; Zhang, X. Estimating surface water area changes using time-series Landsat data in the qingjiang river basin, China. *J. Appl. Remote Sens* 2012, *6*. [CrossRef]

16. Sun, F.; Sun, W.; Chen, J.; Gong, P. Comparison and improvement of methods for identifying waterbodies in remotely sensed imagery. *Int. J. Remote Sens* 2012, *33*, 6854–6875.

17. Water Body Extraction from Multi-Source Satellite Images. Available online: http://citeseerx.ist.psu.edu/viewdoc/download?doi=10.1.1.301.8033&rep=rep1&type=pdf (accessed on 21 June 2003).

18. Xu, H. Modification of normalised difference water index (NDWI) to enhance open water features in remotely sensed imagery. *Int. J. Remote Sens* 2006, *27*,

3025–3033.

19. Water Body Extraction And Change Detection Based on Multi-Temporal SAR Images. Available online: http://adsabs.harvard.edu/abs/2009SPIE.7498E..96Z (accessed on 21 January 2014).
20. Zhou, H.; Hong, J.; Huang, Q. Landscape and water quality change detection in urban wetland: A post-classification comparison method with IKONOS data. *Procedia Environ. Sci* 2011, *10*, 1726–1731.
21. Tang, Z.; Ou, W.; Dai, Y.; Xin, Y. Extraction of water body based on Landsat TM5 imagery—A case study in the Yangtze river. *Adv. Inf. Comm. Technol* 2013, *393*, 416–420.
22. Li, W.; Du, Z.; Ling, F.; Zhou, D.; Wang, H.; Gui, Y.; Sun, B.; Zhang, X. A comparison of land surface water mapping using the normalized difference water index from TM, ETM+ and ALI. *Remote Sens* 2013, *5*, 5530–5549.
23. McFeeters, S.K. Using the normalized difference water index (NDWI) within a geographic information system to detect swimming pools for mosquito abatement: A practical approach. *Remote Sens* 2013, *5*, 3544–3561.
24. McFeeters, S.K. The use of the normalized difference water index (NDWI) in the delineation of open water features.*Int. J. Remote Sens* 1996, *17*, 1425–1432.
25. Alesheikh, A.A.; Ghorbanali, A.; Nouri, N. Coastline change detection using remote sensing. *Int. J. Environ. Sci. Technol* 2007, *4*, 61–66.
26. Lopez-Caloca, A.; Tapia-Silva, F.O.; Escalante-Ramirez, B. Lake Chapala change detection using time series. *Remote Sens. Agric. Ecosyst. Hydrol* 2008, *7104*, 1–11.
27. El-Asmar, H.M.; Hereher, M.E. Change detection of the coastal zone east of the Nile Delta using remote sensing.*Environ. Earth Sci* 2011, *62*, 769–777.
28. Xu, Y.B.; Lai, X.J.; Zhou, C.G. Water surface change detection and analysis of bottomland submersion-emersion of wetlands in Poyang Lake reserve using ENVISAT ASAR data. *China Environ. Sci* 2010, *30*, 57–63.
29. Delju, A.H.; Ceylan, A.; Piguet, E.; Rebetez, M. Observed climate variability and change in Urmia lake basin, Iran.*Theor. Appl. Climatol* 2012, *111*, 285–296.
30. Eimanifar, A.; Mohebbi, F. Urmia lake (northwest Iran): A brief review. *Saline Syst* 2007, *3*, 1–8.
31. Death of Urmia Lake, A Silent Disaster Investigating of Causes, Results Solutions of Urmia Lake Drying. Available online: http://conference.khuisf.ac.ir/DorsaPax/userfiles/file/pazhohesh/zamin%20mashad/127.pdf (accessed on 21 January 2014).
32. Sima, S.; Ahmadalipour, A.; Tajrishy, M. Mapping surface temperature in a hyper-saline lake and investigating the effect of temperature distribution on the lake evaporation. *Remote Sens. Environ* 2013, *136*, 374–385.
33. Ghaheri, M.; Baghal-Vayjooee, M.; Naziri, J. Lake Urmia, Iran: A summary review. *Int. J. Salt Lake Res* 1999, *8*, 19–22.

34. Schroeder, T.A.; Cohen, W.B.; Song, C.; Canty, M.J.; Yang, Z. Radiometric correction of multi-temporal Landsat data for characterization of early successional forest patterns in western Oregon. *Remote Sens. Environ* 2006, *103*, 16–26.

35. Chavez, P.S., Jr. Image-based atmospheric corrections—Revisited and improved. *Photogramm. Eng. Remote Sens* 1996,*62*, 1025–1036.

36. Wilson, E.H.; Sader, S.A. Detection of forest harvest type using multiple dates of Landsat TM imagery. *Remote Sens. Environ* 2002, *80*, 385–396.

37. Shen, L.; Li, C. Water Body Extraction from Landsat ETM+ Imagery Using Adaboost Algorithm. Proceedings of 18th International Conference on Geoinformatics, Beijing, China, 18–20 June 2010; pp. 1–4.

38. Rouse, J.W.; Haas, R.H.; Schell, J.A.; Deering, D.W. Monitoring Vegetation Systems in the Great Plains with ERTS (Earth Resources Technology Satellite). Proceedings of Third Earth Resources Technology Satellite Symposium, Greenbelt, ON, Canada, 10–14 December 1973; Volume SP-351. pp. 309–317.

39. Feyisa, G.L.; Meilby, H.; Fensholt, R.; Proud, S.R. Automated water extraction index: A new technique for surface water mapping using Landsat imagery. *Remote Sens. Environ* 2014, *140*, 23–35.

40. Gao, B.C. NDWI—A normalized difference water index for remote sensing of vegetation liquid water from space.*Remote Sens. Environ* 1996, *58*, 257–266.

41. Phua, M.H.; Tsuyuki, S.; Soo Lee, J.; Ghani, M.A.A. Simultaneous detection of burned areas of multiple fires in the tropics using multisensor remote-sensing data. *Int. J. Remote Sens* 2012, *33*, 4312–4333.

2

CHAPTER

THE USE OF SF_6 AND GIS TO STUDY FARFIELD MODELING OF OCEAN OUTFALL PLUMES

Frederick Bloetscher, Fernando J. Pleitez, Thomas Romah, Ahmed Albasri , Chase Dickinson, Husayn El Sharif, Kevin Matthews, Trung-Duong Nguyen, Lucrece Riche, Frank Youngman, Thomas P. Carsey, Jack Stamates, John R. Proni

Keywords: Ocean Outfall, Treated Sewage, Effluent Disposal, Farfield Modelling, Sulphur Hexafluoride

INTRODUCTION

Six utilities in the Palm Beach, Broward and Miami-Dade counties were discharging 396 million gallons of wastewater per day (MGD) in 2005. In 2009, the Florida Legislature passed a bill to cease outfall operations by 2025. In 2011, this deadline was modified slightly. The proposed solution was to reuse the wastewater. However, how over 300 MGD of brackish water was to be reused was not answered. Instead more questions have been raised about the actual impact of the outfalls as the bearing of policies implemented for the coastal ocean may have profound impacts on other activities. As such, decisions are closely linked to actions to be undertaken with respect to water and energy supplies to the rapidly growing population of southeast Florida. These costs are expected to excel \$20 billion dollars and require over 2 Giga Watts of new power (proposed as nuclear reactors on Biscayne Bay [1]).

The State of Florida is presently faced with critical decisions regarding the future of its coastal environment as a result of terrestrial activities. Coastal ecosystems, including both inshore estuaries and offshore reefs, sustain the State of Florida's

Citation: Bloetscher, F. , Pleitez, F. , Romah, T. , Albasri, A. , Dickinson, C. , Sharif, H. , Matthews, K. , Nguyen, T. , Riche, L. , Youngman, F. , Carsey, T. , Stamates, J. and Proni, J. (2014) The Use of SF_6 and GIS to Study Farfield Modeling of Ocean Outfall Plumes. Journal of Environmental Protection, 5, 1037-1052, doi:10.4236/jep.2014.511103.

natural resource driven economy [2]. Both must be protected to continue providing economic opportunities critical to the State. Increased nutrient loads could promote the growth of algae on the reefs, harming their habitat, their dependent wildlife and affecting tourist and commercial activities. Anthropogenic activities have been linked to the overgrowth of macroalgae and/or benthic cyanobacteria blooms, which cause loss of coral and sponge cover and changes in reef fish demographics [3]. In 1994-1995, blooms of codium ishmocladum were reported in reefs off of the Broward and Palm Beach Counties [4], as well as Caulerpa brachypus C. brachypus var. parvifolia in 2001 [5] [6]. Lyngbya spp blooms were observed off of Broward County in 2003 [7]. Blooms of Cladophora liniformis, Enteromorpha prolifera, and Centroceras clavulatum were observed in the spring of 2007 [8].

While it is asserted that a significant fraction of the nutrients impacting the coastal ecosystem are anthropogenic [9] [10], the current state of understanding concerning point and non-point sources of pollution, as well as natural process, such as upwelling, is incomplete [11]. Prior efforts by the authors has indicated that the canals and upwelling are significant courses of nutrient deposition [12]-[19]. Other sources could include far-field processes, such as coastal pollutants from the Florida Keys, the Caribbean, the Mississippi River, the south-west Florida shelf and atmospheric deposition, depending on the prevailing oceanographic conditions and location. Differentiating which, if any of these discharges may be the source(s) of nutrients is nearly impossible. Despite these numerous sources of nutrients, the outfalls have long been a target for divers, as the cause of the algal blooms, despite the lack of data to implicate the outfalls [18] [20].

South Florida has six discharge outfalls (see **Figure 1**). Each of these outfalls discharges secondary treated wastewater at a depth of 100 feet below the surface (see **Table 1**). Prior researchers have developed equations to describe the nearfield dilution characteristics through the efforts from the Southeast Florida Ocean Outfall Experiments (SEFLOE) conducted from 1984-1994 [20]-[22]. As a result, the southern four outfalls are among the most studied outfalls in the world.

The SEFLOE found that the secondary treated wastewater is less dense than seawater and hence rises as a buoyant jet above the outfall (see **Figure 2**—acoustic photograph of the Broward outfall). The region of the rising plume, where mixing is caused predominantly by buoyancy effects, is known as the nearfield region. Dilution exceeds 20:1 for all four outfalls before reaching the surface during the 10th percentile current condition [20]. Once the plume reaches the surface, it spreads out at the sea surface (top 1/3) as a less salty plume, above the denser seawater, as predicted by Richardson [23]. The surface region is termed the farfield, the result of further mixing that occurs due to ambient ocean currents [24]. Vertical transport is usually considered to be small, as diffusion effects are orders of magnitude smaller than advective horizontal transport [25].

Tracking nutrients from the ocean outfalls in the farfield is difficult as the nutrient concentrations are very low. For example, the total nitrogen concentration is typically less than 20 mg/L in the discharge. If it receives a 20:1 or more dilution in the rise of the plume to the surface, the total nitrogen concentration will be below 1 mg/L, which is hard to track for any distance at the surface. During the SEFLOE effort, only entorocuccus was detectable at 800 m, while no other parameter was detectable at 400 m [20]. Given that the plume rises to the surface, and tracking nutrients is difficult at the surface due to dilution, impacts to the reefs on the bottom are not anticipated in the nearfield vicinity of the outfalls.

A variety of investigators have made suggestions on modelling outfall plumes. Koh and Brooks [26] equated outfalls to underwater chimneys with an emphasis on particulate deposition.

Table 1. Outfall characteristics

Correlation	**0.93**
Maximum Error (ppt)	1193.5
Mean Squared Error (ppt^2)	191.7
Mean Absolute Error (ppt)	4.86

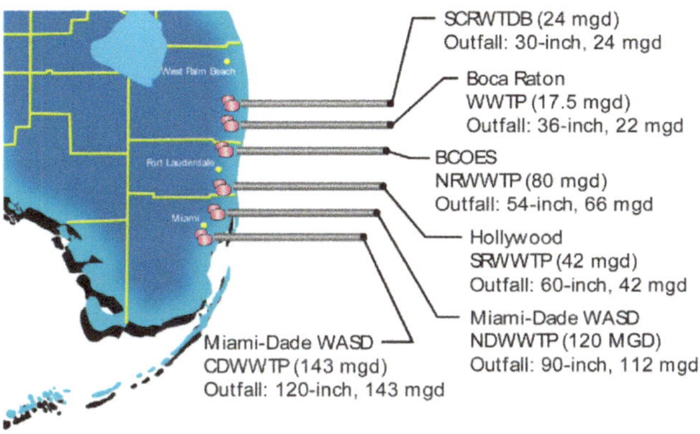

Figure 1. Location of ocean outfalls in SE Florida.

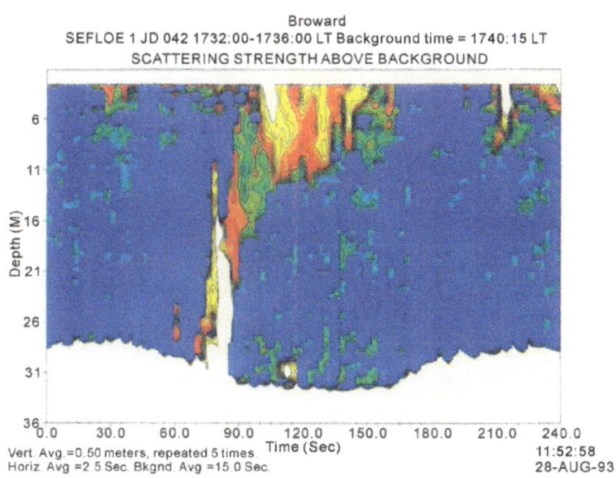

Figure 2. Acoustic photograph of Broward outfall (typical [20]).

They note that "it is practically never feasible to provide such complete treatment that an outfall is not necessary, nor is it common to depend entirely on an outfall with no treatment." As a result, modelling the fate and transport is assumed to follow a Gaussian dispersion, which continually deforms under applied shear stress, to generate a time-averaged pollutant concentration relationship that can be modelled using a normal, Gaussian, distribution curve along its longitudinal axis [26] [27].

There has been much work on nearfield dilution, but far less on the far field. Initial plume models were developed for the nearfield rise, based on thermal principles by Trent and Welty (1973), and subsequently improved on by others using buoyancy. Eulerian integral models simulate buoyant jets based on a set of control volumes fixed in space [28]-[33]. Lagrangian models have also been used by treating buoyant jets as a series of non-interfering moving elements [34]-[36]. Other efforts have been undertaken by Grace [37] and Carvalho et al. [38]. Economopoulou and Economopoulou [39] [40] developed a graphical solution for outfalls that are useful in the nearfield. Others developed or used software, for example, VISJET [41], Delft3D and the Cornell mixing zone expert system (CORMIX [42]) and Visual Plumes [43] [44]. Huang et al. [21] [22] presented a probabilistic approach to develop initial dilution criteria or standards for ocean outfall designs in the nearfield as a part of SEFLOE, while. Lye and Sharp [45] used a Monte Carlo Simulation to predict the ability of outfall to meet compliance requirements in the nearfield.

With respect to farfield outfall studies, a series of investigators attempted to model the farfield without tracers [46] [47]. Several found tracers to be not viable in Florida [48]-[52]. Gray [53] investigated the impact on Australia's outfalls on fish populations, but detected no numerical impacts on fish larvae, which is not inconsistent with the finding of Bloetscher et al. [18] for Florida's outfalls. Outfall plumes that do not follow the flow regime of the Florida outfalls or their currents include plumes modelled by many others [54]-[66]. As a result, there are no prior studies that occur in conditions similar to SE Florida, and due to the costs of conducting tracer experiments, a limited data sample was available for this effort, fully acknowledging that studying the dispersion of effluent from coastal ocean sewage outfalls has continued to be a difficult problem to study in situ because the conditions are constantly changing, as Jones, et al. [67] noted.

METHODS

Given the highly diluted environments in the nearfield, defining farfield effects requires the ability to measure very small quantities of highly diluted contributions on the bottom of the ocean floor. Therefore, to resolve the problem of detecting plume movement in the farfield, NOAA proposed a conservative tracer to prevent impacts on the reef community whenever the terrestrial plume touched the bottom. The indicator chemical that could be detected in-situ at very low levels was sulphur hexafluoride (SF_6). SF_6 was suggested as the tracer due to NOAA's success in using it in the open ocean. Its usefulness in the coastal ocean had not been tested until the Hollywood experiment discussed here, in 2004 [68]. The detection limit of SF_6 is less than 1 part per trillion by volume (pptv) using headspace analysis, and two orders of magnitude lower still using trap and purge techniques [69], which has made it possible to perform tracer studies over large space and temporal scales. The continuous release was performed to create a quasi-stationary tracer field in the ocean. Wannikhof et al. [68] notes this application adds to the utility of the tracer, which had only previously been used as a tracer of gas exchange processes in the ocean [70]-[72], lakes [73] [74], and in rivers [75]; as a surface water dispersion tracer [76], as a tag for surface water in Lagrangian studies ([77]-[79]; and to determine mixing in the deep ocean [80] [81]. As Wannikhof et al. [68] noted success with the Hollywood tracer in June 2004, a second tracer study of the Boynton outfall was conducted in 2007. The intention was to use these data points to develop a model of the farfield plume to evaluate the potential effects of the outfalls on benthic communities.

The Hollywood outfall is located approximately 2.5 km offshore and approximately 8.5 km south of Port Everglades. The outfall terminates at a depth of approximately 30 m, and discharges roughly 46 MGD of treated waste water through a 60-inch diameter armoured concrete pipe. A conservative tracer sulphur hexafluoride gas (SF_6) was injected via bubbling into the wastewater stream. SF_6 was used because it is a good tracer to study point source releases as it has low background levels in the environment. It is nontoxic, stable, and measurable at low concentrations [68]. The injection of SF_6 began on June 4, 2004 (10 a.m. EDT) and ended on June 9, 2004 (3 p.m. EDT), as described in Wannikhof et al. [68].

The ambient current velocity of the ocean during the sampling of the data was available from two different locations. The first was a current profiler, located 5 km downstream of the outfall at 11 m depth. Here, the current was predominantly northward, with little vertical shear and with a flow of approximately 20 $cm \cdot s^{-1}$. Another measurement of northward velocity was taken 12 km offshore just north (downstream) of the study region, with a flow of 80 $cm \cdot s^{-1}$. For the development of this model 20 $cm \cdot s^{-1}$ was used. In the offshore, there is no noticeable difference in salinity [82] [83].

The continuous injection of SF_6 into the outfall pipes was performed to create an environment of elevated SF_6 concentrations along the predominant flow path of the plume, to facilitate tracking of the farfield SF_6 levels. The sampling survey of the ocean waters commenced three days after the initial dosing on the Hollywood outfall, in a continuous mode using a towed "fish" containing a Seabird Micro Cat C (onductivity)-T (emperature) recorder and submersible water pump and samples as described in Wannikhof et al. [68]. The same basic procedures were used in February 2007 for the second SF_6 tracer experiment conducted by NOAA and the Florida Atlantic University (FAU).

Garvine [84] suggests that the plumes may be modelled and characterized by using averaged conditions to develop a simpler model, as opposed to partial or stochastic differential equations, or may be specified only probabilistically. Using this theory, the constituent was assumed effluent, to be released in steady state to the ocean surface at a rate (M) and at a location (x_0, y_0, z_0), as defined by Huang et al. (1996), where the effects of advection are assumed to be significantly larger than the amount of dispersion in the x direction. Pire-Schmidt et al. [13] used limited data to develop a useful predictive model of an inlet plume, based on the same tracer study methods using the 2007 data. The plume was successfully modelled with a Gaussian plume model that appears to mimic the response. It was noted that the tracer concentrated in a series of boluses that migrated north of the inlet [13], which is not expected as the inlet generally pulses with the tidal cycle.

The dispersion coefficients σ_y and σ_z are the standard deviations of the horizontal and vertical Gaussian distribution respectively [27]. As such, the dispersion coefficients represent the area under a Gaussian curve that encompasses 68% of the variance present in the distribution. The most common procedure for estimating the dispersion coefficients was introduced by Pasquill [85] and modified by Gifford [86]. In ocean plume modelling, dispersion coefficients are the driving force behind the advection-dispersion equation. This equation can be solved using a finite element method that is forward in time and central in space for dispersion. Using finite elements, the solution of a 2D diffusion equation for an instantaneous point source of mass per unit length M/L, released at time $t = t_0$ and location $(x, y) = (x_o, y_o)$ in an unbounded domain is:

$$C(x,y,z) = \frac{M}{5\pi(t-t_0)\sqrt{\sigma_y \sigma_z}} \exp\left[-\frac{(y-y_0)}{4\sigma_y(x-x_0)} - \frac{(z-z_0)}{4\sigma_z(x-x_0)}\right] \quad (1)$$

The value of σ_y and σ_z are based on units of ($m^2 \cdot s^{-1}$). Equation (1) can be used to predict the concentration at a distance x, y, and z away from the point source given all parameters. Finding these parameters however, is an art of its own requiring professional expertise, mathematical formulation, and computer modelling. Shear and horizontal dispersion as discussed by Okubo [87]-[91] were reviewed, but the conditions did not appear to be replicated for the outfalls. This work will therefore only use the principles of ADE to model the plume equation in the x-z plane. The x-y plane will be modelled using data-fitting software to derive an empirical equation based on the Hollywood outfall data.

Creating the Base Map of Measured Concentrations

Using the same assumptions as Pire-Schmidt et al. [13], and Pire-Schmidt [12], the process involved taking the data sets, and mining them for information. As noted during the 2007 SF_6 event [13], two sources of bias exist in the sample data. The first is the SF_6 adhering to sampling hardware on eastward (outbound) sampling tracks, which results in artificially increased SF_6 concentrations on eastward tracks. The seconds is the reflection of the SF_6 pollutant from the southeast Florida coast, which lies to the west of the outfall location. Pire-Schmidt [12] outlined how this bias was addressed, but generally the following assumptions were used:

- The SF_6 adherence issue biases maximum SF_6 concentrations towards the east.
- The coastal reflection issue biases maximum SF_6 concentrations towards the west.
- The biases are generally of equal magnitude.
- Rotating east-biased sample points westward and west-biased sample points eastward will sufficiently remove biases.
- A coordinate of the centreline of the SF_6 tracer plume in the Hollywood outfall

experiment is generally located at the point of highest concentration, along a given latitude (after sample points are bias-corrected).

- The centreline of the SF_6 tracer plume in the Hollywood outfall experiment has its origin (passes through) the actual Hollywood outfall location (i.e. the origin of the Hollywood tracer centreline is not significantly affected by the hydrodynamic phenomena at Port Everglades).
- Beyond 13 km north of the Hollywood outfall origin, the tracer plume is completely mixed in the water column.

Spatial Mapping of Experimental Data

The primary tool used to generate spatial interpolation maps was ESRI Arc Map v.10.0 software, using its Geostatistical Analyst Tool. Data sets originally provided in a spreadsheet form were imported as points into the Arc-Map software environment. Spatial data sets were also projected to the NAD 1983 HARN State Plane Florida East FIPS 0901 coordinate system.

To map the SF_6 concentrations, the Kernel Smoothing interpolation, with the barriers method, under the ArcMap Geostatistical Analyst tool, was used. The Kernel Smoothing method is a variant of a first-order local polynomial interpolation [92] [93] and allows for the incorporation of barrier features (i.e. the Florida coastline), into the interpolation. This interpolation method was selected as results appeared to capture the theoretical north-south anisotropy of the tracer plume better than other available interpolation methods. The biases noted above were corrected by separated the incoming (west biased) and outgoing (east biased) data sets and rotating them through a common origin. Initial interpolation algorithms were applied to the rotated concentration data using Inverse Distance Weighting (IDW) and Kriging, with Kriging being the most typically used algorithm for plume modelling, as determined during the literature review. The predictions created by the IDW algorithms produced results that when viewed using a dozen classification bins, highly indicated the presence of the tent pole effect around the data points. Kernel smoothing was applied to achieve the best Root Square Mean Error (RSME).

To map the seafloor depths in the Hollywood outfall study area, the Ordinary Kriging interpolation, under the ArcMap Geostatistical Analyst tool was used. Bathymetry is an important factor due to the reflection caused by the seabed. After the generation of interpolation maps for SF_6 concentrations and seafloor depths in the Hollywood region, results were exported to a rectangular grid of 2368 data points, covering an area of approximately 8.5×10^8 square meters, reflecting the actual data gathered during the experiment.

Creating a Model

Data on current velocity was gathered from bottom instruments. Coastal inlets also were investigated to see if tidal periods had an effect on determining the centreline equation from interpolation. The reason why inlets are of concern is that during tidal periods, large quantities of water are channelled through the inlet as a gradient change. Specifically, the velocity component of the tide is of concern to this project. The underlying assumption is that advection, which is driven by velocity, will cause the spread of the tracer in the direction of the velocity. When considering South Florida, the initial velocity travels parallel to the coast in a northern direction. When peak tide periods are present, the possibility of an additional velocity component acting transversely to the primary advection current can significantly alter the direction of the centreline plume. To determine if tidal influence could be an influencing factor on the centreline and concentrations, an evaluation of the location of test sites to proximity of inlets was made.

Model in the X-Plane

As noted above, it was suggested that under perfect conditions, the behaviour of the plume is such that it can be characterized as having a Gaussian distribution with respect to its longitudinal axis, with the highest peak at the centreline, while exponentially decaying along that same axis. The first step was to develop the centreline, taking into account that the reflection is caused by the plume being trapped at the surface, hitting the shallow ground and/or the westbound coast, and reflecting off those surfaces, creating a higher concentration than what is actually present. The highest concentrations were mapped by the latitude and longitude. Ultimately, the coastline reflection effect was handled by deleting the westbound samples, and assuming that the concentration west of the centreline (towards the shoreline) would have a concentration equal to or greater than that of the centreline, for the same reasons as outlined in Pire-Schmidt [12] and Pire-Schmidt et al. [13]. This implies that the effects of the reflection sources are proportional to depth; hence, depth is a parameter in the model. To determine the magnitude of the effects due to image sources, one can simply reflect the left side of the Gaussian distribution to generate a graph that would exponentially increase the concentration at shallower depths.

X-Y Plane

Conceptually, the concentrations in the x-y plane were found by modifying the "ideal" plume to account for the different factors that alter the dispersion/diffusion. In practice, concentrations were found by calculating the concentration, using the data-fitting modelling software Eureqa v.0.95 at each grid point, given the depth and the centreline concentration associated with the grid point. The Eureqa v.0.95 application was developed by the Cornell Creative Machines Lab [94] as a tool that searches mathematical relationships between input and output variables. Assumptions used for modelling purposes were as follows:

- The initial dilution S, is the dilution (S_o) found at the end of the near-field regime (L_b), which is assumed to be 40:1 based on prior modelling, using and average current speed of 0.2 m/s.
- The average current velocity u_a, was 20 cm/s or 0.2 m/s (based on estimated, average, current meter data).

- Vertical diffusion has minimal effect on the plume until the plume and ambient waters are nearly identical, and therefore can be ignored in the initial phases of the farfield. For analysis purposes, this is assumed to occur when S >100:1 and beyond 13 km from the outfall discharge point.
- Initial plume thickness remains at a thickness of 0.3 of the total depth for some undefined distance (<13 km), but becomes a complete mix system after that point.
- Results of SEFLOE II are used where applicable.

The actual data points were used in Eureqa. Identification of a mathematical relationship between these variables assisted with the characterization of pollutant plume dispersion in the lateral (east-west) direction, without the need for determining horizontal dispersion coefficients directly. The relationship search was simplified by allowing seafloor depth to serve as a surrogate for east-west distance of a point from the Florida coast, based on the high positive correlation between the shallowness of seafloor depth and nearness to the southeast Florida coast. The ideal equation minimizes complexity, while still exhibiting optimal performance in various statistical error and correlation measures, which obfuscates the need to determine the exact relationship between a variable and the system. The concentration at a point (x,y) on the grid changed by $C_{x,y} - C_{xo,yo}$ regardless of the mechanism.

X-Z Plane

Evaluating the x-z plane of the outfall plume requires information about the general behaviour of an effluent under buoyancy-dominated conditions. At the outfall pipe, the jet stream will bend upwards within a short distance from the outlet due to the difference in buoyancy between the effluent and the surrounding environment. Once it hits the surface, the plume is advected with the ambient velocity field at the speed u of the surface plume layer [21]. The subsequent dilution/dispersion of outfall plumes is essentially governed by buoyant spreading and oceanic turbulent diffusion. Results from the SEFLOE field tests indicated that the dilution/dispersion of the Hollywood outfall plume was dominated by buoyant spreading over a range of several hundred meters from the outfall [20] [21]. It is also known that the plume will continue to spread in 2D in the x-z plane until it is fully mixed along its water column. Based on cast data during the 2007 experiment, 13 km downstream of the outfall, the water column appears to be completely mixed. Since the concentration at 13 km is known at the centreline (y = 0), the z-dispersion coefficient can be back calculated.

RESULTS AND DISCUSSION

Hollywood Model

The centreline model was developed as follows based on the data shown in **Figure 3**:

$$C(x, 0, 0) = C_0 \exp[-0.061x] \qquad (2)$$

Figure 3 shows the plot of this equation versus the data points. Figure 4 shows the

Eureqa plot for the equation to describe the horizontal diffusion:

$$C(x,0,0) = C_x Gaus\left[\frac{2.51}{x} - 0.451 - 0.00845x\right] \quad (3)$$

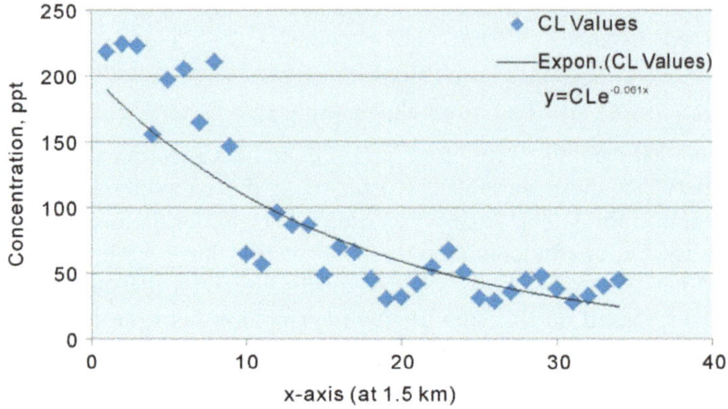

Figure 3. Exponential regression line of CL values for Hollywood outfall.

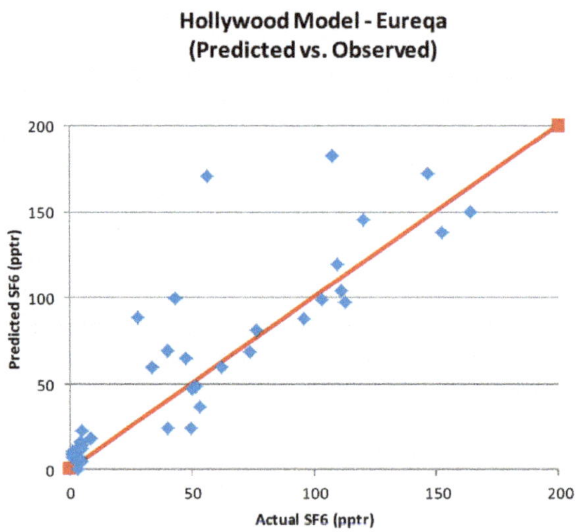

Figure 4. Predicted vs. observed plot

Both predicted and observed values of SF_6 concentrations were plotted along a 45° line to provide a graphical illustration of how the model is performing. A review of the graph indicates that the plotted points generally follow the path of the 45° line. This suggests that the model is performing relatively well. More points appear on the left of the 45° line, which is an indication that the model is slightly overestimating SF_6 concentrations.

A line plot was derived for the analysis (see **Figure 5**). The graph shows how the predicted values (red), matches those of the observed values (blue), indicating a good fit.

The software program MATLAB was used to generate a boxplot of the residual errors (see **Figure 6**). The plot showed four points which could be considered as outliers. Removing these four points reduces the RSME value from 27.77 to 13.13 and the MAE value from 17.08 to 10.70. The statistical performance measures associated with this formulation are presented in **Table 2**.

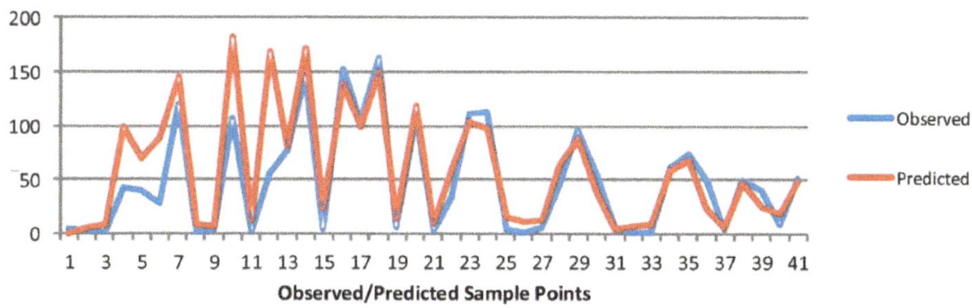

Figure 5. Line plot of predicted and observed values for the Hollywood outfall.

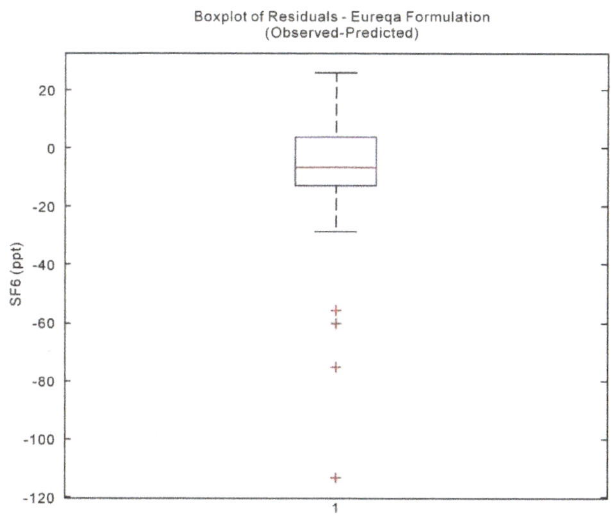

Figure 6. Boxplot of residuals.

X-Z Plane

Using the fact that based on the Hollywood cast samples, at 13 km north of the Hollywood outfall the SF_6 had completely mixed along the water column, the dispersion coefficient in the z direction was back calculated. Additionally, the concentration values at the centreline were plotted against distance and an exponential regression line of best fit was drawn. A trial and error method was employed to find the dispersion coefficient in the z direction as it was known that at 13 km, the ocean plume was completely mixed in the water column. These centreline concentrations were the inputs used to obtain the following expression for the x-z plane:

$$C(x,z) = C_0 \exp\left[\frac{u(z-z_0)^2}{4D_z(x-x_0)}\right] \tag{4}$$

Figure 7 shows a comparison of the SF_6 concentrations from the Hollywood outfall plume obtained using two different methods: the Kernel Smoothing method and the mathematical model. Both tend to follow the same pattern as far as surface concentration distribution is concerned. Additionally, the magnitudes of the concentrations seem to be consistent with one another. The mathematical model however displays a series of boluses that appear along the longitudinal axis, with decreasing magnitude. It is uncertain what causes these boluses, but they may be caused by the different depths present at the ocean bed, since the Kriging method does not take depth into account when formulating its graph. The stars represent the location of the ocean outfall.

An error analysis was performed for the calculation of the concentration plume for the Hollywood outfall in the x and y directions. No error analysis was done for the z direction due to limited sampled data points. A grid system was used for comparison. **Table 3** summarizes how the model correlates to the raw data points.

Table 2. Error measures for selected horizontal dispersion formulation

Correlation	**0.93**
Maximum Error (ppt)	1193.5
Mean Squared Error (ppt^2)	191.7
Mean Absolute Error (ppt)	4.86

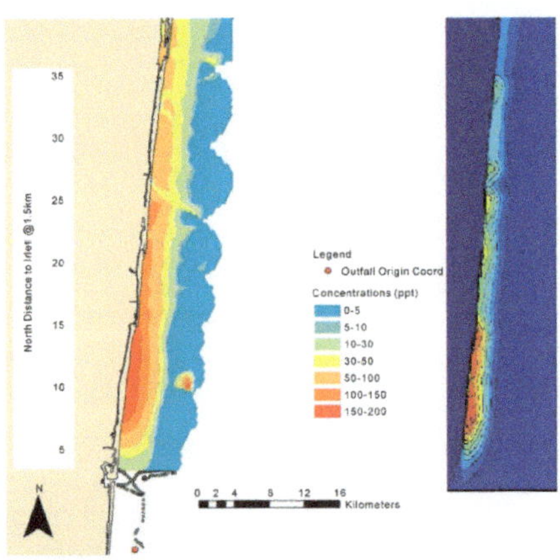

Figure 7. Kernel data smoothing vs. mathematical relationship.

Table 3. Summary statistics for model: Hollywood model (Eureqa) error analysis

RMSE	**27.7653**
MAE	17.0842
AE	700.4532
Correl (ρ)	0.878

CONCLUSIONS

The objective of the plume modelling study was to 1) determine if a farfield tracer could be used for data gathering and 2) determine if this data would permit modelling and calibration of a farfield, coastal ocean plume that adequately describes the advection and dispersion of a near-coastal conservative pollutant plume in the southeast Florida region. The following are the conclusions that can be gained from the experiment:

- SF_6 can successfully be used for monitoring terrestrial discharges to the coastal ocean in the farfield. The tracer captured data ≈ 10 to 60 km downstream.
- From 20 to 60 km northward of the outfall, maximum concentrations decreased less than twofold.
- To track in the far farfield, the SF_6 required may overwhelm detection instruments in the nearfield.
- The complete mix of terrestrial discharges occur less than 13 km downstream of the discharge terminus.
- Terrestrial activities, such as those at the port, is a complicating factor for coastal ocean discharges, especially if they are less than 13 km from the terminus, as their impact on complete mixing is unclear.

The development of the plume model in this study used data gathered from the experiment conducted by Wanninkh of et al. [68] on the Hollywood outfall as well as data other NOAA operations [83] [95].

The results of the modelling showed that the data could be translated into a model. The magnitudes of the concentrations seem to be consistent with one another. The error analysis performed on the Hollywood model (Eureqa method) indicated that the model is performing exceptionally well for that region. Four performance measures were used to evaluate the model; Root Mean Squared Error (RMSE), Mean Absolute Error (MAE), Absolute Error (AE) and Correlation Coefficient (ρ). An analysis was also carried out with the gamma-curve method using the same procedures. The models showed a slight inclination towards overestimating the SF_6 concentrations.

There were several issues that were discovered throughout the experiment that need to be dealt with in the future. A complicating factor was Port Everglades. The tidal considerations create the potential for the change from a continuous flow stream to a pulsed stream, as a result of SF_6 moving to the port during incoming tides. Hence the presence of the boluses. No samples were taken to identify the variation in SF_6 in the port, but a repeated experiment should be undertaken to study this issue. Finally, wind is an issue not addressed during the experiment. The wind was shoreward. As a result, the off-gassing SF_6 could have been reincorporated into the surface waves, creating a false footprint of the plume, which is one reason the west side of the data was ignored. Hence the SF_6 should not be used as the only tracer in future studies intended to trace a specific point source.

Varying physical conditions in the ocean mean that the experiments should be repeated to demonstrate veracity and robustness of the model. A future experiment, under different current speeds is suggested. However, if such an experiment is conducted, the following should be addressed:

- For quantitative interpretation, current (ADCP) measurements and wind data must be collected in conjunction with the SF_6 injection to determine the coastal current conditions.
- A second, conservative, non-gaseous tracer should be included along with the SF_6 to get an additional confirmation of the nutrient sources.
- Reconnaissance of discharge pipelines should be undertaken to insure that there are no leaks and that appurtenances do not leak SF_6 gas.
- Background information on significant current regime disruptors, such as Port Everglades, should be gathered and characterized along with other discharges.

One major finding was that dilution in the farfield (60 km) was extensive (greater than 10,000:1), and within 5 km well over 1000:1. Therefore the outfalls are unlikely to be a major source of contamination to the coast ecosystem. As a result, the bill passed by Florida Legislature to cease outfall operations by 2025 is unlikely to cause any significant change in the nutrient impacts in the coastal environment. Despite the modification of the deadline, the costs to the residents of SE Florida are considerable, as are the new energy demands.

REFERENCES

1. Bloetscher, F. and Muniz, A. (2012) Where Is the Power to Treat All the Water? Potential Utility Driven Solutions to the Coming Power-Water Conflict. Florida Water Resource Journal, 64, 32-46.
2. Johns, G., Leeworthy, V.R., Bell, F.W. and Bonn, M.A. (2003) Socioeconomic Study of Reefs in Southeast Florida. Final Report. Hazen and Sawyer, Hollywood.
3. Lapointe, B.E. and Bedford, B.J. (2010) Ecology and Nutrition of Invasive Caulerpa brachypus f. parvifolia blOoms on Coral Reefs off Southeast Florida, USA. Harmful Algae, 9, 1-12. http://dx.doi.org/10.1016/j.hal.2009.06.001
4. Lapointe, B.E. (1997) Nutrient Thresholds for Bottom-Up Control of Macroalgal Blooms on Coral Reefs in Jamaica and Southeast Florida. Limnol. Oceanogr, 42, 1119-1131.http://dx.doi.org/10.4319/lo.1997.42.5_part_2.1119
5. Lapointe, B.E., Barile, P.J., Littler, M.M., Littler, D.S., Bedford, B.J. and Gasque, C. (2005) Macroalgal Blooms on Southeast Florida Coral Reefs. I. Nutrient Stoichiometry of the Invasive Green Alga Codium isthmocladum in the Wider Caribbean Indicates Nutrient Enrichment. Harmful Algae, 4, 1092-1105. http://dx.doi.org/10.1016/j.hal.2005.06.004

6. LaPointe, B.E., Barile, P.J., Littler, M.M. and Littler, D.S. (2005) Macroalgal Blooms on Southeast Florida Coral Reefs II. Cross-shelf Discrimination of Nitrogen Sources Indicates Widespread Assimilation of Sewage Nitrogen. Harmful Algae, 4, 1106-1122.http://dx.doi.org/10.1016/j.hal.2005.06.002

7. Paul, V.J., Thacker, R., Banks, K. and Golubic, S. (2005) Benthic Cyanobacterial Bloom Impacts on the Reefs of South Florida (Broward County, USA). Coral Reefs, 24, 693-697.http://dx.doi.org/10.1007/s00338-005-0061-x

8. Banks, K.W., Riegl, B.M., Shinn, E.A., Piller, W.E. and Dodge, R.E. (2007) Geomorphology of the Southeast Florida Continental Reef Track (Miami-Dade, Broward, and Palm Beach Counties, USA). Coral Reefs, 26, 617-633.http://dx.doi.org/10.1007/s00338-007-0231-0

9. Lapointe B.E. and Bedford, B.J. (2010) Ecology and Nutrition of Invasive Caulerpa brachypus f. parvifolia Blooms on Coral Reefs off Southeast Florida, USA. Harmful Algae, 9, 1-12. http://dx.doi.org/10.1016/j.hal.2009.06.001

10. Zhang, J-Z., Kelble, C.R., Fischer, C.J. and Moore, L. (2008) Hurricane Katrina Induced Nutrient Runoff from an Agricultural Area to Coastal Waters in Biscayne Bay, Florida. Estuarine, Coastal and Shelf Science, 84, 209-218.http://www.sciencedirect.com/science/article/pii/S0272771409002960

11. Collier, C., Ruzicka, R., Banks, K., Barbieri, L., Beal, J., Bingham, D., Bohnsack, J., Brooke, S., Nancy, C., Dodge, R., Fisher, L., Gadbois, N., Gilliam, D., Gregg, L., Kellison, T. and Kosmynin, V. (2008) The State of Coral Reef Ecosystems of Southeast Florida. In The State If Coral Reef Ecosystems of the United States and Pacific Freely Associated States: 2008. NOAA.http://ccmaserver.nos.noaa.gov/ecosystems/coralreef/coral2008/welcome.html

12. Pire-Schmidt, J., Bloetscher, F., Meeroff, D.M., Carsey, T., Stamates, J., Sullivan, K. and Proni, J.R. (2012) Farfield Modeling of the Boynton Inlet Plume. Environmental Management and Sustainable Development, 1, 74-89. [Paper reference:4]

13. Pire-Schmidt, J. (2009) Farfield Modeling of the Boynton Inlet Plume Using Sulfur Hexafluoride as Tracer. A Thesis Submitted to the Faculty of the College of Engineering and Computer Science in Partial Fulfillment of the Requirements for the Degree of Master of Science, Florida Atlantic University, Boca Raton. [Paper reference:5]

14. Carsey, T., Stamates, J., Amornthammarong, N., Bishop, J., Bloetscher, F., Brown, C., Craynock, J., Cummings, S., Dammann, P., Davis, J., Featherstone, C., Fischer, C., Goodwin, K., Meeroff, D., Proni, J.R., Sinigalliano, C., Swart, P. and Zhang, J-Z. (2012) Boynton Inlet 48 Hou Sampling Intensives: June and September, 2007. Ocean Chemistry Division, Atlantic Oceanographic and Meteorological Laboratory, Miami.

15. Carsey, T., Casanova, H., Drayer, C., Featherstone, C., Fischer, C., Goodwin, K., Proni, J.R., Saied, A., Sinigalliano, C., Stamates, J., Swart, P. and Zhang, J.Z. (2010) FACE Outfalls Survey Cruise—October 6-19, 2006, NOAA Technical Report, OAR AOML-38. AOML, Miami.

16. Stamates, S.J. (2013) Boynton Inlet Flow Measurement Study. NOAA Technical Report, OAR-AOML-43, 13 p.

17. Fergen, R.E., Davis, P.A. and Bloetscher, F. (2004) Ammonia Dynamics on the Ocean Environment. Florida Water Resource Journal, 56, 33-36.

18. Bloetscher, F., Meeroff, D.E. and Plummer, J.D. (2010) Environmental Reviews and Case Studies: Evaluation of Costal Ocean Discharges, Environmental Impacts, and Regional Solutions in Southeast Florida. Environmental Practice, 12, 285-303.http://dx.doi.org/10.1017/S1466046610000360 [Paper reference:2]

19. Bloetscher, F. and Meeroff, D.E. (Florida Atlantic University) (2007) Boynton Inlet Sampling Event; NOAA Dye Study Coordination. FAU, Boca Raton.

20. Hazen, S. (1994) SEFLOE II Final Report: Broward County Office of Environmental Services North Regional Wastewater Treatment Plant, City of Hollywood Utilities Department Southern Region Wastewater Treatment Plant, Miami-Dade Water and Sewer Department North District Wastewater Treatment Plant, Miami-Date Water and Sewer Department Central District Wastewater Treatment Plant. Hollywood. Submitted to National Oceanic and Atmospheric Administration. [Paper reference:6]

21. Huang, H., Proni, J.R. and Tsai, J.J. (1994) Probabilistic Approach to Initial Dilution of Ocean Outfalls. Water Environment Research, 66, 787-793. [Paper reference:3]

22. Huang, H., Fergen, R.E., Proni, J.R. and Tsai, J.J. (1998) Initial Dilution Equations for Buoyancy-Dominated Jets in Currents. Journal of Hydraulic Engineering, 124, 105-108.http://dx.doi.org/10.1061/(ASCE)0733-9429(1998)124:1(105) [Paper reference:2]

23. Richardson, L.F. (1926) Atmospheric Diffusion Shown on a Distance-Neighbor Graph. Proceedings of the Royal Society A, 110, 709-737.http://dx.doi.org/10.1098/rspa.1926.0043

24. Proni, J., Adler, M., Amornthammarong, N., Bishop, J., Bloetscher, F., Carsey, T., Craynock, J., Cummings, S., Dammann, P., Drayer, C., Featherstone, C., Goodwin, K., Meeroff, D., Peltola, E., Pierrot, D., Sinigalliano, C., Swart, P., Stamates, J., Sullivan, K., Wanninkhof, R. and Zhang, J.Z. (2011) Final Report: An Investigation of the South Central Regional Wastewater Treatment Plan Ocean Outfall and Coastal Environment. Ocean Chemistry Division, Atlantic Oceanographic and Meteorological Laboratory, Miami, 58 p.

25. Chin, D.A. (2000) Water Resources Engineering. Prentice-Hall, Upper Saddle River.

26. Koh, R.C.Y. and Brooks, N.H. (1975) Fluid Mechanics of Waste-Water Disposal in the Ocean. Annual Review of Fluid Mechanics, 7, 187-211.http://www.annualreviews.org/aronline [Paper reference:2]

27. Lewis, R. (1997) Dispersion in Estuaries and Coastal Water. John Wiley & Sons, New York. [Paper reference:2]

28. Hirst, E. (1971) Buoyant Jets Discharged to Quiescent Stratified Ambient.

Journal of Geophysical Research, 76, 7375-7384. http://dx.doi.org/10.1029/JC076i030p07375

29. Schatzmann, M. (1979) An Integral Model of Plume Rise. Atmospheric Environment, 13, 721-731. http://dx.doi.org/10.1016/0004-6981(79)90202-6
30. Delvigne, G.A.L. (1980) Round Buoyant Jet with Three-Dimensional Trajectory in Ambient Flow. Pub. No. 22H, Delft.
31. Gu, R. and Stefan, H.G. (1988) Analysis of Turbulent Buoyant Jet in Density-Stratified Water. Journal of Environmental Engineering, 114, 878-897.
32. Bemporad, G.A. (1994) Simulation of Round Buoyant Jet in Stratified Flowing Environment. Journal of Hydraulic Engineering, 120, 529-543.http://dx.doi.org/10.1061/(ASCE)0733-9429(1994)120:5(529)
33. Chu, V.H. and Lee, J.H.W. (1996) General Integral Formulation of Turbulent Buoyant Jets in Cross-Flow. Journal of Hydraulic Engineering, 122, 27-34.http://dx.doi.org/10.1061/(ASCE)0733-9429(1996)122:1(27)
34. Frick, W.E. (1984) Non-Empirical Closure of the Plume Equations. Atmospheric Environment, 18, 653-662.http://dx.doi.org/10.1016/0004-6981(84)90252-X
35. Lee, J.H.W. and Cheung, V. (1990) Generalized Lagrangian Model for Buoyant Jets in Current. Journal of Environmental Engineering, 116, 1085-1105.http://dx.doi.org/10.1061/(ASCE)0733-9372(1990)116:6(1085)
36. Frick, W.E., Baumgartner, D.J. and Fox, C.G. (1994) Improved Prediction of Bending Plumes. Journal of Hydraulic Research, 32, 935-950.
37. Grace, R.A. (2005) Marine Outfall Performance. I: Introduction and Flow Restoration. Journal of Performance of Constructed Facilities, 19,347-358.
38. Carvalho, J.L.B., Feitosa, R.C., Rosman, P.C.C. and Roberts, P.J.W. (2004) Bacterial Decay Model for Coastal Outfall Plumes. Journal of Coastal Research, 1524-1528.
39. Economopoulou, M.A. and Economopoulos, A.P. (2001) Graphical Sizing and Analysis of Ocean Outfalls with Buoyant Plumes. Journal of Environmental Engineering, 127, 3-12.http://dx.doi.org/10.1061/(ASCE)0733-9372(2001)127:1(3)
40. Economopoulou, M.A., Economopoulos, A.A. and Economopoulos, A.P. (2003) Sensitivity Analysis and Comparative Performance of Outfalls with Single Buoyant Plumes. Journal of Environmental Engineering, 129, 168-178.
41. Cheung, S.K.B., Leung, D.Y.L., Wang, W., Lee, J.H.W. and Cheung, V. (2000) VISJET—A Computer Ocean Outfall Modelling System. 2000 Proceedings of Computer Graphics International, Geneva, 19-24 June 2000, 75-80.
42. Bleninger, T. and Jirka, G.H. (2004) Nearand Far-Field Model Coupling Methodology for Wastewater Discharges. In: Lee, J.H.W. and Lam K.L., Eds., Proccedings of 4th International Symposium on Environmental Hydraulics and 14th Congress of Asia and Pacific Division, International Association of Hydraulic Engineering and Research, Hong Kong, China, 15-18 December 2004, . http://outfalls.ifh.uni-karlsruhe.de/download/near-far-field-coupling-final.pdf

43. Frick, W.E., Roberts, P.J.W., Davis, L.R., Keyes, J., Baumgartner, D.J. and George, K.P. (2003) Dilution Models for Effluent Discharges, 4th Edition (Visual Plumes), United States National Exposure EPA/600/R-03/025, Environmental Protection Agency Research Laboratory, Research Triangle Park NC 27709.

44. Frick, W.E. (2004) Visual Plumes Mixing Zone Modeling Software. Environmental Modeling & Software, 19, 645-654. http://dx.doi.org/10.1016/j.envsoft.2003.08.018

45. Lye, L.M. and Sharp, J.J. (2002) Methods of Compliance Evaluation for Ocean Outfall Design and Analysis. Environmental Management, 30, 536-546.http://dx.doi.org/10.1007/s00267-002-2642-6

46. Bogle, G.V., Volioulis, I.A. and Meiorin, L. (1993) Estimation of Far-Field Dilution in Ocean Waste Discharges. Journal of Waterway, Port, Coastal, and Ocean Engineering, 119, 15-29. http://dx.doi.org/10.1061/(ASCE)0733-950X(1993)119:1(15)

47. Roberts, P. (1999) Modeling Mamala Bay Outfall Plumes. II: Farfield. Journal of Hydraulic Engineering, 125, 574-583. http://dx.doi.org/10.1061/(ASCE)0733-9429(1999)125:6(574)

48. Loh, P.C., Fujioka, R.S. and Lau, L.S. (1979) Recovery, Survival and Dissemination of Human Enteric Viruses in Ocean Waters Receiving Sewage in Hawaii. Water, Air, and Soil Pollution, 12, 197-217.

49. Lin, J., Xie, L., Pietrafesa, L.J., Ramus, J.S. and Paerl, H.W. (2007) Water Quality Gradients across Albemarle-Pamlico Estuarine System: Seasonal Variations and Model Applications. Journal of Coastal Research, 23, 213-229. http://dx.doi.org/10.2112/05-0507.1

50. Elliott, A.H. (1998) Prediction of Illness Risk near Ocean Outfalls Using Frequency Distributions of Bacterial Concentrations. Water Research, 32, 3182-3187.

51. Chang, N.B., Yeh, S.C. and Chang, C.H. (2011) Optimal Expansion of a Coastal Wastewater Treatment and Ocean Outfall System under Uncertainty (II): Optimisation Analysis. Civil Engineering and Environmental Systems, 28, 39-59.http://dx.doi.org/10.1080/10286600903243138

52. Washburn, L., Stone, S. and MacIntyre, S. (1999) Dispersion of Produced Water in a Coastal Environment and Its Biological Implications. Continental Shelf Research, 19, 57-78. http://dx.doi.org/10.1016/S0278-4343(98)00068-5

53. Gray, C.A. (1997) Field Assessment of Numerical Impacts of Coastal Sewage Disposal on Fish Larvae Relative to Natural Variability. Environmental Biology of Fishes, 50, 415-434.http://dx.doi.org/10.1023/A:1007365232465

54. Ozgokmen, T.M., Fischer, P.F., Duan, J. and Iliescu, T. (2006) Three-Dimensional Turbulent Bottom Density Currents from a High-Order Nonhydrostatic Spectral Element Model. Journal of Physical Oceanography, 34, 2006-2026.

55. Nageswara, S.V. (2006) Identification of Simple Product-Form Plumes Using Networks of Sensors with Random Errors. Computer Science and Mathematics

Division, Oak Ridge National Laboratory, Oak Ridge, TN 37831.raons@ornl.gov

56. Yapa, P.D. and Li, Z. (1997) Simulation of Oil Spills from Underwater Accidents I: Model Development. Journal of Hydraulic Research, 35, 673-688.http://dx.doi.org/10.1080/00221689709498401
57. Koh, R.C.Y. (1982) Initial Sedimentation of Waste Particulates Discharged from Ocean Outfalls. Environmental Science & Technology, 16, 757-763.http://dx.doi.org/10.1021/es00105a006
58. Thomassin, B.A., Gourbesville, P., Gout, B. and Arnoux, A. (1998) Impact of Industrial and Urban Sewage Outflows off Tropical High Islands Is an Important Environmental Problem for Many Developing Countries of the Indopacific and Caribbean Regions. OCEANS '98 Conference Proceedings, 28 September-1 October 1998.
59. Frick, W.E., Khangaonkar, T., Sigleo, A.C. and Yang, Z. (2007) Estuarine-Ocean Exchange in a North Pacific Estuary: Comparison of Steady State and Dynamic Models. Estuarine, Coastal and Shelf Science, 74, 1-11.
60. Blumberg, A.F. (1993) Modeling Transport Processed in the Coastal Ocean. Journal of Marine Environmental Engineering, 1, 31-52.
61. Blumberg, A.F., Ji, Z.G. and Ziegler, C.K. (1996) Modeling Outfall Plume Behavior Using Farfield Circulation Model. Journal of Hydraulic Engineering, 122, 610-616.http://dx.doi.org/10.1061/(ASCE)0733-9429(1996)122:11(610)
62. Hunt, C.D., Mansfield, A.D., Mickelson, M.J., Albro, C.S., Geyer, W.R. and Roberts, P.J.W. (2010) Plume Tracking and Dilution of Effluent from the Boston Sewage Outfall. Marine Environmental Research, 70, 150-161.http://dx.doi.org/10.1016/j.marenvres.2010.04.005
63. Purnama, A. and Al-Barwani, H.H. (2006) Spreading of Brine Waste Discharges into the Gulf of Oman. Desalination, 195, 26-31.
64. Al-Barwani, H.H. and Purnama, A. (2007) Simulating Brine Plumes Discharged into the Seawaters. Desalination, 221, 608-613. http://dx.doi.org/10.1016/j.desal.2007.02.060
65. Bleninger, T. and Jirka, G.H. (2008) Modeling and Environmentally Sound Management of Brine Discharges from Desalination Plants. Desalination, 221, 585-597.http://dx.doi.org/10.1016/j.desal.2007.02.059
66. Ahmed, M., Shayyab, W.H., Hoey, D. and Al-Handaly, J. (2001) Brine Disposal from Reverse Osmosis Desalination Plants in Oman and the United Arab Emirates. Desalination, 133, 135-147.http://dx.doi.org/10.1016/S0011-9164(01)80004-7
67. Jones, B.H., Barnett, A. and Robertson, G.L. (2001) Towed Mapping of the Effluent Plume from a Coastal Ocean Outfall. Proceedings of MTS/IEEE International Conference Oceans 2001, MTS 0-933957-29-7, Biloxi, 5-8 November 2001, 1985-1989.
68. Wanninkhof, R., Sullivan, K.F., Dammann, W.P., Proni, J.R., Bloetscher, F., Soloviev, A.V. and Carsey, T. (2005) Farfield Tracing of a Point Source Discharge Plume in the Coastal Ocean Using Sulfur Hexafluoride. Environmental Science

Technology, 39, 8883-8890.http://dx.doi.org/10.1021/es048126+ [Paper reference:7]

69. Wanninkhof, R., Ledwell, J.R. and Watson, A.J. (1991) Analysis of Sulfur Hexafluoride in Seawater. Journal of Geophysical Research: Oceans, 96, 8733-8740.http://dx.doi.org/10.1029/91JC00104

70. Wanninkhof, R., Asher, W., Weppernig, R., Chen, H., Schlosser, P., Langdon, C. and Sambrotto, R. (1993) Gas Transfer Experiment on Georges Bank Using Two Volatile Deliberate Tracers. Journal of Geophysical Research: Oceans, 98, 20237-20248.http://dx.doi.org/10.1029/93JC01844

71. Nightingale, P.D., Liss, P.S. and Schlosser, P. (2000) Measurements of Air-Sea Gas Transfer during an Open Ocean Algal Bloom. Geophysical Research Letters, 27, 2117-2120. http://dx.doi.org/10.1029/2000GL011541

72. Watson, A.J., Upstill-Goddard, R.C. and Liss, P.S. (1991) Air-Sea Exchange in Rough and Stormy Seas Measured by a Dual Tracer Technique. Nature, 349, 145-147.http://dx.doi.org/10.1038/349145a0

73. Wanninkhof, R., Ledwell, J.R. and Broecker, W.S. (1985) Gas Exchange-Wind Speed Relationship Measured with Sulfur Hexafluoride on a Lake. Science, 227, 1224-1226.http://dx.doi.org/10.1126/science.227.4691.1224

74. Clark, J.F., Schlosser, P., Wanninkhof, R., Simpson, H.J., Schuster, W.S.F. and Ho, D.T. (1995) Gas Transfer Velocities for SF6 and 3He in a Small Pond at Low Wind Speeds. Geophysical Research Letters, 22, 93-96.http://dx.doi.org/10.1029/94GL02410

75. Caplow, T., Schlosser, P., Ho, D.T. and Enriquez, R. (2004) Effect of Tides on Solute Flushing from a Strait: Imaging Flow and Transport in the East River with SF6. Environmental Science & Technology, 38, 4562-4571.http://dx.doi.org/10.1021/es035248d

76. Ho, D., Schlosser, P. and Caplow, T. (2002) Determination of Longitudinal Dispersion Coefficient and Net Advection in the Tidal Hudson River with a Large-Scale, High Resolution SF6 Tracer Release Experiment. Environmental Science & Technology, 36, 3234-3241.

77. Coale, K.H., Johnson, K.S., Chavez, F.P., Buesseler, K.O., Barber, R.T., Brzezinski, M.A., Cochlan, W.P., Millero, F.J., Falkowski, P.G., Bauer, J.E., Wanninkhof, R.H., Kudela, R.M., Altabet, M.A., Hales, B.E., Takahashi, T., Landry, M.R., Bidigare, R.R., Chase, Z., Strutton, P.G., Friederich, G., Gorbunov, M.Y., Lance, V.P., Hilting, A.K., Hiscock, M.R., Demerest, M., Hiscock, W.T., Sullivan, K.A., Tanner, S.J., Gordon, R.M., Hunter, C.L., Elrod, V.A., Fitzwater, S.E., Tozzi, S., Koblizek, M., Roberts, A.E., Herndon, J., Timothy, D., Brown, S.L., Selph, K.E., Sheridan, C.C., Twining, B.S. and Johnson, Z.I. (2004) Southern Ocean Iron Enrichment Experiment: Carbon Cycling in Highand Low-Si Waters. Science, 304, 408-414. http://dx.doi.org/10.1126/science.1089778

78. Martin, J.H., Coale, K.H., Johnson, K.S., Fitzwater, S.E., Gordon, R.M., Tanner, S.J., et al. (1994) Testing the Iron Hypothesis in Ecosystems of the Equatorial Pacific Ocean. Nature, 371, 123-129. http://dx.doi.org/10.1038/371123a0

79. Jahne, B., Heinz, G. and Dietrich, W. (1987) Measurement of the Diffusion Coefficients of Sparingly Soluble Gases in Water. Journal of Geophysical Research: Oceans, 92, 10767-10776. http://dx.doi.org/10.1029/JC092iC10p10767

80. Ledwell, J.R., Watson, A.J. and Law, C.S. (1998) Mixing of a Tracer in the Pycnocline. Journal of Geophysical Research: Oceans, 103, 21499-21529.http://dx.doi.org/10.1029/98JC01738

81. Watson, A.J., Messias, M.J., Fogelqvist, E., Van Scoy, K.A., Johannessen, T., Oliver, K.I.C., Stevens, D.P., Rey, F., Tanhua, T., Olsson, K.A., Carse, F., Simonsen, K., Ledwell, J.R., Jansen, E., Cooper, D.J., Kruepke, J.A. and Guilyardi, E. (1999) Mixing and Convection in the Greenland Sea from a Tracer-Release Experiment. Nature, 40, 902-904. http://dx.doi.org/10.1038/44807

82. Bloetscher, F. and Meeroff, D.E. (2007) Implications of Nutrients in Coastal Environments. NAEP Conference Proceedings, Orlando.

83. Carsey, T., Featherstone, C., Goodwin, K., Sinigalliano, C., Stamates, J., Zhang, J.Z., Proni, J.R., Bishop, J., Brown, C., Adler, M., Blackwelder, P. and Alsayegh, H. (2011) Boynton-Delray Coastal Water Quality Monitoring Program, NOAA Technical Report, OAR AOML-39. Atlantic Oceanographic and Meteorological Laboratory, Miami. [Paper reference:2]

84. Garvine, R.W. (1998) Penetration of Buoyant Coastal Discharge onto the Continental Shelf: A Numerical Model Experiment. Journal of Physical Oceanography, 29, 1892-1909.

85. Pasquill, F. (1961) The Estimation of the Dispersion of Windborne Material. Meteorological Magazin, 90, 33-49.

86. Gifford Jr., F.A. (1961) Use of Routine Observations for Estimating Atmospheric Dispersion. Nuclear Safety, 2, 47-57.

87. Okubo, A. (1969) Ocean Mixing. Management Information Services, Detroit.

88. Okubo, A. (1971) Ocean Diffusion Diagrams. Deep-Sea Research, 18, 789-802

89. Okubo, A. and Karweit, M. (1964) Diffusion from a Continuous Source in a Uniform Shear Flow. Limnology and Oceanography, 14, 514-520.http://dx.doi.org/10.4319/lo.1969.14.4.0514

90. Okubo, A. (1966) A Note on Horizontal Diffusion from an Instantaneous Source in a Nonuniform Flow. Journal of the Oceanographical Society of Japan, 22, 35-40.

91. Okubo, A. (1968) Some Remarks on the Importance of the "Shear Effect" on Horizontal Diffusion. Journal of the Oceanographical Society of Japan, 24, 60-69

92. ESRI (2008) Transcript: Understanding the ArcGIS Desktop Applications: What Is ArcMap? ESRI.

93. ESRI (2010) ArcGIS 10 Help. ArcGIS ArcMap Version 10. ESRI.

94. Cornell Creative Machine Lab (2012) Eureqa.http://creativemachines.cornell.edu/eureqa

95. NOAA: National Oceanographic and Atmospheric Agency (2008) Interim

Report South Central Regional Waste Water Treatment Plant. Florida Area Coast Environment (FACE) NOAA-AOML, Miami.

3

CHAPTER

SUBSURFACE OCEAN SIGNALS FROM AN ORBITING POLARIZATION LIDAR

James H. Churnside, Brandi J. McCarty and Xiaomei Lu

Keywords: CALIPSO, CALIOP, ocean lidar, phytoplankton, polarization, lidar, ocean color, chlorophyll

INTRODUCTION

Airborne lidar measurements, a wavelength of 532 nm have proven to be an effective tool for probing the upper ocean. Studies include detection of fish and zooplankton [1–5], thin plankton layers [6], internal waves propagating on the pycnocline [7,8], and bubbles [9,10]. Extending these capabilities to a satellite-based lidar would provide the depth dimension that is missing in satellite-based ocean-color measurements.

The Cloud-Aerosol Lidar with Orthogonal Polarization (CALIOP) is the primary instrument on the Cloud-Aerosol Lidar and Infrared Pathfinder Satellite Observation (CALIPSO) satellite. This lidar has many of the same characteristics as the airborne lidars, including the 532 nm laser wavelength, a laser pulse length of 10–20 ns, and dual-polarization receivers [11]. These characteristics suggest that the CALIOP might be useful for oceanographic studies, even though its mission is to investigate clouds and aerosols in the atmosphere. Our objective in this study was to determine whether or not there was a subsurface component to the CALIOP return over the ocean. There have been several investigations of CALIOP subsurface signals. Rodier et al. [12] estimated

Citation: Churnside, J.H.; McCarty, B.J.; Lu, X. Subsurface Ocean Signals from an Orbiting Polarization Lidar. Remote Sens. 2013, 5, 3457-3475, doi:10.3390/rs5073457.

the subsurface lidar signal from the depth-integrated, cross-polarized lidar signal, and compared that with surface chlorophyll concentration on a global scale. Shi and Wang [13] compared the total signal in the nominal surface bin (reported altitude of −7 m) and the first subsurface bin (nominal altitude −37 m) in the Gulf of Martaban in the eastern Indian Ocean before and after passage of Cyclone Nargis in 2008. They used the relative values in these two depth bins to argue for vertical mixing of sediments by the cyclone. Barton and Jasinski [14] used the total return integrated over the nominal surface bin and all subsurface bins to estimate total subsurface scattering in Tampa Bay, Florida. This estimate was based on corrections for atmospheric attenuation and for surface reflections, including specular reflections and scattering from surface foam. The surface reflection components were based on models involving the wind speed.

Despite these investigations, there are several difficulties in applying CALIOP data to subsurface scattering. The primary difficulty is the processing of the signals prior to downlink. The data is filtered to condense the signal profiles in range, ultimately decreasing the downlinked data volume. The lidar returns are low-pass filtered using a three-pole Bessel filter with a bandwidth of 2 MHz and digitized at a rate of 10 MHz, corresponding to the initial 15 m range sampling. Two successive samples are averaged to provide a range resolution of 30 m in air or 22.5 m in water. Another factor is that the incidence angle was 0.3° until November of 2007 and 3° after this date. Airborne lidars typically operate with an incidence angle of around 15° to reduce the specular reflection from the surface. The CALIOP has a very strong specular component, and this component has been suggested as a technique to infer wind speed [15]. One other difficulty is a slow transient recovery time of the photomultiplier tubes used in the detection of the 532 nm channel. This detector feature can mask weak subsurface returns [16].

We considered several characteristics of lidar returns that included a subsurface contribution to determine whether the CALIOP could, in fact, respond to subsurface scattering. The specular return from the sea surface preserves polarization, but subsurface scattering depolarizes initially polarized light [17–19]. Therefore, we concentrated on a comparison of the CALIOP 532 channels, the cross-polarized return and the total, or unpolarized, return.

The null hypothesis for this study was that there was no contribution to the signal from below the surface of the ocean. This hypothesis implies no depolarization of the return near the surface, since the surface itself is a polarization-preserving specular reflector. It further implies that the shape of the cross-polarized return would be identical to the shape of the unpolarized return, since the only source of depolarization would be incomplete polarization separation in the lidar. Finally, the null hypothesis implies that there would be no correlation between any characteristics of the lidar return and subsurface properties, such as phytoplankton concentration. Conversely, the null hypothesis would be refuted if there is a significant depolarization at ranges corresponding to below the surface, since this could only come from subsurface scattering. Alternatively, it would be refuted if the broad surface return in the cross-polarized receiver was delayed or further broadened relative to the unpolarized return, since this could only come from scattering occurring after the surface reflection. If the distribution of these properties were correlated with near-surface chlorophyll concentration, used as a proxy for scatterer density, this would provide additional evidence to refute the null hypothesis.

We restricted the analysis to regions of relatively clear, open-ocean waters. In very turbid water, the lidar signal attenuates rapidly. The near-surface return would be partially depolarized by subsurface return, but the cross-polarized pulse delay and broadening would be very small. In addition, the delay and broadening would be affected more by penetration depth than the concentration of subsurface scatterers, so one might expect a negative correlation with chlorophyll concentration for these cases.

MATERIALS AND METHODS

The primary data set included version 3.01, level 1b CALIOP total attenuated backscatter, βT, and cross-polarized attenuated backscatter, βX, for the year 2010 and for latitudes between 60°S and 60°N [20]. Only the 532 nm wavelength was used, because the 1064 nm light is strongly absorbed by sea water. Daytime data was removed to minimize noise from background light. To avoid turbid coastal waters, the associated land/water mask was used to exclude all but deep-ocean data (>120 m). The level 2 cloud layer product was used to remove profiles contaminated by clouds. Similarly, the aerosol optical depths, AOD as available in the level 2 aerosol layer product and estimated from the CALIOP profiles, were used to remove profiles with AOD > 0.2, minimizing possible contamination by aerosols. Lidar shots were also removed from the data set if maximum signals were so small ($\beta T < 0.1\ m^{-1}\cdot sr^{-1}$ or $\beta X < 0.001\ m^{-1}\cdot sr^{-1}$) that the resulting profiles were very noisy. The lidar calibration is described in detail [11], but is briefly reviewed here. The 532 nm total return was calibrated using backscatter from altitudes between 30 and 34 km above mean sea level, where the atmosphere is purely molecular and the expected return can be accurately estimated. The calibration coefficient was measured during the night-time segment of each orbit and averaged over 1485 km of ground track. The 532 nm cross-polarized channel calibration coefficient was calculated from that of the total channel using a factor that was periodically measured by inserting a pseudo-depolarizer into the optical path before the polarization beam splitter.

The CALIOP data is reported at a vertical resolution of 30 m in the altitude region between −0.5 and 8.3 km above the surface, but the exact altitude changes with variations in satellite orbit. A shift of several range bins may be necessary to account for these differences. It is because of this uncertainty in altitude that data is reported to 0.5 km below the surface, so to ensure that the atmosphere was sampled to the surface. A suggested shift in altitude bins is provided with the level 1b data based on estimated orbital parameters, and we applied this shift to the data used here. For each polarization channel, an additional shift of one range bin in either direction was applied if the peak of the attenuated backscatter was one bin above or below the nominal surface bin. If the peak was more than one bin from the nominal surface bin in either channel, that shot was not used in the analysis. For each of the remaining lidar shots and for each channel, the nominal surface sample and the two adjacent samples were selected to calculate the three parameters of a Gaussian curve. These parameters are the amplitude, A, the altitude of the peak, R, and the width, W:

$$\beta(z) = A\exp\left[-\frac{(z-R)^2}{W^2}\right].$$

where z is the altitude in air. The parameters were calculated using values for z of 22.8 m, −7.1 m, and −37.1 m and the corresponding data values for each shot. The nominal surface bin at −7.1 m in air corresponds to a depth of 5.3 m and the first subsurface bin at −37.1 m corresponds to a depth of 27.9 m. We then calculated the differences ΔR = RT − RX and ΔW = WX − WT. The altitude difference between the two channels was used rather than the absolute altitude in the cross-polarized channel, because the absolute altitude of the surface (as defined in the lidar altitude bins) is not known precisely [16]. Use of the difference removes this uncertainty. Similarly, the width difference removes any effects that would be common to both channels, such as a temporal spreading of the surface return by waves. Note that these differences were defined such that both are positive if the cross-polarized return is delayed and broadened when compared with the total. These differences, ΔR and ΔW, were averaged into one-degree latitude and longitude bins by month.

The depolarization was defined as:

$$d(z) = \frac{\beta_X(z)}{\beta_T(z) - \beta_X(z)}.$$

and the values in the first subsurface bin (z = −27.9 m) were averaged in the same way as ΔR and ΔW.

A total of 9.6 million lidar shots passed all of the criteria and were used in the analysis. The number in each one degree by one degree by one-month average ranged from 0 to 555. From these, 139,242 contained at least one shot, with an average number of 69.

An additional dataset used in the analysis was chlorophyll-a concentration, C, derived from the Moderate Resolution Imaging Spectroradiometer (MODIS) on the Aqua satellite [21]. Chlorophyll concentration was estimated using the three-band OC3 algorithm using a polynomial fit in the logarithm of the ratio of blue to green remote-sensing reflectance. In this algorithm, the blue remote-sensing reflectance is the larger of that at 443 nm or 489 nm, and the green at 550 nm. Monthly averaged, level 3 data with 4 km resolution was averaged to the same one-degree latitude and longitude bins as the CALIOP data.

An observed correspondence in the spatial patterns of the lidar data and the chlorophyll concentration was investigated quantitatively using Spearman's rank correlation of the one degree by one degree monthly averages.

Wind speed for each CALIOP profile was obtained from the level 2b data from the Advanced Microwave Scanning Radiometer—Earth Observing System (AMSR-E) sensor on NASA's Aqua satellite [22]. While the spatial resolution of the AMSR-E wind product is much larger than the CALIOP footprint, the root-mean-square difference between that and the wind speed inferred directly from CALIOP is only 1.2 m·s^{-1} for clear skies [15], which is sufficient for our purposes.

RESULTS AND DISCUSSION

A typical profile (Figure 1a) shows the delay and broadening of the cross-polarized return relative to the total return as described in the introduction. This example was averaged over 100 lidar shots to show the general features without noise. The standard deviations of the 100 values at each depth were around 20% of the corresponding means for the total return, and almost 100% for the cross-polarized return. Much of this variability is probably a result of surface roughness variations, but a detailed investigation is beyond the scope of this paper. The amplitude, AT, of the Gaussian fit to the total surface return peaks at 2.2 m above the surface, while the amplitude, A_X, of the fit to the cross-polarized return peaks at 0.2 m below the surface, hence $\Delta R = 2.4 \pm 0.2$ m. The width, W_T, of the total return is 22.8 m, while the width of the cross-polarized return, W_X, is 24.8 m, such that $\Delta W = 2.0 \pm 0.4$ m. The variability in the single-shot parameter values is also large, but the resulting uncertainties in average ΔR and ΔW for this example are 9% and 18%, respectively. The depolarization of the same averaged profile (Figure 1b) shows a maximum in the depolarization at a depth of 28 m. This is consistent with subsurface scattering. Figure 1 also shows the effects of the relatively slow transient response of the photomultiplier tubes for both channels.

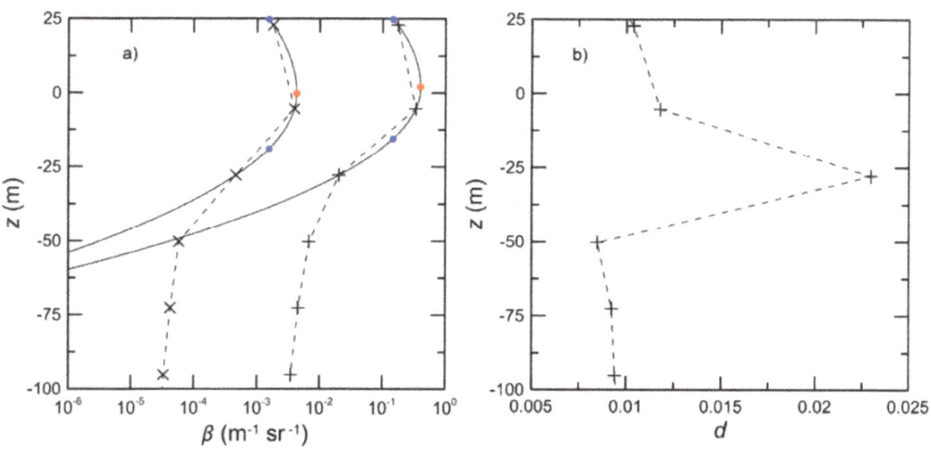

Figure 1. (**a**) Typical depth profiles of lidar attenuated backscatter β. Dashed lines connect the values for the total (+) and cross-polarized (×) samples, averaged over 100 shots. Solid lines are the corresponding Gaussian fit to the near surface samples. Red points mark the points where z = R and β = A for each curve and the blue points mark the points where z = R ± W. (**b**) Depolarization, d, for the same averaged profile.

For all months of 2010, the global averages (60°S to 60°N) of both ΔR and ΔW (Figure 2) were significantly greater than zero. The number of lidar shots contributing to those monthly averages ranged from 0.4 million in August to 1.6 million in March.

Despite the shot-to-shot variability, the large number of shots in each average resulted in a standard deviation of each average that was less than 0.01 of the corresponding average value, and the averages were all significantly greater than zero. The subsurface depolarization is also shown in Figure 2. The figure shows that the range difference, ΔR, was always significantly greater than the increase in width, ΔW. It is interesting that ΔR and ΔW track each other fairly well, apart from July, August, and September. For these months, data from the southern hemisphere was relatively sparse, and this may have been an influencing factor. The northern hemisphere for these months has large regions with high chlorophyll concentration near the surface. In all months, however, the global average depolarization at 28 m depth tracks the increase in width.

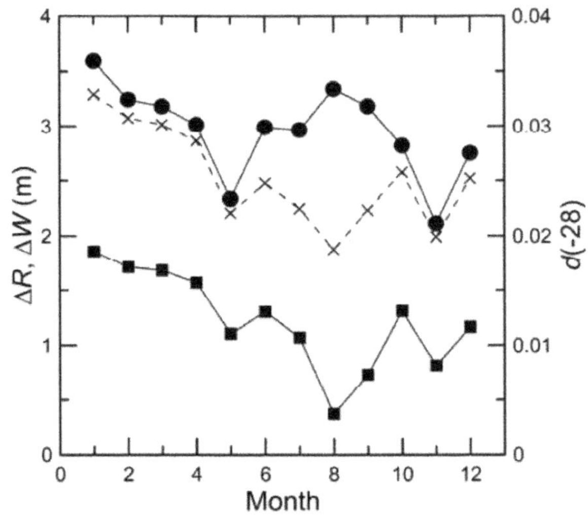

Figure 2. Global average difference between total and cross-polarized range ΔR (solid circles), average difference between total and cross-polarized surface return width ΔW(solid squares), and average depolarization of the first subsurface return at z = −28 m,d(−28) (× connected by dashed line) by month of 2010.

We were also interested in the spatial distribution of the three lidar parameters, ΔR, ΔW, and d(−28). The example of January (Figure 3) presents the parameters after averaging with a 3° filter to make the patterns more visible. There is a rough correspondence between the global patterns of all three of the fitted CALIOP parameters and surface-chlorophyll concentration derived from MODIS data. The highest chlorophyll concentration in January is below 40°S, and this region clearly shows higher values for all three lidar parameters. A region of high chlorophyll above 30°N is clearly seen in the depolarization, is less visible in the range difference, and even less visible in the width difference. The equatorial upwelling zones, which appear as regions of increased chlorophyll concentration, are also seen in the depolarization and range difference, but not in the width difference. There is also a correspondence on smaller scales. Note, for example, the region of higher chlorophyll in the northwest Indian Ocean that is matched by increased range difference and increased depolarization.

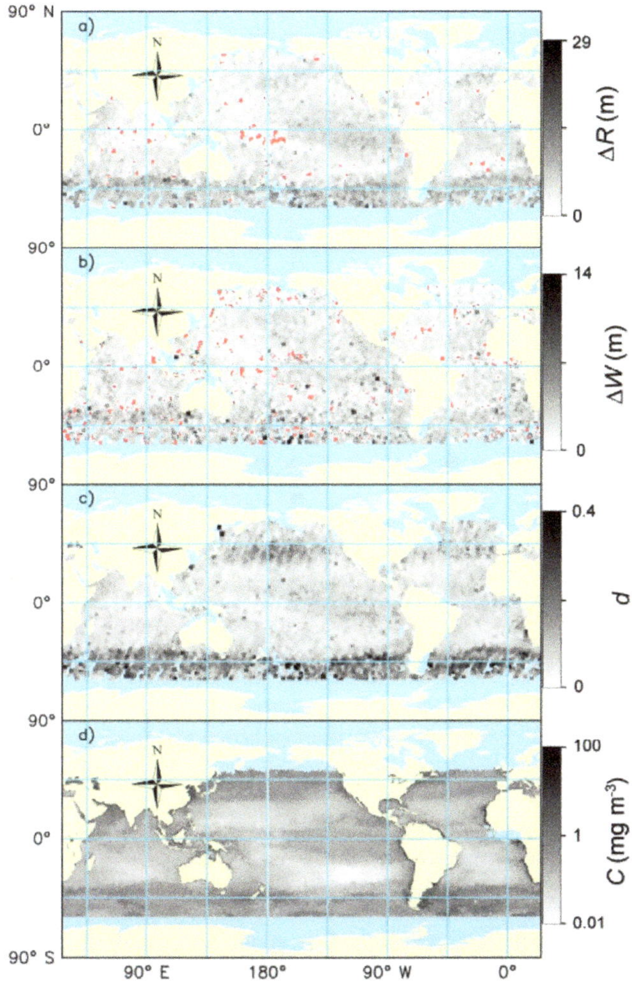

Figure 3. Maps of (**a**) ΔR, (**b**) **Δ**W, (**c**) d(−28), and (**d**) C for January 2010. Yellow is land, light blue is ocean area with no lidar data. Red points denote negative values.

The correlations between monthly averages of all three of the lidar parameters and chlorophyll concentration (Figure 4) are all significant, with statistical p-values $< 10^{-6}$. The number of data pairs in each of these correlations varies between 6,500 (August) and 19,000 (March), with an average of 11,000. These correlations suggest that phytoplankton might be an important component of the subsurface returns. In restricting the analysis to deep-ocean CALIOP data, we expect that most, although not all, of the data would be from Case 1 waters where the optical properties, including depolarized backscatter, are largely determined by phytoplankton. Bubbles may be present, but would be concentrated near the surface and would not significantly depolarize the lidar return. Concentrations of suspended sediments are generally low away from the coasts.

Zooplankton are likely to be present where there are phytoplankton, and these could contribute to the observed correlations. However, we would expect concentrations of zooplankton to be lower than the phytoplankton on which they feed. Particulate organic matter (POM) is also likely to be present where there are phytoplankton, and this could also contribute to the observed correlations. The success of ocean color measurements in measuring chlorophyll concentration in Case 1 waters would suggest that, even if POM is a significant source of backscatter, it can be used to infer chlorophyll.

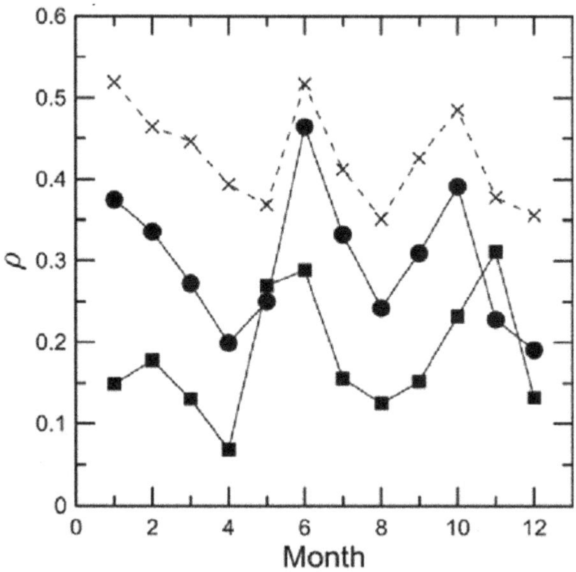

Figure 4. Correlation coefficient ρ between 1° by 1° average chlorophyll concentration,C, and average difference between total and cross-polarized range, ΔR, (solid circles), average difference between total and cross-polarized surface return width, ΔW, (solid squares), and average depolarization of the first subsurface return, d(−28) (× connected by dashed line) by month of 2010.

On the other hand, the correlations are generally <0.5, so the lidar parameters are not completely determined by the near-surface chlorophyll concentration inferred from MODIS. Part of this difference may be related to differences in phytoplankton populations. The depolarization from a phytoplankton cell depends on its size and shape in ways that are unrelated to its chlorophyll content. Part may be related to non-uniform depth distributions of phytoplankton. The presence of subsurface layers will affect remote-sensing reflectance and cross-polarized lidar return differently.

The spatial distribution of these correlations (Figure 5) was calculated using sets of 10° by 10° by 12-month data. There are fewer data pairs used for each correlation in this analysis, ranging from 17 to 905, with an average of 420. The general features of these maps are consistent with the global averages. The depolarization has the highest correlations, with an average of 0.31, and the fewest regions with no significant correlation. Conversely, the width has the most regions with no significant correlation and the lowest average correlation at 0.20. A detailed analysis of the regional differences is beyond the scope of this paper, but there some interesting features. One region that deserves further study is the western tropical Pacific, which has relatively high correlation values. Another is the region from 0°–10°N and 40°–50°W, where all three of the lidar parameters are negatively correlated with chlorophyll concentration and which includes the mouth of the Amazon River. This is consistent with our argument that subsurface returns would not be detectable in waters with high attenuation like the Amazon River plume. Precise determination of how much attenuation could be accommodated is difficult, however, since detection also depends on the level of surface reflection, the amount of subsurface scattering, and the depolarization introduced by that scattering.

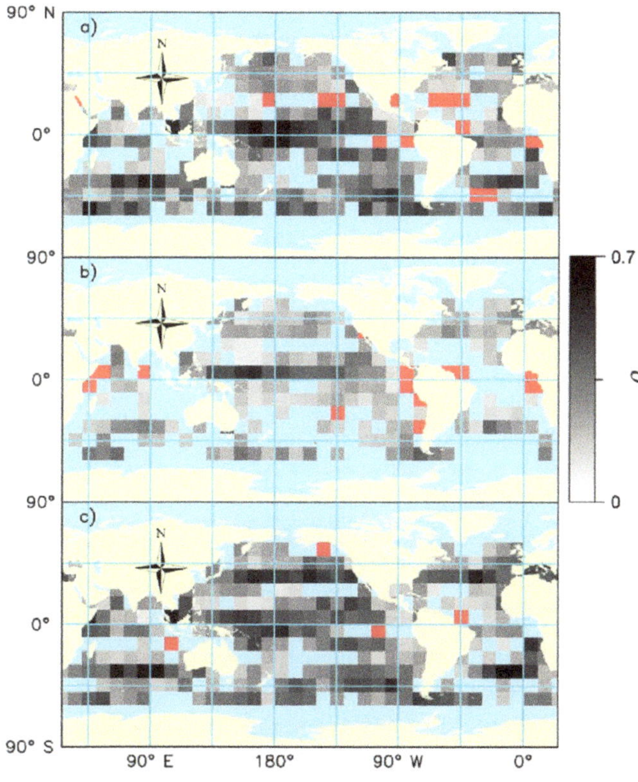

Figure 5. Correlation coefficient ρ between C and (**a**) ΔR, (**b**) ΔW, and (**c**) d(−28) for all months in 10° by 10° regions. Red regions had negative correlations. Light blue regions have latitude > 60° or had no significant ($p < 0.05$) correlation.

Atmospheric Interference

While the correlations between the lidar parameters and chlorophyll concentration suggest that subsurface scattering is present in the lidar return in addition to the specular surface reflection, it was necessary to rule out the possibility of interference by atmospheric aerosols. That is, depolarization by forward scattering in the atmosphere coupled with a correlation between chlorophyll concentration and AOD might produce correlations between depolarization and chlorophyll concentration that are not representative of subsurface scattering. Note that the question is of depolarization by narrow-angle forward scattering in the atmosphere before or after reflection from the sea surface. This is different from the issue of depolarization by backscattering from non-spherical aerosols, since the return from aerosol backscattering does not contribute to the subsurface signal.

For the small aerosol optical depths used in the analysis, this depolarization can be estimated assuming a single aerosol scattering event. This means that we will consider those photons that are scattered in the forward direction by an aerosol particle and specularly reflected from the sea surface back to the receiver or are specularly reflected from the sea surface and then scattered back to the receiver by an aerosol particle. From symmetry arguments, we can evaluate the aerosol-surface case and use the same result for the surface-aerosol case.

Consider a linearly-polarized beam scattered by an aerosol particle. The scattered field can be expressed in terms of the incident field by the matrix expression:

$$\mathbf{E}_s = \frac{\exp(ikr)}{-ikr}\mathbf{Q}^{-1}\mathbf{SQE}_0 \ .$$

where **Es** is the scattered field vector, k is the optical wavenumber, r is the distance to the observation plane, **Q** is the rotation matrix between the initial plane of polarization and the scattering plane, and **S** is the scattering matrix. For linearly polarized light, we can let:

$$\mathbf{E}_0 = \begin{pmatrix} E_0 \\ 0 \end{pmatrix}.$$

For scattering at an azimuthal angle of φ with respect to the plane of polarization:

$$\mathbf{Q} = \begin{pmatrix} \cos\phi & -\sin\phi \\ \sin\phi & \cos\phi \end{pmatrix}.$$

The scattering matrix for a large, spherical aerosol particle is given by:

$$\mathbf{S} = \begin{pmatrix} S_1 & 0 \\ 0 & S_2 \end{pmatrix}.$$

where the scattering amplitudes, S1 and S2, depend on the characteristics of the particle and on the scattering angle. The off-diagonal elements in the scattering matrix are zero for spherical aerosols when the plane of polarization and the scattering plane coincide. The initial coordinate system is such that the polarization and scattering planes coincide, the scattering matrix is applied, and the inverse rotation matrix restores the original coordinate system.

For a distribution of large, spherical aerosol particles, we can assume that S2 = S1cos(θ), where θ is the scattering angle. We will also let cos(θ) = 1 − δ. Because of the receiver field of view (130 μrad), δ will generally be small, but this condition is not required in the following development. Averaging over all azimuthal scattering angles, the cross-polarized irradiance can be expressed as:

$$I_X = \frac{I_0}{8k^2r^2}|S_1|^2\delta^2 .$$

The scattered irradiance in the co-polarization is:

$$I_{CO} = \frac{I_0}{8k^2r^2}|S_1|^2\left[3 + 5(1-\delta)^2\right].$$

The depolarization of the scattered light is:

$$d_A = \frac{\langle I_X\rangle}{\langle I_{CO}\rangle}.$$

where the angle brackets denote the average over all scattering angles within the field of view of the lidar and over height. We will approximate the field of view by denoting a maximum δ given by:

$$\delta_{max}(z) = 1-\cos\left[\tan^{-1}\left(\frac{47\text{ m}}{z}\right)\right].$$

where z is altitude, and the value of 47 m was obtained from the product of the field of view half angle (65 μrad) and the average satellite altitude (720 km). While this maximum value goes to unity at the surface, it is less than 0.1 for z > 100 m.

To illustrate, we can approximate the distribution of scattering angles by the Henyey-Greenstein function [23]:

$$|S_1(\delta)|^2 = \frac{1}{2}\frac{1-g^2}{\left[(1-g)^2+2g\delta\right]^{\frac{3}{2}}}.$$

where the asymmetry parameter g is the average value of cos(θ) at the optical wavelength of interest. Winter and Chýlek [24] calculated a value for sea-salt aerosol of g = 0.75. Fiebig and Ogren [25] reported values for g that were generally between 0.58 and 0.64 at one coastal monitoring site and between 0.62 and 0.67 at another. Andrews et al. [26] reported values between 0.59 and 0.72 for "wet" aerosols in Oklahoma. Moorthy et al. [27] reported values between 0.58 and 0.7 over the Indian Ocean. These are not exhaustive, but suggest that the value will generally be somewhere between 0.6 and 0.75 over the ocean.

As a first approximation, we will assume that the aerosols are distributed uniformly in height from the surface to some height H0. Figure 6 presents the depolarization for the expected range of g. We conclude that dA is not very sensitive to the vertical distribution of aerosols for heights above 200 m, and that will almost certainly be less than 0.005 under realistic conditions. Note that this is the depolarization of the scattered light. The aerosol optical depth of the data selected is clustered around 0.05, so only around 10% of the light backscattered from the ocean is also scattered by aerosols during the round trip path through the atmosphere. The actual measured depolarization would then be approximately a factor of ten less than that of the scattered light, or less than 5×10^{-4}. The average value of the measured depolarization values in the monthly one-degree bins, on the other hand, was 0.036 and only 1.8% of the values were less than ten times the maximum estimated aerosol-induced depolarization. Aerosol scattering, then, does not contribute significantly to the cross-polarized return from the CALIOP data used in this analysis.

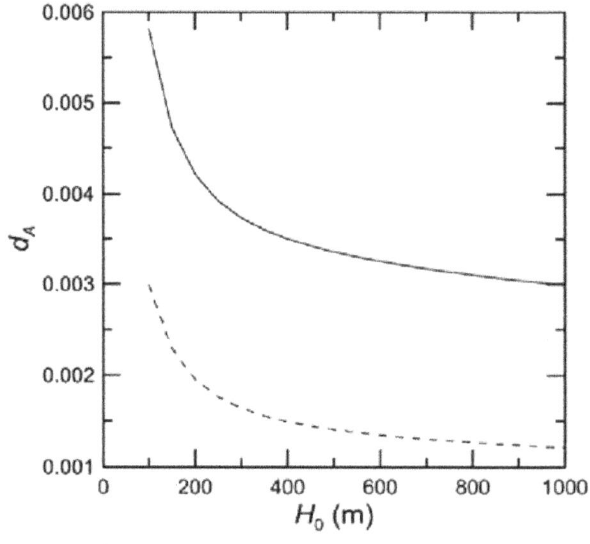

Figure 6. Plot of depolarization by atmospheric aerosols, d_A, as a function of the surface aerosol layer depth, H_0, for asymmetry parameter values of 0.6 (solid line) and 0.75 (dashed line).

As a further check, we recalculated the correlations using only cases with very low winds (<5 m·s^{-1}) and very clear atmospheric conditions (AOD < 0.025). This reduced the number of data pairs in each of the correlation calculations drastically, with the average dropping from 11,000 in the full data set to 2000 in thc reduce data set. The correlations between chlorophyll concentration and the three lidar parameters are reduced as a result, but the reductions are relatively small. For d, the change was $\rho = 0.43 \pm 0.06$ to 0.32 ± 0.08 for the average, ± the standard deviation of the 12 monthly values. For ΔR, the change was $\rho = 0.30 \pm 0.08$ to 0.26 ± 0.07 and for ΔW, it was $\rho = 0.18 \pm 0.08$ to 0.13 ± 0.09.

Receiver Limitations

One limit on the depth to which the CALIOP signal may be attributed to subsurface scattering is the ability to distinguish it from the tail of the impulse response of the photomultipliers used in the 532 nm receivers, and is evident in Figure 1a. The impulse response of each receiver was carefully measured from orbit for both polarizations using land-surface returns collected in 2006 and 2007 [28], and these impulse response functions can be approximated by:

$$f_T = \exp\left[-\left(\frac{z}{15}\right)^2\right] + 0.015\exp\left(\frac{z}{118}\right)$$

$$f_C = \exp\left[-\left(\frac{z}{15}\right)^2\right] + 0.014\exp\left(\frac{z}{110}\right)$$

for the total and cross-polarized channels, respectively. Note that z is negative below the surface as in Figure 1a. These functions include both a Gaussian approximation to the effects of the Bessel filter and the exponentially-decaying contribution from signal-induced fluorescence in the photomultipliers [29]. The photomultiplier decay depth of over 100 m is slower than the attenuation of the lidar return in even the clearest waters, since the round-trip attenuation depth would be only about 10 m for pure sea water, which has an attenuation coefficient of 0.05 m^{-1} at the lidar wavelength. Thus, the temporal response of the photomultiplier will dominate the return at large depths.

To investigate the limits imposed on depth penetration by the transient response, we considered the following scenario. The attenuated backscatter from the water column from an ideal detector was taken to be:

$$\beta_{WT} = 3.98\times10^{-4}\exp(0.168z)$$

where the water volume backscatter coefficient and attenuation are the average of reported values for pure sea water and clear ocean from the Petzold measurements [30]. The attenuation was calculated from absorption and backscatter using the Lee formula [31,32] for the diffuse attenuation coefficient. The corresponding cross-polarized return, β_{WX}, was taken to be 10% of the total return based on clear-ocean measurements with an airborne lidar [19]. The depth profile of the total surface return was taken as:

$$\beta_{ST} = \frac{\gamma}{\int f_t dz} f_t \,.$$

where γ is the integrated surface reflection coefficient from [15]. For the cross-polarized component, β_{SX}, we note that the surface return is a specular reflection that does not create depolarization. This implies that the cross-polarized surface reflection is the result of specular reflection of light that is depolarized by atmospheric aerosols. For this, we assumed a depolarization of 5×10^{-4}, as discussed in Section 3.1.The water volume returns for ideal detectors and surface returns for the CALIOP detectors are plotted for this case in Figure 7 with the assumption of an ideal receiver with unlimited range resolution.

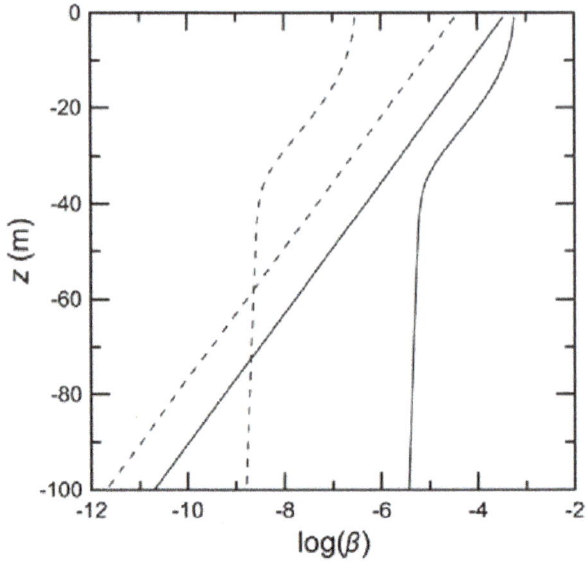

Figure 7. Depth profiles of lidar attenuated backscatter, β, for volume backscatter using an ideal detector (β_{WT} and β_{WX}, straight lines) and surface reflection using the Cloud-Aerosol Lidar with Orthogonal Polarization (CALIOP) detectors (β_{ST} and β_{SX}, curved lines). Solid lines are total return (β_{WT} and β_{ST}), and dashed lines are cross-polarized return (β_{WX} and β_{SX}).

The surface reflection is greater than the volume scattering at all depths for the total return. Enough subsurface scattering to be greater than the surface reflection would require a volume backscattering coefficient typical of very turbid waters. Very turbid waters also have a high attenuation, and only the subsurface scattering very close to the surface could contribute. This would be indistinguishable from the specular surface reflection, so there is no information about subsurface scattering in the total channel. The situation for the cross-polarized return is very different. Here the subsurface scattering is greater than the surface return from the surface to a depth of 58 m. This is fairly robust. This depth decreases, but only to 54 m, if we make conditions much less favourable by halving wind speed, doubling surface depolarization, halving water volume depolarization, or halving water volume backscatter coefficient.

Another way of looking at this issue is to compare the total depth-integrated subsurface return with the surface reflection. For uniform water properties, the total depth-integrated subsurface return is given by:

$$I_{WT} = \frac{\psi(\pi)b}{2K_D},$$

where Ψ is the scattering phase function at a scattering angle of π radians, b is the scattering coefficient, and K_D, the diffuse attenuation coefficient, describes the attenuation of the lidar signal for the CALIOP geometry [32]. For the purposes of comparison, we will use simple models based on chlorophyll concentration [33]. At 530 nm, we have:

$$b = 0.311C^{0.62},$$

and

$$K_D = 0.0518 + 0.0467C^{0.67}.$$

We can separate the scattering coefficient into sea water and particulate contributions and apply the appropriate phase-function values from Petzold [30] to get:

$$I_{WT} = \frac{2.94\times10^{-4} + 3.15\times10^{-3}\left(0.311C^{0.62} - 2.5\times10^{-3}\right)}{0.104+0.0934C^{0.67}}.$$

To get the corresponding cross-polarized signal, we will assume a constant depolarization of 0.1.

The relationship between the integrated surface reflection and wind can be inverted to obtain the wind speed at which the surface and subsurface returns would be equal. At wind speeds above this value, we would expect to see a significant return from below the surface. At lower wind speeds, we would not. Figure 8 shows these regimes for the total and cross-polarized signals, assuming a surface depolarization of 0.005. This figure clearly shows that, under most ocean conditions, the total return from the lidar will be dominated by the surface reflection and the cross-polarized return will be dominated by subsurface scattering.

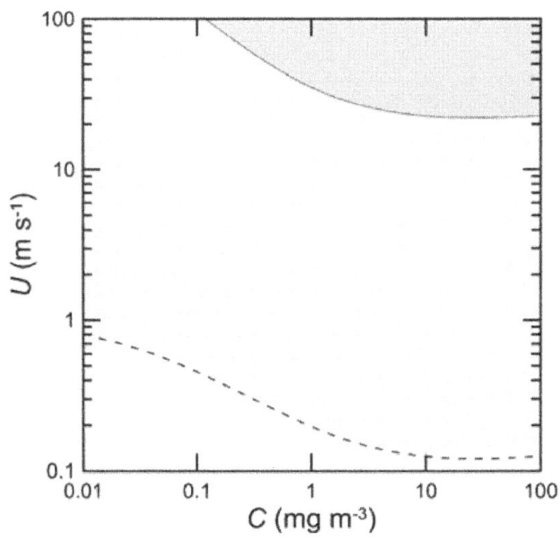

Figure 8. Wind speed, U, where the integrated subsurface signal will be greater than the integrated surface return for total (above the solid line) and cross-polarized (above the dashed line) channels as functions of chlorophyll concentration, C.

Examples

As an example of detection of subsurface scattering layers with CALIOP, we looked for single-shot lidar profiles that, in addition to our other selection criteria, had a depolarization greater than 0.1 at the depth of 50 m. At the lidar depth bin above this, a peak in depolarization can result from the delay and spreading of the near-surface return and does not necessarily imply a peak in scattering. At the depth bin below, the lidar returns can be contaminated by the photomultiplier response. This suggests that the 50 m depth bin is most likely to show a clear signature of a subsurface scattering layer. Therefore, we selected two lidar profiles that were near hydrographic stations reported in the World Ocean Database [34] which included profiles of chlorophyll inferred from fluorescence measurements. The first profile (Figure 9) was taken on 15 July 2010 in the Pacific Ocean 275 km off the coast of California (35.553°N, 124.740°W) on the R/V Point Sur by the US Navy Postgraduate School. The second profile was taken on 28 May 2010 in the Atlantic Ocean 150 km off the coast of Delaware (38.562°N, 73.207°W) on the R/V Delaware II by the Woods Hole Oceanographic Institution under the Ecosystem Monitoring Program (ECOMON). In both cases, there is a maximum in the chlorophyll concentration between 40 and 50 m in depth, which roughly corresponds to the peak in the depolarization. It is interesting that the layer with higher chlorophyll concentration produced more depolarization, but techniques for quantitative retrieval of chlorophyll concentration have yet to be developed.

These examples also provide some insight into the water clarity required to detect layers at depths near 50 m. For the example in the Pacific, the average chlorophyll concentration in the top 30 m was 0.17 mg·m^{-3}. The integrated two-way transmission to this depth, was 0.019. The example in the Atlantic had an average chlorophyll concentration in the top 30 m of 0.19 mg·m^{-3} and a two-way transmission of 0.017.

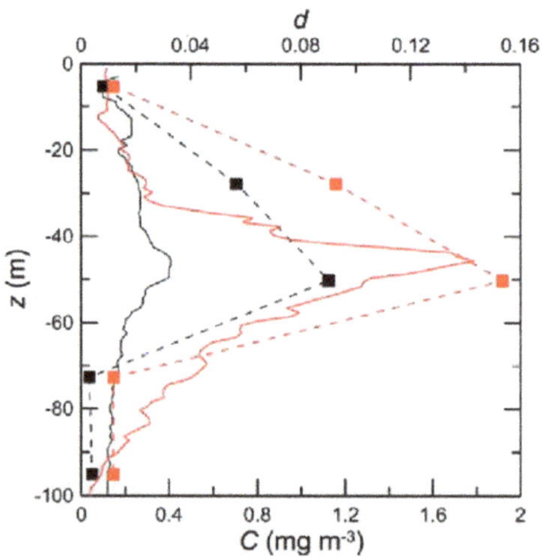

Figure 9. Depth profiles of lidar depolarization, d, (squares connected by dashed line) and chlorophyll concentration, C, (lines) from May 2010 in the NW Atlantic Ocean (black) and July 2010 in the NE Pacific (red).

CONCLUSIONS

The primary conclusion of this study is that subsurface particles can be detected by the CALIOP. Characteristics of the scattering from these particles were shown to be related to chlorophyll concentration near the surface. An increased depolarization (>0.1) at a depth of 50 m was observed under certain conditions. In the two examples presented, this depolarization peak corresponds to a peak in the chlorophyll profile, and the example with the higher chlorophyll concentration also had a greater depolarization. This implies that rough profiles of subsurface scattering should be possible to a depth of 50 m. Retrieval of these profiles to infer properties of the water column will require more study.

There are some important limitations, however. The total return is dominated by the strong surface reflection; so useful information will generally only be in the cross-polarized channel. Returns from depths greater than 50 m are not possible under any conditions as a result of the temporal response of the detectors. The correlations are less than unity, so the subsurface scattering is not completely determined by surface chlorophyll. This might be because of changes of chlorophyll concentration with depth and differences in phytoplankton taxa that produce different scattering properties for the same chlorophyll concentration.

The primary effects of subsurface scattering on the broad cross-polarized surface return are a delay of the peak of the return relative to the unpolarized return, a broadening of the temporal pulse, and a depolarization in the return from a nominal depth of 28 m. These are all positively correlated with surface chlorophyll concentration, and are likely to be at least partly a result of scattering by phytoplankton. These correlations are not artifacts of aerosol scattering in the atmosphere and are not artifacts of the temporal response of the detectors. Of the three effects, depolarization seemed to provide the best indicator of subsurface scattering.

ACKNOWLEDGMENTS

This work was partially supported by NASA under project number NNH10AO57I. CALIOP data were obtained from NASA Langley Research Center Atmospheric Science Data Center. MODIS AQUA chlorophyll concentrations were obtained from NASA Goddard Space Flight Center. Chlorophyll profiles were provided by the NOAA World Ocean Database. We would like to thank Yongxiang Hu for helpful discussions.

REFERENCES

1. Brown, E.D.; Churnside, J.H.; Collins, R.L.; Veenstra, T.; Wilson, J.J.; Abnett, K. Remote sensing of capelin and other biological features in the north Pacific using lidar and video technology. ICES J. Mar. Sci **2002**, 59, 1120–1130.
2. Churnside, J.H.; Demer, D.A.; Mahmoudi, B. A comparison of lidar and echosounder measurements of fish schools in the Gulf of Mexico. ICES J. Mar. Sci **2003**, 60, 147–154.
3. Churnside, J.H.; Thorne, R.E. Comparison of airborne lidar measurements with

420 kHz echo-sounder measurements of zooplankton. Appl. Opt **2005**, 44, 5504–5511.

4. Churnside, J.H.; Sharov, A.F.; Richter, R.A. Aerial surveys of fish in estuaries: A case study in Chesapeake Bay. ICES J. Mar. Sci **2011**, 68, 239–244.
5. Churnside, J.H.; Brown, E.D.; Parker-Stetter, S.; Horne, J.K.; Hunt, G.L.; Hillgruber, N.; Sigler, M.F.; Vollenweider, J.J. Airborne remote sensing of a biological hot spot in the southeastern Bering Sea. Remote Sens **2011**, 3, 621–637.
6. Churnside, J.H.; Donaghay, P.L. Thin scattering layers observed by airborne lidar. ICES J. Mar. Sci **2009**, 66, 778–789.
7. Churnside, J.H.; Ostrovsky, L.A. Lidar observation of a strongly nonlinear internal wave train in the Gulf of Alaska.Int. J. Remote Sens **2005**, 26, 167–177.
8. Churnside, J.H.; Marchbanks, R.D.; Lee, J.H.; Shaw, J.A.; Weidemann, A.; Donaghay, P.L. Airborne lidar detection and characterization of internal waves in a shallow fjord. J. Appl. Remote Sens **2012**, 6, 063611–063615.
9. Krekova, M.M.; Krekov, G.M.; Shamanaev, V.S. Influence of air bubbles in seawater on the formation of lidar returns.J. Atmos. Ocean. Tech **2004**, 21, 819–824.
10. Churnside, J.H. Lidar signature from bubbles in the sea. Opt. Express **2010**, 18, 8294–8299.
11. Hunt, W.H.; Winker, D.M.; Vaughan, M.A.; Powell, K.A.; Lucker, P.L.; Weimer, C. CALIPSO lidar description and performance assessment. J. Atmos. Ocean. Tech **2009**, 26, 1214–1228.
12. Rodier, S.; Zhai, P.; Josset, D.; Hu, Y.; Vaughan, M. CALIPSO Lidar Measurements for Ocean Subsurface Studies. Proceedings of the 34th International Symposium on Remote Sensing of the Environment, Sydney, NSW, Australia, 10–15 April 2011.
13. Shi, W.; Wang, M. Three-dimensional observations from MODIS and CALIPSO for ocean responses to Cyclone Nargis in the Gulf of Martaban. Geophys. Res. Lett **2008**, 35, L21603.
14. Barton, J.S.; Jasinski, M.F. Sensitivity of depth-integrated satellite lidar to subaqueous scattering. Remote Sens **2011**, 3, 1492–1515.
15. Hu, Y.; Stamnes, K.; Vaughan, M.; Pelon, J.; Weimer, C.; Wu, D.; Cisewsk, M.; Sun, W.; Yang, P.; Lin, B.; et al. Sea surface wind speed estimation from space-based lidar measurements. Atmos. Chem. Phys. Discuss **2008**, 8, 2771–2793.
16. Hu, Y.; Powell, K.; Vaughan, M.; Tepte, C.; Weimer, C.; Beherenfeld, M.; Young, S.; Winker, D.; Hostetler, C.; Hunt, W.;et al. Elevation information in tail (EIT) technique for lidar altimetry. Opt. Express **2007**, 15, 14504–14515.
17. Krekov, G.M.; Krekova, M.M.; Shamanaev, V.S. Laser sensing of a subsurface oceanic layer. II. Polarization characteristics of signals. Appl. Opt **1998**, 37, 1596–1601.
18. Vasilkov, A.P.; Goldin, Y.A.; Gureev, B.A.; Hoge, F.E.; Swift, R.N.; Wright, C.W. Airborne polarized lidar detection of scattering layers in the ocean. Appl.

Opt **2001**, 40, 4353–4364.

19. Churnside, J.H. Polarization effects on oceanographic lidar. Opt. Express **2008**, 16, 1196–1207.
20. NASA Langley Research Center. CALIPSO Search and Subsetting Web Application. Available online: https://www-calipso.larc.nasa.gov/search/ (accessed on 2 December 2012).
21. Feldman, G.C.; McClain, C.R. Ocean Color Web, MODIS Reprocessing #3; Kuring, N., Bailey, S.W., Eds.; NASA Goddard Space Flight Center: Greenbelt, MD, USA, 2013; Available online: http://oceancolor.gsfc.nasa.gov/ (accessed on 18 April 2013).
22. Wentz, F.; Meissner, T. AMSR-E/Aqua Level 2b Global Swath Ocean Products Derived from Wentz Algorithm V002; National Snow and Ice Data Center: Boulder, CO, USA, 2004.
23. Henyey, L.C.; Greenstein, J.L. Diffuse radiation in the galaxy. Astrophys. J **1941**, 93, 70–83.
24. Winter, B.; Chýlek, P. Contribution of sea salt aerosol to the planetary clear-sky albedo. Tellus B **1997**, 49, 72–79.
25. Fiebig, M.; Ogren, J.A. Retrieval and climatology of the aerosol asymmetry parameter in the NOAA aerosol monitoring network. J. Geophys. Res.-Atmos **2006**, 111, D21204.
26. Andrews, E.; Sheridan, P.J.; Fiebig, M.; McComiskey, A.; Ogren, J.A.; Arnott, P.; Covert, D.; Elleman, R.; Gasparini, R.; Collins, D.; et al. Comparison of methods for deriving aerosol asymmetry parameter. J. Geophys. Res.-Atmos **2006**, 111, D05S04.
27. Moorthy, K.K.; Nair, V.S.; Babu, S.S.; Satheesh, S.K. Spatial and vertical heterogeneities in aerosol properties over oceanic regions around India: Implications for radiative forcing. Q. J. Roy. Meteor. Soc **2009**, 135, 2131–2145.
28. Li, J.; Hu, Y.; Huang, J.; Stamnes, K.; Yi, Y.; Stamnes, S. A new method for retrieval of the extinction coefficient of water clouds by using the tail of the CALIOP signal. Atmos. Chem. Phys. Discuss **2011**, 11, 2903–2916.
29. Zhao, Y. Signal-induced fluorescence in photomultipliers in differential absorption lidar systems. Appl. Opt **1999**, 38, 1639–4648.
30. Mobley, C.D. Light and Water: Radiative Transfer in Natural Waters; Academic Press: San Diego, CA, USA, 1994; p. 592.
31. Lee, Z.-P.; Darecki, M.; Carder, K.L.; Davis, C.O.; Stramski, D.; Rhea, W.J. Diffuse attenuation coefficient of downwelling irradiance: An evaluation of remote sensing methods. J. Geophys. Res **2005**, 110, C02017.
32. Lee, J.H.; Churnside, J.H.; Marchbanks, R.D.; Donaghay, P.L.; Sullivan, J.M. Oceanographic lidar profiles compared with estimates from in situ optical measurements. Appl. Opt **2013**, 52, 786–794.
33. Morel, A. Optical modeling of the upper ocean in relation to its biogenous matter content (Case I waters). J. Geophys. Res **1988**, 93, 10749–10768.

34. WODselect Maintained by NOAA's National Oceanographic Data Center. Available online: http://www.nodc.noaa.gov/OC5/SELECT/dbsearch/dbsearch.html (accessed on 14 June 2013).

4

CHAPTER

SPECTRAL MODELS FOR IMPROVING ATMOSPHERIC CORRECTION RESULTS IN COMPLEX OCEANIC WATERS

Palanisamy Shanmugam, Muniyandi Tholkapiyan

Keywords: Ocean Color, Spectral Models, MODIS-Aqua, Algal Blooms, Arabian Sea, India

INTRODUCTION

Ocean color instruments measure the spectrum of sunlight reflected from the ocean-atmosphere system at several visible and near-infrared (NIR) wavebands. The major portion of these signals is from the contribution of scattering by molecules and particles (aerosols) in the atmosphere (~80% - 90%). The process of removing the atmospherically scattered signal in order to retrieve the ocean color signal (i.e., water-leaving radiance L_w) is referred to as atmospheric correction [1]. The current SeaDAS atmospheric correction ("SAC" for brevity) algorithm for producing the global ocean color products from SeaWiFS and MODIS uses the algorithm of Gordon and Wang [2]. It specifically uses two near infrared (NIR) bands centered at MODIS 748 and 869 nm (SeaWiFS 765 and 865 nm) to determine aerosol type and estimate the atmospheric effects in the visible by extrapolating the aerosol effect from the NIR into visible bands. Though the basic assumption of negligible waterleaving radiance (L_w) in the NIR for deriving aerosol properties was replaced by Stumpf et al. [3] and then updated by Bailey et al. [4], the SAC algorithm still produces low or negatively biased L_w at short wavelengths in areas with absorbing aerosols or high biomass concentrations or sediment levels typical of productive coastal regions.

Citation: P. Shanmugam and M. Tholkapiyan, "Spectral Models for Improving Atmospheric Correction Results in Complex Oceanic Waters around India," Advances in Remote Sensing, Vol. 2 No. 2, 2013, pp. 51-57, doi:10.4236/ars.2013.22007.

The principal difficulty in these waters is: 1) the adequacy of aerosol models is presently difficult to judge. This holds true for mineral dust aerosols (from deserts) because their absorption can seriously reduce $L_a + L_{ra}$ (atmospheric signal due to aerosol and Rayleigh scattering) in the visible and their vertical distribution profoundly influences the TOA radiance in the visible (especially in the blue) [1,5]. Moreover, the absorption properties are very difficult to determine on the basis of the observations of $L_a + L_{ra}$ in the NIR which means that in-situ measurements in the visible are required for correction. However, in the visible (especially blue) L_w can be significant, and cannot be estimated a priori [1]. 2) The errors involved in extrapolating aerosol reflectance to visible bands are considerably higher and this may require a more realistic method [6]. 3) Assumption of negligible water-leaving radiance in the NIR is not valid in optically complex waters, as high sediment loads and phytoplankton will contribute NIR radiance [6-8]. 4) Although modifications to point 3 (i.e., "non-zero L_w in the NIR" based on in-water models and spatial homogeneity assumption) [6,9] show noticeable improvements in turbid waters, the problems are still encountered in highly bloomed waters and could be attributed to the inadequacy of aerosol models and inadequate/inappropriate in-water models. Thus, an accurate retrieval of ocean color has become a prolonging issue and requires an alternate and robust solution.

In this study, spectral models are developed to improve the results of atmospheric correction with standard methods and their results are validated with in-situ data sets. Preliminary results are also presented for a case study using MODIS-Aqua data over highly turbid and productive waters in the Arabian Sea.

DATA AND METHODS

In-Situ Data

In-situ data radiometric data at key wavelengths and pigment (Chlorophyll) concentrations for different regions were obtained from the NASA bio-Optical Marine Algorithm Data set (NOMAD) [10]. These data are global, high quality bio-optical datasets collected over a wide range of optical properties, trophic status, and geographical locations in open ocean waters, estuaries, and coastal waters. Despite the NOMAD data encompass large samples, only few cruises data (between 2003 and 2007) were considered because satellite sampling for a given station was masked by clouds or affected by sensor digitization problems.

Satellite Data

MODIS-Aqua Level 1A data data were obtained from the US NASA GSFC (Goddard Space Flight Center), and processed using SeaDAS5 to generate Level 2 data (water-leaving radiance L_w). Many of these data were collected coincidently with the above field data and few were acquired in highly bloomed waters of the Arabian Sea. For the reason that L_w data at the short wavelengths are profoundly deteriorated by the SAC algorithm, all Level 2 L_w data were reprocessed based on the new spectral models and then converted to remote sensing reflectance (R_{rs} (l)) using the standard procedures available with SeaDAS. In theory, the spectral models developed in this study can be applied to the entire global ocean color dataset. In practice, however, due to large volume of MODIS data, a few coastal and offshore regions were selected to demonstrate their accuracy and reliability.

SPECTRAL MODELS

To avoid known problems with atmospheric correction at the short wavelengths (note the inaccuracy of the atmospheric correction grows for decreasing wavelengths so that the 412 nm waveband is the most affected one [11], a correction is carried out in two steps based on the definition of water into two types: relatively clear waters (Chl < 0.5 mg/m^3) and optically complex waters (bloom, plume and colored waters where Chl > 0.5 mg/m^3 and negative L_w (412) as produced by the SAC algorithm). Accordingly, two separate sets of spectral models are developed considering the fact that the water-leaving radiance for clear waters is weakly variable with a smooth decrease towards the shorter wavelengths (412 nm) with increasing pigment concentration which is in contrast with that for optically complex waters. These models take into account a change (distortion) in spectral shape characteristics in the blue wavelengths (412 and 443 nm) for estimating the atmospheric correction distortion and reconstructing the L_w spectra in the visible domain (412 - 555 nm). For optically complex waters (Chl > 0.5 mg/m^3; negative L_w (412)), the spectral models rely on the normalized spectral difference between L_w (412) and L_w (443) multiplied by its wavelength ratio factor which is then added with measurements at each wavelength, as shown in Equation (1), where $\lambda_1 = 412$, $\lambda_2 = 443$ and $\kappa = 0.85$. To compensate for a maximum distortion at 412 nm and ensure spectral continuity, the normalized spectral values (sum) of L_w (412) and L_w (443) and coefficient κ are used in the above equations. These models ensure a maximal shift in L_w (toward positive) when its distortion is elevated and a minimal shift in L_w when it is undistorted.

$$\left.\begin{aligned}
n_L_w(412) &= \left[\left(\left(L_w(412)+L_w(443)\right)/2\right)+\left(\left(L_w(443)-L_w(412)\right)/2\right)\right]\times[\lambda_1/\lambda_2]\times\kappa \\
n_L_w(443) &= \left[\left(L_w(443)\right)+\left(\left(L_w(443)-L_w(412)\right)/2\right)\right]\times[\lambda_1/\lambda_2] \\
n_L_w(488) &= \left[\left(L_w(488)\right)+\left(\left(L_w(443)-L_w(412)\right)/2\right)\right]\times[\lambda_1/\lambda_2] \\
n_L_w(531) &= \left[\left(L_w(531)\right)+\left(\left(L_w(443)-L_w(412)\right)/2\right)\right]\times[\lambda_1/\lambda_2] \\
n_L_w(551) &= \left[\left(L_w(551)\right)+\left(\left(L_w(443)-L_w(412)\right)/2\right)\right]\times[\lambda_1/\lambda_2]
\end{aligned}\right\} \quad (1)$$

$$\left.\begin{aligned}
n_L_w(412) &= L_w(412)+\left[(551-489)/(551-412)\right]\times\left[(412/551)^2-\left(L_w(551)/L_w(489)\right)^{0.5}\right] \\
n_L_w(443) &= L_w(443)+\left[(551-489)/(551-412)\right]\times\left[(412/551)^2-\left(L_w(551)/L_w(489)\right)^{0.5}\right] \\
n_L_w(489) &= L_w(489)+\left[(551-489)/(551-412)\right]\times\left[(412/551)^2-\left(L_w(551)/L_w(489)\right)^{0.5}\right] \\
n_L_w(531) &= L_w(531)+\left[(551-489)/(551-412)\right]\times\left[(412/551)^2-\left(L_w(551)/L_w(489)\right)^{0.5}\right] \\
n_L_w(551) &= L_w(551)+\left[(551-489)/(551-412)\right]\times\left[(412/551)^2-\left(L_w(551)/L_w(489)\right)^{0.5}\right]
\end{aligned}\right\} \quad (2)$$

For relatively clear waters (Chl < 0.5 mg/m³), a correction is achieved based on a set of spectral models (one for each band) to obtain the best spectral shape of L_w (412 - 551 nm). These models that readjust the spectral shape of the L_w from the SAC algorithm are defined in Equation (2).

The newly derived water-leaving radiance values will be helpful to reduce errors in bio-geochemical products such as chlorophyll concentration in optically complex waters (**Figure 1**). For example, the OC3 algorithm [12] estimates Chl a from a cubic polynomial formulation by using the maximum band ratio determined as the greater of the $R_{rs}(443)/R_{rs}(551)$ or $R_{rs}(488)/R_{rs}(551)$. It is then feasible to obtain error in the atmospheric correction and in the retrievals of pigment concentrations in both clear and optically complex waters.

RESULTS

The comparison between original (SAC) and reconstructed R_{rs} (spectral models) is shown in **Figure 2** for bloomed and turbid waters in the Arabian Sea, using several MODIS-Aqua data collected during the year 2007. It reveals dramatic anomalous differences between these two spectra, i.e., R_{rs} produced by the SAC algorithm are significantly biased low and negative at short wavelengths (412 and 443 nm) and this resulted in a large distortion of the radiance structures across the visible spectrum (left). In particular, most of the distorted spectra were observed in bloomed waters. The effect of the spectral models on raising radiance values particularly at 412 and 443 is clearly seen in **Figure 2** (right). The physically realistic shape of R_{rs}(l) is also typical for waters with phytoplankton blooms in the Arabian Sea [13].

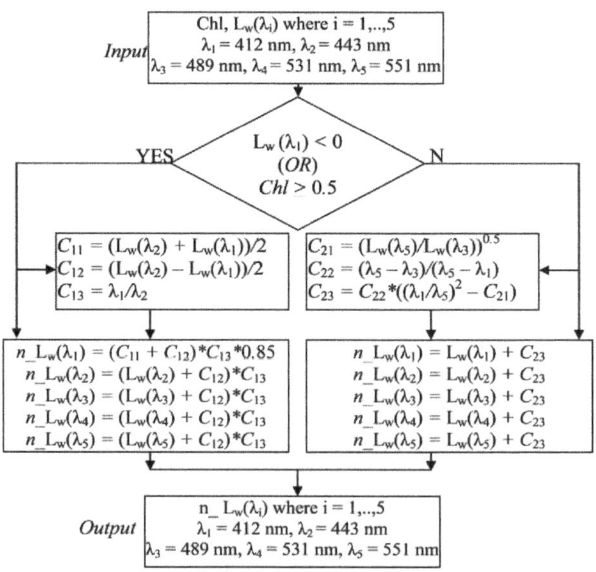

Figure 1. Flowchart showing an iterative method to reconstruct the atmospherically distorted spectra. n_Lw is the reconstructed (new) water-leaving radiance.

Because the quality of the atmospheric correction affects the quality of the R_{rs}(l) which is key to the estimation of pigments and other geophysical products, satellite-derived remote sensing reflectances were compared with those from the in-situ measurements in both clear and optically complex waters. **Figure 3** shows comparisons between the original (SAC) R_{rs}(l) and reconstructed R_{rs}(l) and in-situ R_{rs}(l) at the selected five wavelengths. Although there is no remarkable difference, the results in these plots show a reasonably good matchup between reconstructed and in-situ reflectance measurements. Most of the corrected R_{rs}(l) values are aligned closer to the 1:1 line as compared with those from the SAC algorithm. To gain further insight into the differences between the satellite R_{rs}(l) and in-situ R_{rs}(l), the mean relative error (MRE), root means square error (RMSE), slope (S), intercept (I), correlation coefficient (R^2) and bias were calculated using the standard statistical procedures and the results are summarized in Tables 1(a) and (b). Notably, these statistics are better for the reconstructed R_{rs}(l) than for the SAC algorithm (see relatively lower MRE, RMSE and I, higher S and R^2, and almost same biases). This clearly indicates significant improvements over the SAC algorithm.

The effect of atmospheric correction errors on pigment retrievals was also assessed using the OC3 algorithm as applied to both the original R_{rs}(l) and reconstructed R_{rs}(l) (Figures 4(a) and (b)). As noted earlier, the uncorrected R_{rs}(l) showed noticeable overestimations in Chl with respect to in situ measurements in bloomed/coastal waters [13], although producing Chl consistent with the in-situ data in clear waters.

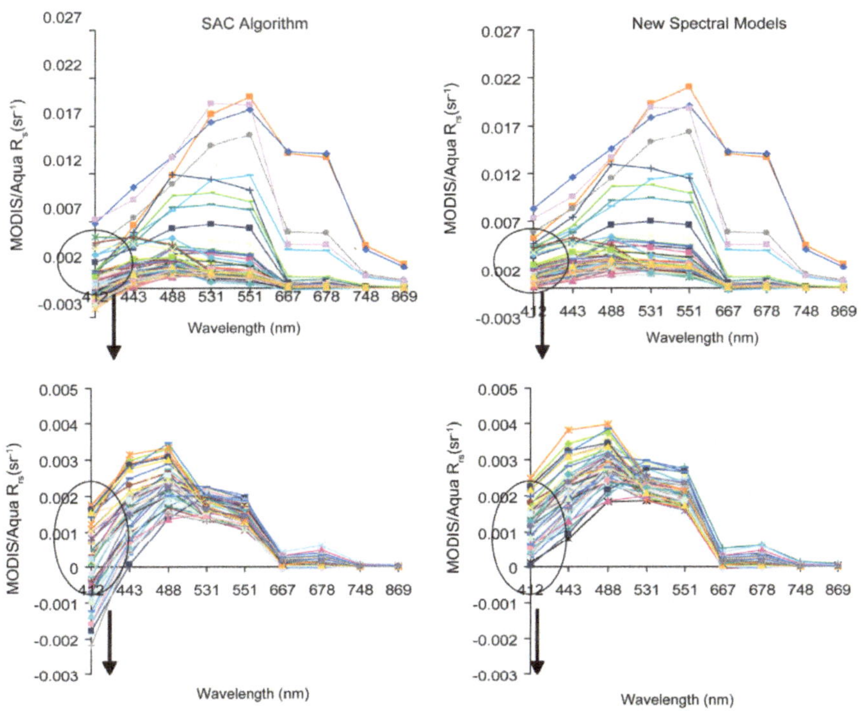

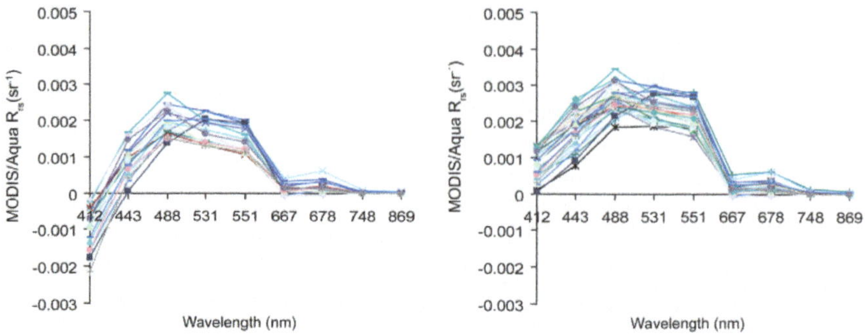

Figure 2. Examples of the remote sensing reflectance (R_{rs}) spectra obtained from several MODIS-Aqua data using the SAC algorithm (left) and the spectral models (right). MODIS-Aqua data from bloomed and coastal waters in the Arabian Sea are: A2007006065500, A2007007073500, A2007008064500, A2007009073000, A2007017095500, A2007018072000, A2007021075- 000, A2007022065500, A2007023074000, A2007024064500, A2007025073000, A2007026063500, A2007028075500, and A2007- 029070500.

By contrast, the validation matchups between in-situ Chl and satellite Chl from the reconstructed R_{rs}(l) are very excellent and demonstrate good agreement in both clear and optically complex waters (lower MRE, RMSE, I and bias, and higher S and R^2). The averaged minimum, mean and maximum Chl values for this case are also in good agreement with in-situ data (**Table 1**(b)), with its mean value being 1.65 times lower than that produced by the former case.

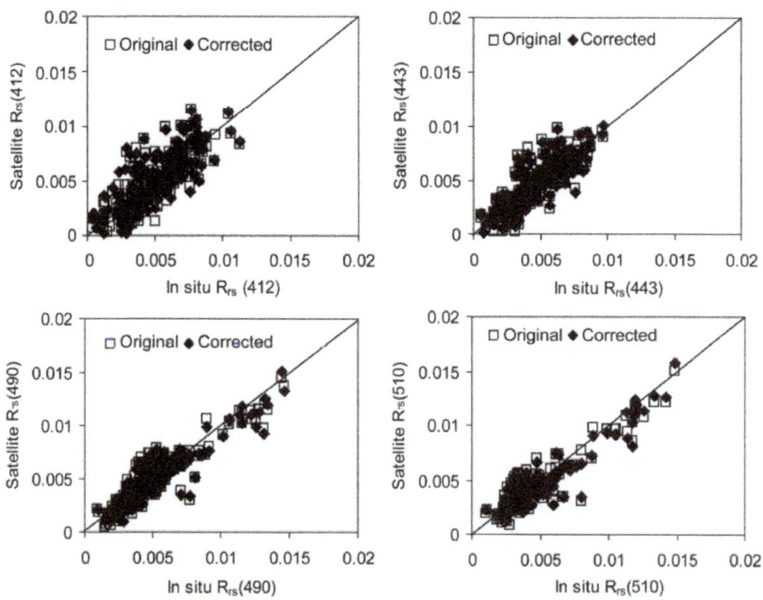

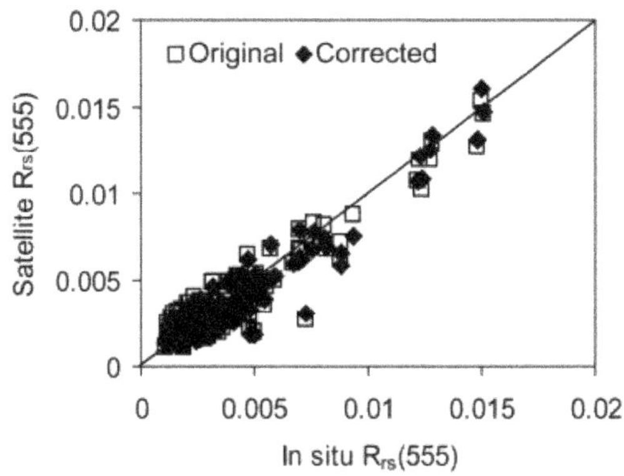

Figure 3. MODIS-derived remote sensing reflectance (R_{rs} in sr^{-1}) values compared with in-situ measurements for the wavelengths of 412, 443, 490, 510 and 555 nm. The in-situ data collected in both clear and optically complex waters were obtained from the NOMAD database.

Shown in Figures 4(c) and (d) are two hypothetical cases from the bloom water (c) and clear water (d). It should be noted that when Chl > 5 mg/m^3, the situation is even worse (also see **Figure 4**(b)) and an excessive overcorrection is often inevitable with the SAC algorithm because of its excessive aerosol correction. This overcorrection or reduction in L_w eventually appears as an elevated pigment concentrations as observed in Figures 4(a) and (b). On the contrary, the spectral models have potential to not only improve L_w retrievals (thus constituents' concentrations) in optically complex waters but also reliably reproduce these spectra in clear waters.

Optically complex waters are consistently turbid due to the presence of suspended sediments and algal blooms and have large NIR water-leaving radiances. The enhanced water signal interferes with the SAC algorithm to produce unacceptable errors in retrieved water-leaving radiance and pigment concentrations. This leaves only two chaises for users who need to derive variables such as seasonal mean concentrations or annual primary production: either (1) to completely discord the spectrally distorted L_w data or (2) to use them as is, in error. In either case the results will be biased. Our spectral models reduce errors in estimated water-leaving radiance for both optically complex and clear waters. The number of negative water-leaving radiance pixels drops significantly. Furthermore, data from many coastal areas with quality masks due to atmospheric correction failure can be recovered. Consequently, the spatial patterns in the masked areas, even if the accuracy in the estimated waterleaving radiance or pigment concentrations is doubtful, can be derived. The improvement in the water-leaving radiance data will be helpful to improve the accuracy bio-geo-physical products. This scheme is easy to use for any operational processing of the satellite ocean color data on daily basis.

CONCLUSION

In conclusion, the new models show promise in reconstructing the spectrally distorted water-leaving radiance in both clear and optically complex waters. The spectral shape/form produced by these models is found to be better consistent with in-situ measurements in the visible domain (412 - 555 nm) that allows more accurate estimations of constituents' concentrations. The minimal errors of the satellite validation matchups imply that these spectral models can be used to improve the various biogeophysical products desired for coastal oceanic waters. Though the spectral models have been described for MODIS/SeaWiFS imagery, it can be applied rather generally to other satellite-based and probably also airborne ocean color sensors (with only two near-infrared bands (i.e., 765 and 865 nm) and no shortwave bands). The directions for further research include a more detailed validation of these models with large in-situ data sets from highly turbid and productive waters, and an attempt to tune it to have a wider applicability. The effect of absorbing aerosols on the retrieved water-leaving radiance will be assessed and further refinements of the models will be anticipated.

Table 1. (a) and (b) Comparison of algorithm performances using in-situ data sets (e.g., remote sensing reflectance R_{rs} (sr^{-1}) and Chlorophyll concentration (mg/m^3))

(a)

	MRE	RMSE	Slope	R^2	Bias	Sample density
			Seadas R_{rs}			
R_{rs} (412)	0.0298	0.1846	0.5017	0.5882	0.0700	194
R_{rs} (443)	0.0111	0.1836	0.5636	0.5628	0.0261	194
R_{rs} (490)	0.0041	0.1067	0.7947	0.7214	0.0096	194
R_{rs} (510)	0.0084	0.1105	0.8246	0.6890	0.0202	194
R_{rs} (555)	0.0027	0.1231	0.9457	0.7818	0.0069	194
			Corrected R_{rs}			
R_{rs} (412)	0.0098	0.2196	0.5747	0.4735	0.0230	194
R_{rs} (443)	0.0076	0.1553	0.6582	0.6045	0.0177	194
R_{rs} (490)	0.0043	0.1021	0.8309	0.7321	0.0099	194
R_{rs} (510)	0.0085	0.1064	0.8601	0.7011	0.0214	194
R_{rs} (555)	0.0022	0.1186	0.9800	0.7957	0.0074	194
			Chlorophyll			
Old	−0.123	0.2096	0.8308	0.8590	−0.043	194
New	−0.124	0.1900	0.8655	0.8769	−0.041	194

(b)

	In-situ	OC3-Chl from SeaDAS R_{rs}	OC3-Chl from corrected R_{rs}
Minimum	0.076	0.0602	0.079
Maximum	10.62	32.037	19.39
Mean	0.958107	1.405	1.225

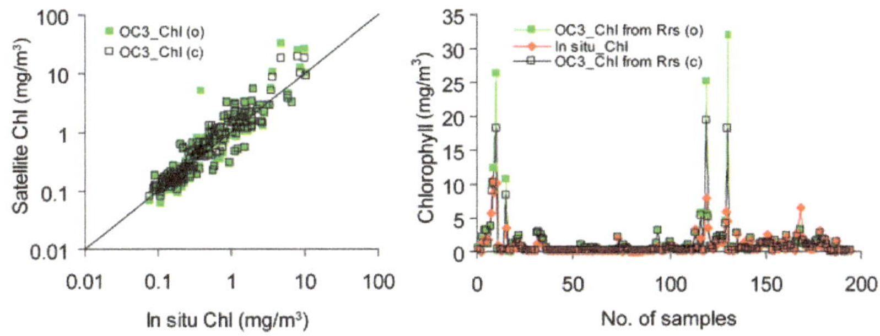

(a)

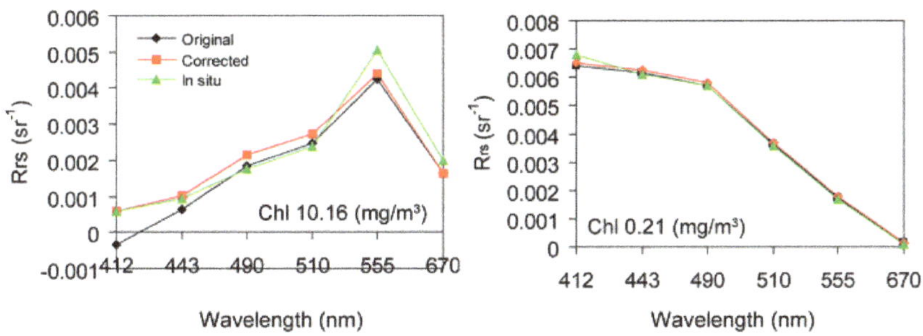

(b)

Figure 4. Scatterplot illustrating the comparison of estimated Chl concentrations by OC3 algorithm (using original R_{rs}(l) from the SAC algorithm and reconstructed R_{rs}(l)) to in-situ Chl concentrations (a), Chlorophyll at different stations obtained from the in-situ measurements and estimated from the OC3 algorithm by using the original and reconstructed R_{rs}(l), (c) and (d) R_{rs} spectra for two typical cases—Algal bloom (c) and clear waters (b).

ACKNOWLEDGEMENTS

This work was supported by the ISRO-IITM cell project (ICSR/ISRO-IITMJPC34/2009). The author gratefully acknowledges the NASA Ocean Biology Processing Group for making available the global, high quality in situ bio-optical (NOMAD) dataset as well as the global coverage MODIS-Aqua data to this study.

REFERENCES

1. H. R. Gordon, "Atmospheric Correction of Ocean Color Imagery in the Earth Observing System Era," Journal of Geophysical Research, Vol. 102, No. D14, 1997, pp. 17081-17106. doi:10.1029/96JD02443
2. H. R. Gordon and M. Wang, "Retrieval of Water-Leaving Radiance, Aerosol Optical Thickness over the Oceans with SeaWiFS: A Preliminary Algorithm," Applied Optics, Vol. 33, No. 3, 1994, pp. 443-452. doi:10.1364/AO.33.000443

3. S. W. Bailey, B. A. Franz and P. J. Werdell, "Updated NIR Water Leaving Radiance Estimation for Ocean Color Data Processing," Optics Express, Vol. 18, No. 7, 2010, pp. 7521-7527. doi:10.1364/OE.18.007521

4. C. Moulin, H. R. Gordon, V. F. Banzon and R. H. Evans, "Assessment of Saharan Dust Absorption in the Visible from SeaWiFS Imagery," Journal of Geophysical Research: Atmospheres, Vol. 106, No. D16, 2001, pp. 18239-18249. doi:10.1029/2000JD900812

5. K. G. Ruddick, F. Ovidio and M. Rijkeboer, "Atmospheric Correction of SeaWiFS Imagery for Turbid Coastal and Inland Waters," Applied Optics, Vol. 39, No. 6, 2000, pp. 897-913.doi:10.1364/AO.39.000897

6. R. A. Arnone, P. Martinolich, R. W. Gould, R. Stumpf and S. Ladner, "Coastal Optical Properties Using SeaWiFS," Proceedings of Ocean Optics XIV Conference, Kailua-Kona, Washington DC, 10-13 November 1998, pp. 10-13.

7. D. A. Siegel, G. M. Wang, S. Maritorena and W. Robinson, "Atmospheric Correction of Satellite Ocean Color Imagery: The Black Pixel Assumption," Applied Optics, Vol. 39, No. 21, 2000, pp. 3582-3591. doi:10.1364/AO.39.003582

8. C. Hu, K. L. Carder and F. E. Muller-Karger, "Atmospheric Correction of SeaWiFS Imagery over Turbid Waters: A Practical Approach," Remote Sensing of Environment, Vol. 74, No. 2, 2000, pp. 195-206. doi:10.1016/S0034-4257(00)00080-8

9. P. J. Werdell and S. W. Bailey, "An Improved In-Situ Bio-Optical Data Set for Ocean Color Algorithm Development, Satellite Data Product Validation," Remote Sensing of Environment, Vol. 98, No. 1, 2005, pp. 122- 140. doi:10.1016/j.rse.2005.07.001

10. A. Morel and B. Gentili, "A Simple Band Ratio Technique to Quantify the Colored: Dissolved and Detrital Organic Material from Ocean Color Remotely Sensed Data," Remote Sensing of Environment, Vol. 113, No. 5, 2009, pp. 998-1011.doi:10.1016/j.rse.2009.01.008

11. J. E. O'Reilly, S. Maritorena, B. G. Mitchell, D. A. Siegel Kendall, L. Carder, S. A. Garver, M. Kahru and C. McClain, "Ocean Color Chlorophyll Algorithms for SeaWiFS," Journal of Geophysical Research: Ocean, Vol. 103, No. C11, 1998, pp. 24937-24953.

12. P. Shanmugam, "A New Bio-Optical Algorithm for the Remote Sensing of Algal Blooms in Complex Ocean Waters," Journal of Geophysical Research, Vol. 116, No. C4, 2001, Article ID: C04016. doi:10.1029/2010JC006796

13. Z. P. Lee, "Remote Sensing of Inherent Optical Properties: Fundamentals, Tests of Algorithms and Applications," International Ocean-Color Coordinating Group, Dartmouth, 2006.

5
CHAPTER

SOIL MOISTURE RETRIEVAL QUANTITATIVELY WITH REMOTELY SENSED DATA AND ITS CRUCIAL FACTORS ANALYSIS

Ji JIAN, Peifen PAN, Yuanyuan CHEN, Wunian YANG

Keywords: Soil moisture, quantitative Remote Sensing, NDVI

INTRODUCTION

The Biospheric Aspects of the Hydrological Cycle (BAHC), one of the core projects of the International Geosphere Biosphere Programme (IGBP) coordinated by the International Council for Science (ICSU), was established to study the role of vegetation in the hydrological interactions between the land surfaces and atmosphere. One objective of BAHC is to determine the biospheric controls of the hydrological cycle through field measurements for the purpose of developing models of energy and water fluxes in the soil-vegetation-atmosphere system at temporal and spatial scales ranging from vegetation patches to General Circulation Model (GCM) grid cells [1,2]. This encouraged us to focus not only on the water itself, but also on its correlation factors such as the vegetation, the ground and so on when studying water environments and water resources, especially for water closely related to vegetation [3–9]. Soil moisture observations are one of these studies. Soil moisture is a key component of the hydrological cycle, controlling the partitioning of precipitation between runoff, evapotranspiration and deep infiltration. As a link between the biosphere and the edaphic zone, soil water plays a crucial role in terrestrial ecosystems by determining plant growth. If the soil water level falls below a species-specific threshold, plants experience water stress, and decreased soil moisture under warmer conditions can

Citation: J. JIAN, P. PAN, Y. CHEN and W. YANG, "Soil Moisture Retrieval Quantitatively with Remotely Sensed Data and Its Crucial Factors Analysis," Journal of Water Resource and Protection, Vol. 1 No. 6, 2009, pp. 439-447. doi:10.4236/jwarp.2009.16053.

inhibit photosynthesis [10]. Soil moisture observations over large areas are increasingly required in a range of environmental applications including meteorology, hydrology, water resource management and climatology. Remote sensing can provide considerable cost-and time-savings when applied to mapping soil moisture over large areas. Various approaches have been developed over the past two decades to infer near-surface soil moisture from remote sensing measurements of surface temperature, radar backscatter and microwave brightness temperature [11–15].

In this paper, the soil moisture in study area was retrieved with Landsat Enhance Thematic Mapper Plus (ETM+) data and Advanced Spaceborne Thermal Emission and Reflection Radiometer (ASTER) data first, then, the influence factors of soil moisture was analyzed. The objective of this paper is to provide a new method for soil moisture monitoring.

METHODS

Study Area

In this paper, the Maoergai area (from latitude 32°20' to 32°40' north and from longitude 103°00' to 103°30' east, shown as **Figure 1**), located in the upper Minjiang river, northern Sichuan province, in the southwest of China, was selected as the study area. The upper Minjiang river area has been famous as a "natural barrier" and the "green lake" of Sichuan province for a long time, where the forest area reached 12 thousand km², almost 50% of the whole 24.5 thousand km², 600 years ago. However, with the destruction of the forest, it had been reduced to only 7.4 thousand km², or about 30% of the whole area, by 1950. Meanwhile, the upper Minjiang is transforming from a pristine region to one with an increasing level of human activity, which leads to a degraded ecosystem in the area. The study area is a specific area in the upper Minjiang River with regard to its vegetation, soil, and mountainous characters, thus, the study result can be popularized in Minjiang river basin.

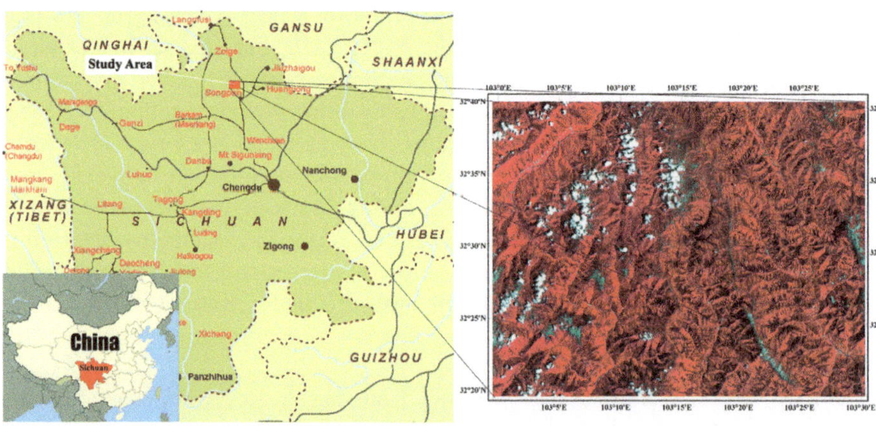

Figure 1. Location of the study area.

Data

In this study, we collected the soil type, groundwater and vegetation distribution data of the study area for the year 2000, which were collected within the gold-eye project subsidized by National High Technology Research and Development Program (HTRDP) [16], 48 ground control points (GCPs) distributed averagely in study area, the moisture of representative bare soils in the study area , the moisture of representative vegetations in the study area , all spectrums of representative soil and representative vegetation in the study area, Landsat ETM+ data of 10 July 2002 and ASTER data of 2 November 2003 of the study area for retrieving soil moisture. The remotely sensed data have been registered to UTM WGS-84 coordinates using the GCPs to allow co-registration of the ground data.

Retrieval and Analysis Processes

The retrieval and analysis processes were divided into 3 parts in this paper: First, pre-process of remotely sensed data, including geometric register, radiometric correction(radiometric calibration, topographic correction and atmospheric correction); Second, retrieval of biophysical parameter for land cover, including vegetation indices, soil brightness, soil wetness, greenness and surface temperature; Third, soil moisture retrieval with Ts/NDVI model and soil moisture analysis with soil type distribution data, vegetation distribution data and groundwater distribution data.

To better understand the soil moisture retrieval and analysis processes, a flow chart is provided in **Figure 2**.

RESULTS

Soil Moisture Retrieval from Landsat ETM+ Data

Surface temperature (T_s, Kelvin), normalized difference vegetation index (NDVI) and the feature special diagram retrieved from ETM+ data of July 2002 in study area were shown in **Figure 3**.

To calculate the relative soil moisture, the dry edge linear equation and wet edge linear equation of T_s / NDVI feature space should be calculated first. The dry and wet edge linear equations are calculated as follows:

Dry edge:

$$T_{s\text{-}max} = -18.356343 * NDVI + 307.57766 \quad r = 0.47030258 \qquad (1)$$

Wet edge:

$$T_{s\text{-}min} = 0.68168266 * NDVI + 288.26377 = 0.30658477 \qquad (2)$$

With the data of soil moisture and experiential formulas [1,2,4,8–10] in study area as a reference, the relative soil moisture of any point in ETM+ data's T_s / NDVI feature space was calculated as Equation 3, where RSM is the relative soil moisture of any point, T is the surface temperature of the point, T_D is the surface temperature of dry edge which is corresponding to the NDVI of the point and T_w is the one of wet edge.

$$RSM = 100 - \frac{T - T_W}{T_D - T_W} \times 59.29378862 \tag{3}$$

Then, with the Equation 3, the relative soil moisture of every pixel can be calculated. With the relative data of soil moisture content data and investigation data in study area as a reference, the practical soil moisture retrieved from ETM+ of July 2002 in study area is shown as **Figure 4**. According to **Figure 4**, the soil moisture was relatively high in July 2002 in study area; most points have a soil moisture between 18% and 25%, while few points have a soil moisture less than 15%. Contrasting the soil moisture map with the terrain and false color maps, we can get that the soil moisture have a relatively high relation with the distribution of vegetation and terrain: the point near to water has a higher soil moisture value, vice versa; the point far from water has a relatively low soil moisture value.

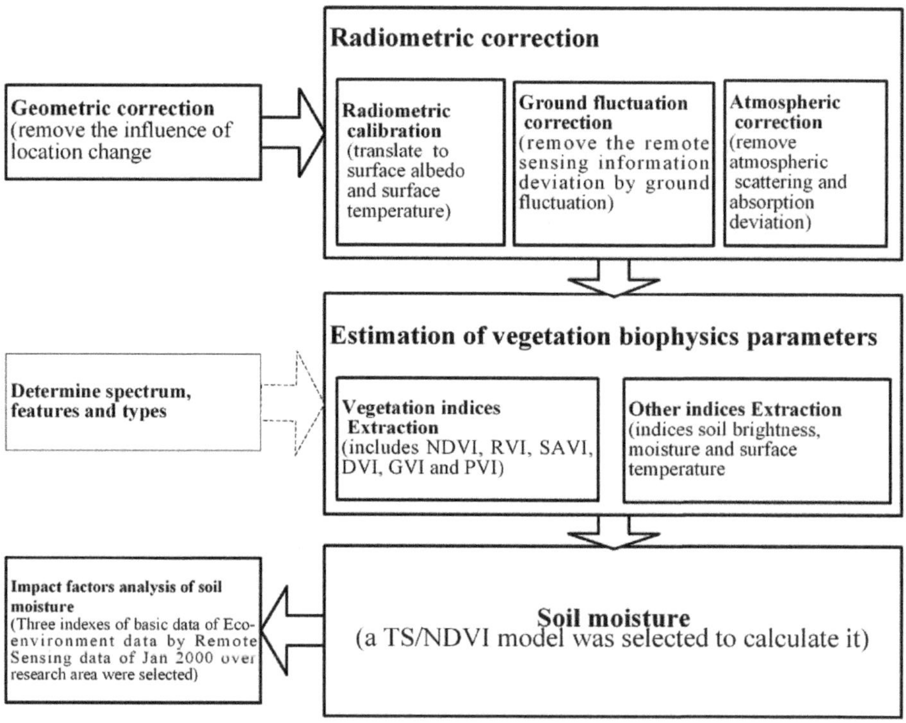

Figure 2. Flow chart of the soil moisture retrieval and analysis.

Figure 3. NDVI, surface temperature and feature space diagram retrieved from ETM+ data.

Soil Moisture Retrieval from ASTER Data

The surface temperature (T_S, Kelvin), the normalized difference vegetation index (NDVI) and the feature special diagram retrieved from ASTER data of November 2003 in study area were shown as **Figure 5**.

As the data process of ETM+ data above, the dry edge linear equation and wet edge linear equation in ASTER T_S/ NDVI feature space were calculated first, shown as Equations (4) and (5).

Dry edge:

$$T_{S\text{-max}} = -35.698122*NDVI+307.18527 \quad r = 0.78242239 \tag{4}$$

Wet edge:

$$T_{S\text{-min}} = 7.9501126*NDVI+271.93125 \quad r = 0.64890765 \tag{5}$$

With the data of soil moisture and experiential Formulas [17] in study area being as a reference, the relative soil moisture of any point in ETM+ data's T_S/ NDVI feature space was calculated as Equation 6. Where RSM is the relative soil moisture of a point, T is the surface temperature of the point, and T_D is the surface temperature of dry edge.

$$RSM = 100 - \frac{T - T_W}{T_D - T_W} \times 79.29378862 \tag{6}$$

Figure 4. Overlaid map of groundwater distribution map with the soil moisture map retrieved from Remotely Sensed data. (White lines are the boundaries of groundwater, red words are groundwater codes).

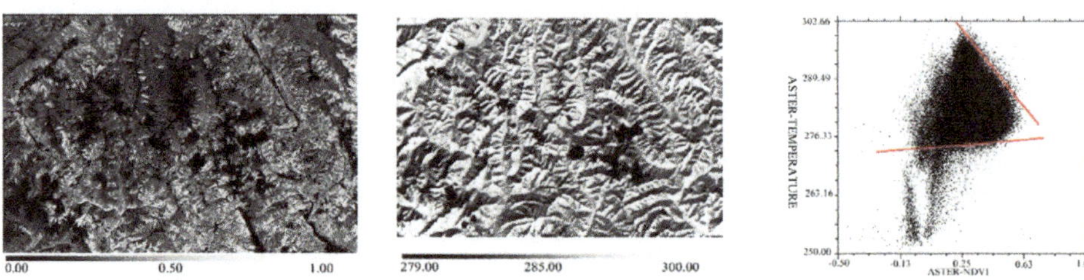

(a) NDVI (b) Surface temperature(K) (c) TS-NDVI feature space

Figure 5. NDVI, surface temperature and feature special diagram retrieved from ASTER data.

Then, with the Equation 6, the relative soil moisture of every pixel can be calculated. With the relative data of soil moisture content data and investigation data in study area being as a reference, the practical soil moisture retrieved from ETM+ of July 2002 in study area was shown as **Figure 6**. According to **Figure 6**, the soil moisture was widely distributed in November 2003 in study area, between 5% and 25%, the reason is mostly like that it was the time when autumn was changing into winter with some bare land and some evergreen woods, leading to a various kinds of soil moisture distribution but most between 10% and 20%, which was consistent with practical situation. Some points have soil moisture of 0, the reason is mostly like that the soil moisture of bare rock and hardened soil in this season was very low. Meanwhile, some points have relatively high soil moisture, the reason is mostly like that snow and clouds exist in ASTER data.

Figure 6. Soil moisture image by ASTER in research area (%).

Contrasting **Figure 4** with **Figure 6**, the soil moisture retrieved from ASTER data of November 2003 in study area has a relatively lower value than that retrieved from ETM+ of July 2002, the reason is that there is less water in winter but more in summer in study area.

DISCUSSION

From the results above, we can see that soil moisture was closely related to soil type, vegetation type and water. In order to make clearer the relationship between soil moisture and soil type, vegetation type and water, the distribution map of soil moisture was overlaid on the distribution map of vegetation type, soil type and water system.

Figure 7 is the overlaid map of the soil type distribution map with the soil moisture distributed map retrieved from ETM+ and ASTER. From the contrast of **Figure** 7(a) and 7(b), we can see that soil moisture distributed map is deeply related to soil type, that is to say, soil moisture is different from one kind soil to another. As a whole, the soil moisture of brown soil (with code of B210) and dark brown soil (with code of B310) is higher, but the soil moisture of terra cinnamon soil' (with code of C212) is lower, the soil moisture of other soil types are mid. However, the soil moisture of the same soil is different with the place change, which is similar in the two maps. Also, the soil moistures of different soils in different seasons are different.

Figure 8 is the overlaid map of the groundwater distribution map with the soil moisture distributed map retrieved from ETM+ and ASTER. From the contrast of **Figure 8**(a) and 8(b), we can see that the distribution of groundwater has some effect on soil moisture distribution, but the influence is weaker than the soil types'. General speaking, the influence of groundwater on soil moisture is stronger in summer than it is in winter; the reason is that the evapotranspiration is stronger in summer.

Figure 9 is the overlaid map of the vegetation distribution map with the soil moisture distributed map retrieved from ETM+ and ASTER. From the contrast of **Figure 9**(a) and 9(b), we can see that the distribution of soil moisture map is highly related to vegetation. Soil moisture varied with vegetation change. General speaking, the soil moisture on evergreen forest land is relatively higher, while cutover land's is relatively lower. With the season change, the relativity of vegetation and soil moisture would change accordingly, because most deciduous forests' evapotranspiration is different in different seasons, leading to the change of the ratio of soil water-holding.

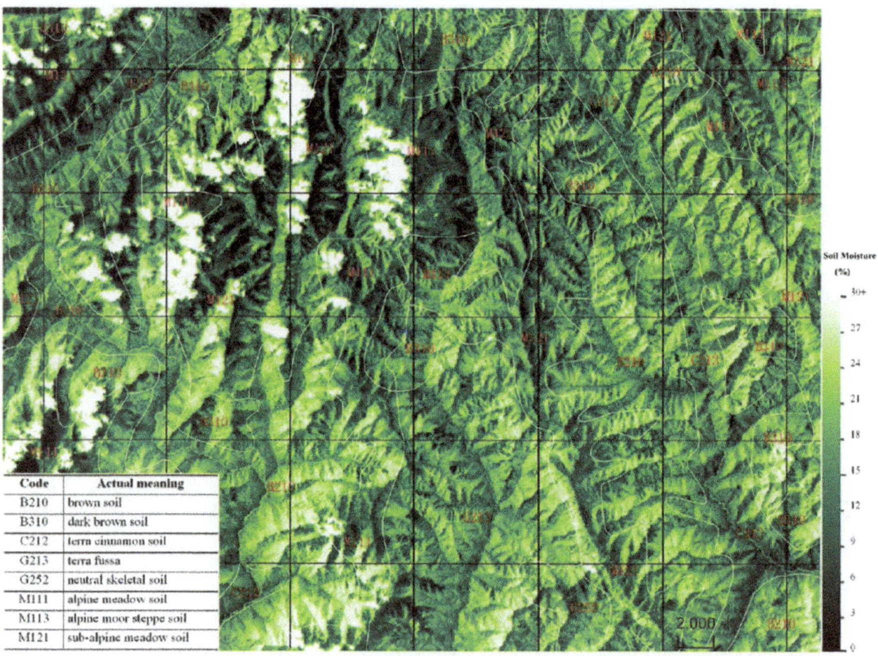

(a).Overlaid map of soil type distribution map with soil moisture map retrieved from ASTER.

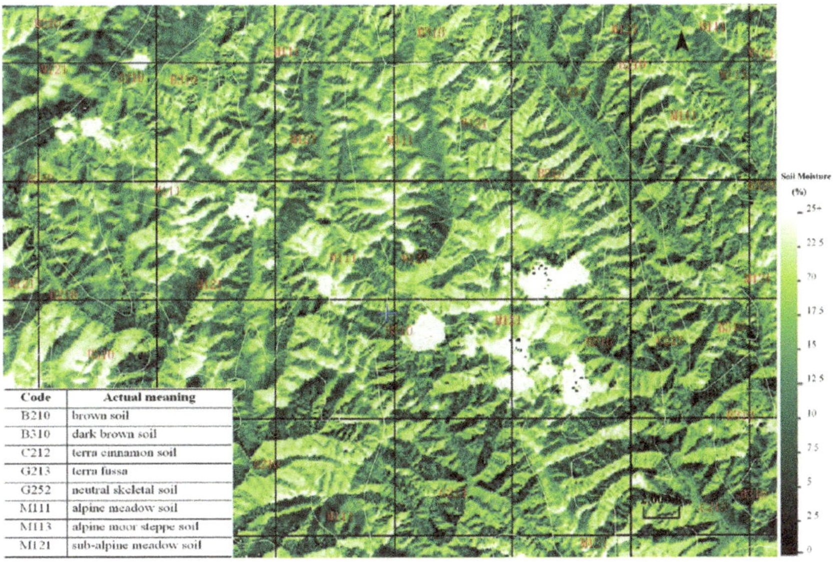

(b).Overlaid map of soil type distribution map with soil moisture map retrieved from ETM.

Figure 7. Overlaid map of the soil type distribution map with the soil moisture map retrieved from Remotely Sensed data. (White lines are the boundaries of soil types, red words are soil codes).

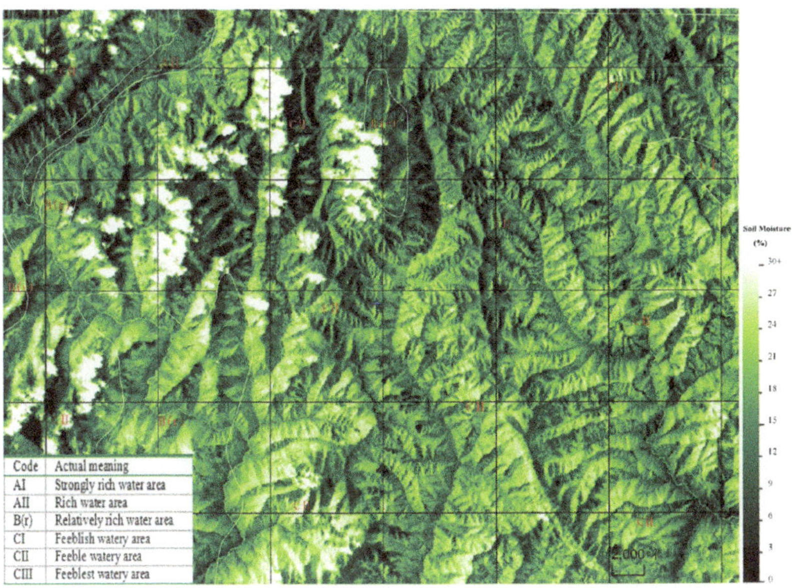

(a).Overlaid map of groundwater distribution map with soil moisture map retrieved from ETM.

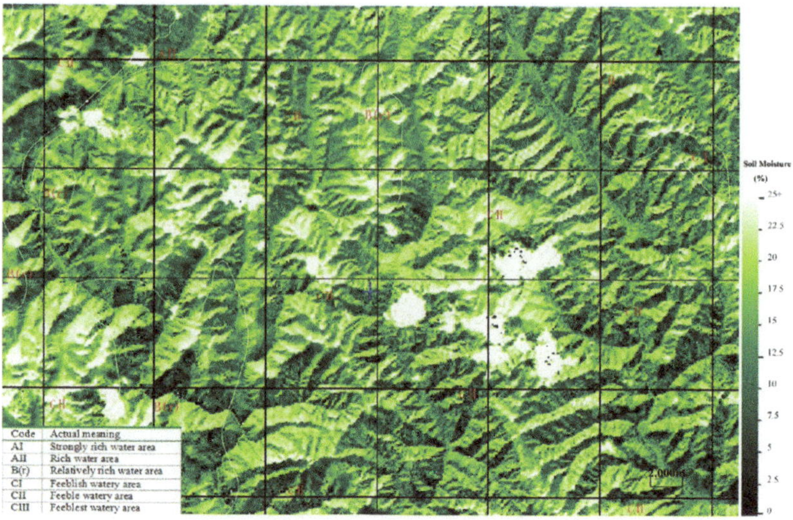

(b).Overlaid map of groundwater distribution map with soil moisture map retrieved from ASTER.

Figure 8. Overlaid map of groundwater distribution map with the soil moisture map retrieved from Remotely Sensed data. (White lines are the boundaries of groundwater, red words are groundwater codes).

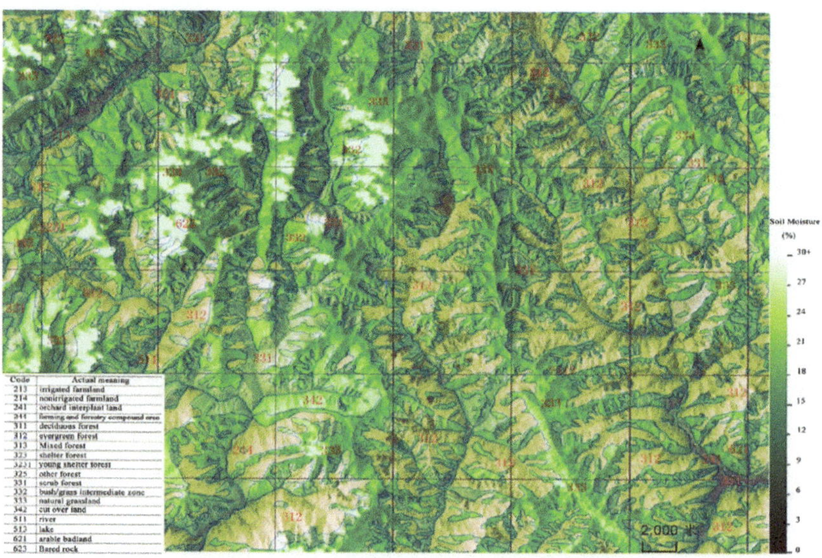

(a).Overlaid map of vegetation distribution map with soil moisture map retrieved from ETM.

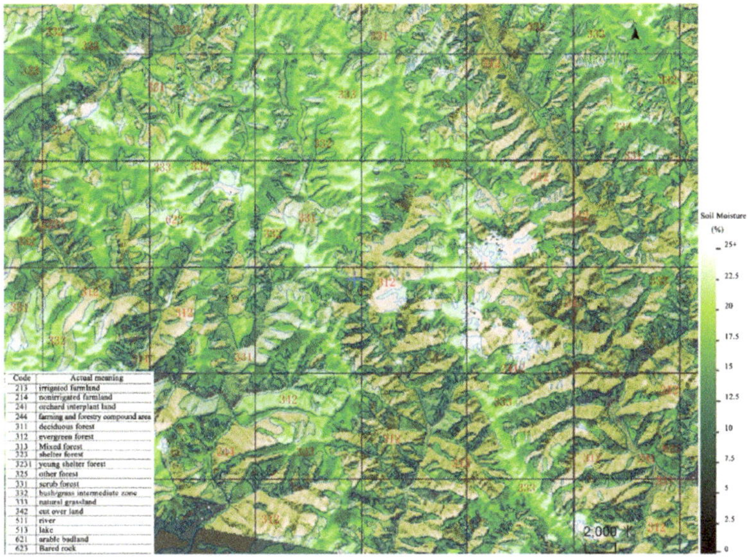

(b).Overlaid map of vegetation distribution map with soil moisture map retrieved from ASTER.

Figure 9. Overlaid map of vegetation distribution map with the soil moisture map retrieved from remotely sensed data. (Blue lines are the boundaries of groundwater, red words are groundwater codes).

From the analysis above, the model to retrieve soil moisture from remotely sensed data is viable, and, the potential influence factors of soil moisture are soil type and vegetation type.

CONCLUSIONS

In this paper, with the ETM+ and ASTER data, the soil moisture maps in study area were retrieved. The retrieved soil moisture maps were in agreement with the soil type and vegetation, which were also the two main factors determining the distribution of soil moisture.

As the remote sensing retrieval model of soil moisture was complex, the study was only at the beginning. In the following work, the authors plan to investigate soil moisture with hyper-spectral remotely sensed data to raise the accuracy and reliability of soil moisture retrieval.

ACKNOWLEDGMENTS

The research was supported by the Science and Technology Bureau of Sichuan Province of China. The authors sincerely thank Hong Jiang, from Nanjing University, Li Peng, from Chengdu University of Technology for their suggestions and help.

REFERENCES

1. R. W. A. Hutjes, P. Kabat, et al., "Biospheric aspects of the hydrological cycle," Journal of Hydrology, Vol. 212–213, pp. 1–21, 1998.
2. Y. Gao and C. Wang, "Biospheric aspects of hydrological cycle: BAHC plan and its research progress," Progress in Geography, Vol. 19, pp. 97–103, 2000.
3. D. Entekhabi, H. Nakamurai, and E. G. Njoku, "Solving the inverse problem for soil moisture and temperature profiles by sequential assimilation of multifrequeney remotely sensed observations," IEEE Transactions on Geoscience and Remote Sensing, Vol. 32, pp. 438–448, 1994.
4. IGPB, WCRD and IHDP, "Abstract volume of challenge of a changing earth," in Proceedings Global Change Opening Science Conference, Amsterdam, Netherland, pp. 10-13, July, 2001.
5. D. Gerten, S. Schaphoff, U. Haberlandt, W. Lucht, and S. Sitch, "Terrestrial vegetation and water balance-hydrological evaluation of a dynamic global vegetation mode," Journal of Hydrology, Vol. 286, pp. 249–270, 2004.
6. N. A. Brunsell, "Characterization of land-surface precipitation feedback regimes with remote sensing," Remote Sensing of Environment, Vol. 112, pp. 200–211, 2006.
7. R. H. French, J. J. Miller, C. Dettling, and J. R. Carr, "Use of remotely sensed data to estimate the flow of water to a playa lake," Journal of Hydrology, Vol. 325, pp. 67–81, 2006.
8. M. S. Feldman, T. Howard, E. McDonald-Buller, G. Mullins, D. T. Allen, A. Webb, and Y. Kimura, "Applications of satellite remote sensing data for estimating dry deposition in eastern Texas," Atmospheric Environment, Vol. 44, pp. 7562–7576, 2007.
9. E. García-Cuesta, I. M. Galván, and A. J. de Castro, "Multilayer perceptron as inverse model in a groundbased remote sensing temperature retrieval problem," Engineering Applications of Artificial Intelligence, Vol. 21, pp. 26–34, 2008.
10. A. Holsten, T. Vetter, K. Vohland, V. Krysanova, "Impact of climate change on soil moisture dynamics in Brandenburg with a focus on nature conservation areas," Ecological Modelling, Vol. 220, pp. 2076–2087, 2009.
11. K. Y. Li, M. T. Coe, N. Ramankutty, and R. D. Jong, "Modeling the hydrological impact of land-use change in West Africa," Journal of Hydrology, Vol. 337, pp. 258– 268, 2007.
12. J. Liu, E. Pattey, M. C. Nolin, J. R. Miller, and O. Ka, "Mapping within-field soil drainage using remote sensing," DEM and apparent soil electrical conductivity. Geoderma, Vol. 143, pp.261–272, 2008.
13. Y. Liu, W. Yang, and X. X. Wang, "Development of a SWAT extension module to simulate riparian wetland hydrologic processes at a watershed scale," Hydrological Processes, Vol. 22, pp. 2901–2915, 2008.
14. A. Loew, "Impact of surface heterogeneity on surface soil moisture retrievals from passive microwave data at the regional scale: The Upper Danube case,"

Remote Sensing of Environment, Vol. 112, pp. 231–248, 2008.

15. M. F. McCabe, E. F. Wood, R. Wójcik, M. Pan, J. Sheffield, H. Gao, and H. Su, "Hydrological consistency using multi-sensor remote sensing data for water and energy cycle studies," Remote Sensing of Environment, Vol. 112, pp. 430–444, 2008.

16. W. N. Yang, J. Jian, Y. X. Li, X. N. Wan, L. Peng, H. H. Liu, H. Y. Shao, X. A. Dai, T. Zeng, and X. M. Wu, "Remote sensing inversion of eco-water resource quantity in international workshop on earth observation and remote sensing applications," 29 June-2 July 2008, Beijing, China(IEEE), pp. 1–6, 2008.

17. J. Jian, "Quantitative investigation of eco-water with remote sensing in the upper of Minjiang river," Doctoral thesis, Chengdu University of Technology, China, 2006.

6
CHAPTER

SATELLITE-OBSERVED BLACK WATER EVENTS OFF SOUTHWEST FLORIDA: IMPLICATIONS FOR CORAL REEF HEALTH IN THE FLORIDA KEYS NATIONAL MARINE SANCTUARY

Jun Zhao, Chuanmin Hu, Brian Lapointe, Nelson Melo, Elizabeth M. Johns and Ryan H. Smith

Keywords: Black water, MODIS, SeaWiFS, water quality, coral reef, Florida Keys

INTRODUCTION

The Florida Keys National Marine Sanctuary (FKNMS) is a marine protected area designated, in 1990, to protect the coral reef ecosystems adjacent to the Florida Keys (Figure 1). Extensive seagrass beds, mangrove-fringed islands, coral reefs, and more than 6,000 species of marine life within the boundaries of the FKNMS make the area outstanding in terms of conservational, recreational, commercial, ecological, historical, scientific, educational, and aesthetic values [1,2]. The FKNMS is not isolated, but is connected to the deep ocean, the land and the atmosphere through ocean circulation, terrestrial discharge, and atmospheric deposition [3]. The FKNMS is located closely to South Florida, which attracts residents and tourists due to its natural assets.

Citation: Zhao, J, Hu, C, Lapointe, B, Melo, N, Johns, E.M, Smith, R.H. Satellite-Observed Black Water Events off Southwest Florida: Implications for Coral Reef Health in the Florida Keys National Marine Sanctuary. Remote Sens. 2013, 5, 415-431, doi:10.3390/rs5010415.

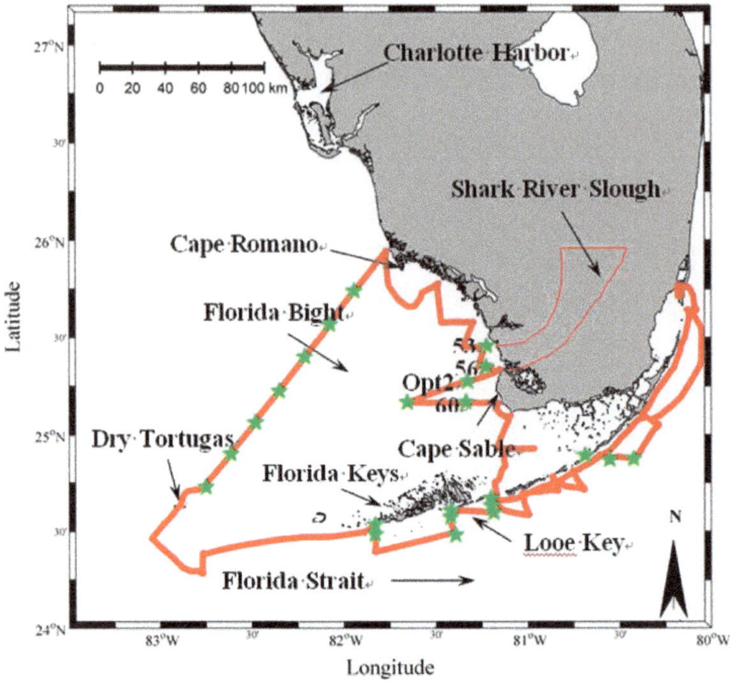

Figure 1. The thick red line shows the ship track from 27 February to 2 March 2012 and the green pentagrams show the station locations where surface water samples were collected. The region outlined by the thin red lines shows the Shark River Slough.

The increasing population and development in the Florida Keys, together with agriculture and other human activities on the mainland of South Florida, imperil the downstream FKNMS ecosystems. Degradation of water quality around the FKNMS has been a major concern for decades. Between 1991 and 1995, extensive cyanobacterial blooms developed in Florida Bay and the FKNMS in response to increased flows of nitrogen-enriched water from the Everglades Agricultural Area via Shark River Slough [4,5]. A Landsat image (Path 15, Row 43) in late May, 1992, showed the presence of highly discoloured and turbid water in western and central Florida Bay, where dense phytoplankton blooms developed in response to nitrogen-enrichment from Shark River Slough [6]. The extensive phytoplankton blooms in Florida Bay led to regional-scale eutrophication in downstream waters of the FKNMS, which included a proliferation of coral diseases and a 38% loss of living coral within the FKNMS [7]. The Comprehensive Everglades Restoration Plan (CERP), a major component of the South Florida Ecosystem Restoration Initiative (SFERI), will be accomplished through a series of projects to modify the freshwater flow to south Florida coastal waters. The operation may further increase nitrogen-rich discharges from Shark River Slough, which could further affect the water quality in the vicinity of the FKNMS. Discoloured coastal water off southwest Florida has often been termed as "black water". It refers to dark coloured water, whereby the blue light is significantly absorbed by high concentrations of phytoplankton and/or coloured dissolved organic matter (CDOM). In January 2002, a black water event was reported off the southwest Florida coast, which lasted through April 2002 [8,9]. The black water between January and April 2002 contained abundant non-toxic (dominated by *Rhizosoliniaceae* diatoms) and toxic (*Karenia brevis*) microscopic phytoplankton. The 2002 event resulted in severe benthic resource decline: 70% of stony coral, 40% of other coral species and almost complete sponge colonies at two reef sites north of Key West, Florida disappeared [10]. Runoff from the Everglades watershed to the southwest Florida Shelf and *K. brevis* blooms, which occurred off Charlotte Harbor (located in the central west of Florida) earlier and migrated to the south along the coast, have been implicated as the causes of the highly publicized black water event in early 2002. Black water has come under scrutiny since then, and several other black water events have been reported [3,11].

Studies using more recent satellite data, which can take precedence over *in situ* surveys and buoy stations in spatial and temporal coverage, have great potential to augment and improve policy making for protecting marine ecosystems [12], e.g., NOAA's existing coral reef decision support systems, including the Integrated Coral Observing Network (ICON)/Coral Reef Early Warning System (CREWS) operated by NOAA's Coral Reef Conservation Program (CRCP).

In early 2012, satellite and *in situ* measurements indicated that another black water event was taking place in the southwest Florida coastal waters. Several questions then arose: Was this event stronger or weaker than before? What caused this event?

In this paper, we will focus on the 2012 black water event off the southwest Florida coast (Figure 1) and compare it with the 2002 event. Our main objectives are:

1) To document the 2012 black water event in the southwest Florida coastal waters using the Moderate-Resolution Imaging Spectroradiometer (MODIS) and the Sea-viewing Wide Field-of-view Sensor (SeaWiFS) measurements, and compare with the 2002 black water event in the same region;

2) To understand what caused the black water event in 2012.

Spectral normalized water-leaving radiance (nL_w) has already been used as an index for determining black water, where low levels of nL_w at 443 nm was observed due to high absorption of phytoplankton and/or CDOM [10]. In this study, we will first delineate the 2012 black water event using satellite-derived nL_w data. Secondly, the 2012 event will be compared with the prolonged 2002 event in terms of spatial and temporal coverage. This is beneficial for environmental change studies, and can provide information on the ecosystem health and variations of the water quality in the southwest Florida coastal waters. Thirdly, the potential sources for the black water will be analysed and discussed. Finally, we will discuss the potential effects of black water on benthic communities.

DATA AND METHODS

In Situ Survey

A flow-through system equipped with sensors to measure chlorophyll and CDOM fluorescence, salinity, and turbidity was positioned on-board the R/V *Walton Smith* (WS1202), off the south Florida coastal waters from 27 February to 2 March 2012. This was one of the routine bimonthly cruise surveys from the South Florida Program (SFP) of NOAA's Atlantic Oceanographic and Meteorological Laboratory (AOML). The ship track is shown in Figure 1. Water was pumped from about 2.3 m below the surface to a chamber where all parameters were measured after de-bubbling. Data was recorded at 10-second intervals, with spikes and outliers removed using a median filter.

Water samples were collected from the flow-through system to determine chlorophyll-a concentrations and CDOM absorption coefficients using NASA recommended protocols, and then used to calibrate the flow-through chlorophyll and CDOM fluorescence into chlorophyll-a concentrations and CDOM absorption coefficients. The former was converted to phytoplankton pigment absorption using the relationship found in this region [13].

Satellite-tracked drifters were deployed during two AOML-SFP cruises conducted between October and December 2011. The drifters were of the CODE/Davis design, with 1 m surface drogues. The data was transmitted to satellites every 30 minutes.

Satellite Data

MODIS and SeaWiFS data at 1 km resolution were downloaded from the NASA Goddard Space Center and processed using the most recent updates in calibration and algorithms embedded in the software package SeaDAS (version 6.3). Normalized water leaving radiance (nL_w) at 443 nm, 488 nm and 547 nm was generated. Enhanced RGB (ERGB, R = 547 nm, G = 488 nm, B = 443 nm) images were composited with nL_w at the three wavelengths, following the same approach as in previous studies [10,11]. The 547-nm channel was used as the red band instead of the conventional 670-nm channel as the $nL_w(670)$ value was nearly zero for most waters, therefore yielding much less information than the value for $nL_w(547)$. The ERGB images provide qualitative visualization on various water types: the brownish/reddish colour represents high concentrations of phytoplankton, the bright colour results from sediment-rich waters or shallow bright bottom, and the darkish colour is caused by high concentrations of phytoplankton and/or CDOM. Shallow, seagrass bottom can also cause a darkish colour, but the location of such discoloured water is static in time and therefore can be easily distinguished from the analysis of the image time series.

The spectral nL_w data was also used to derive the downwelling diffuse attenuation coefficients at 488 nm, based on the QAA algorithm [14] (K_{d}_488_lee, m^{-1}). A recent validation showed that the satellite-based K_{d}_488_lee data was robust, with RMS uncertainties of <15% for the range of 0.03–0.65 m^{-1} in this study region [15].

River Discharge and Nutrient Concentrations

Monthly mean discharge data for Shark River Slough was obtained from the South Florida Water Management District for S12_T, as a summation of S12A, S12B, S12C and S12D. Monthly climatology was calculated for the period from 1990 to 2012. Similarly, the annual mean and climatologic discharge data was calculated for 1984–2012.

Determining nL_w Threshold

In this study, we will delineate the black water coverage using nL_w gradients. As aforementioned, black water patches always show low nL_w values. Thus, the threshold for nL_w, below which the region can be referred to as black water, needs to be quantified.

The methods used here are similar to [16]. Firstly, all ERGB images during the period between January and April of 2002 and 2012 from SeaWiFS and MODIS/Aqua were examined and only those with cloud-free conditions over the black water regions were retained. With this criterion, 29 and 16 images were chosen for SeaWiFS and MODIS/Aqua, respectively. Next a region of interest covering the visually identified black water was outlined manually in ENVI for the chosen ERGB images, where dark colours indicate the black water patches. A gradient image was then generated for each $nL_w(443)$ image, corresponding to the outlined ERGB images. For each pixel, the gradient was defined as the difference from adjacent pixels in a 3 × 3 window. A histogram was produced and the mode of the maximum gradient was acquired, which can be used to differentiate fronts induced by the black water. A mean value for all pixels in the nL_w image associated with the gradient mode was calculated and used as a threshold.

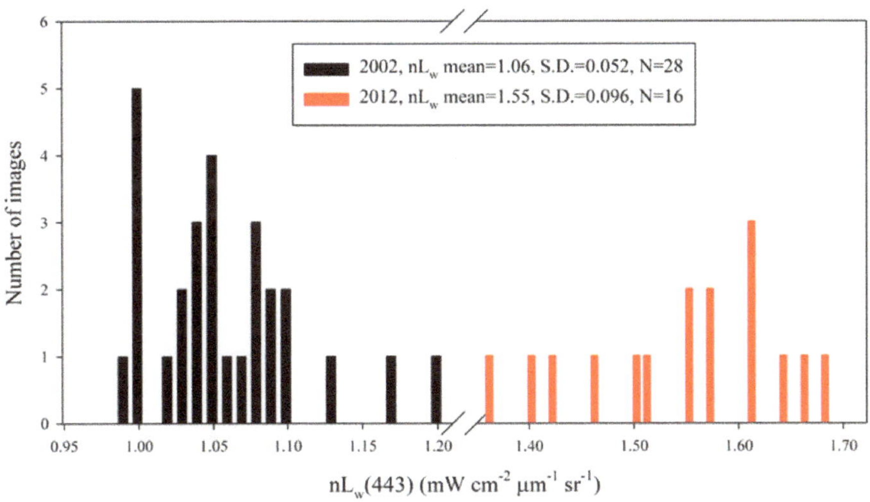

Figure 2. Normalized water-leaving radiance ($nL_w(443)$) thresholds for the 2002 and 2012 black water events during the periods from January to April from SeaWiFS (black bars) and MODIS/Aqua (red bars) cloud-free observations.

Thus, each nL_w image had its own threshold. Finally, all threshold values were pooled together, and the histogram was computed, as well as the mean and the standard deviation, as shown in Figure 2. The thresholds for SeaWiFS ranged from 0.99 to 1.2 $mW \cdot cm^{-2} \cdot \mu m^{-1} \cdot sr^{-1}$ for the 2002 black water event and from 1.36 to 1.68 $mW \cdot cm^{-2} \cdot \mu m^{-1} \cdot sr^{-1}$ for MODIS/Aqua for the 2012 black water event. If a universal threshold were to be derived, it would lead to 1.16 and 1.74 $mW \cdot cm^{-2} \cdot \mu m^{-1} \cdot sr^{-1}$ for the two events, respectively, as a result of the average plus two standard deviations. However, in order to calculate the spatial coverage of the black water events more accurately using the gradient method, the image-specific $nL_w(443)$ thresholds were used.

RESULTS

Figure 3 shows the development and progression of the 2012 black water event. The black water patches are shown by dark color in the ERGB images. In late 2011, a phytoplankton bloom patch, which initiated north of Charlotte Harbor in late September 2011 and was dominated by *K. brevis* as indicated by *in situ* taxonomy data (Figure 3(a–d)), moved southward into the Florida Bight. In early January 2012, the black water patch extended from the region near the coast to the region to the north of the Florida Keys. The patch continued to move southward from late January until early February. At the beginning of February, the black water patch reached the Florida Keys region. From late February to early March the patch retreated slightly toward the southwest coast. In late March, the patch shrank and moved toward the Florida Keys again. The patch nearly disappeared around 12 April. These movement patterns of the black water patch were consistent with the results obtained from satellite-tracked surface drifters deployed between late October and December 2011 in this region (Figure 4). The drifter tracks indicated that the regional flow was generally slow (2.6–7.7 $cm \cdot s^{-1}$) and to the southwest between late October 2011 and early February 2012, and the flow pattern was consistent with the speed and direction of the black water patch observed from satellite images. After early February, the flow moved generally west and then northwest from the beginning of March to late May 2012. This agreed well with the timeline of the dissipation of the black water plume. Furthermore, the flow patterns were in good agreement with previous studies [17]. Water samples collected on 1 and 2 March 2012 showed that the black water patch contained low (between 10^3 and 10^4 $cells \cdot L^{-1}$) to high (between 10^5 and $10^6 cells \cdot L^{-1}$) concentrations of *K. brevis*, as annotated in Figure 3(i).

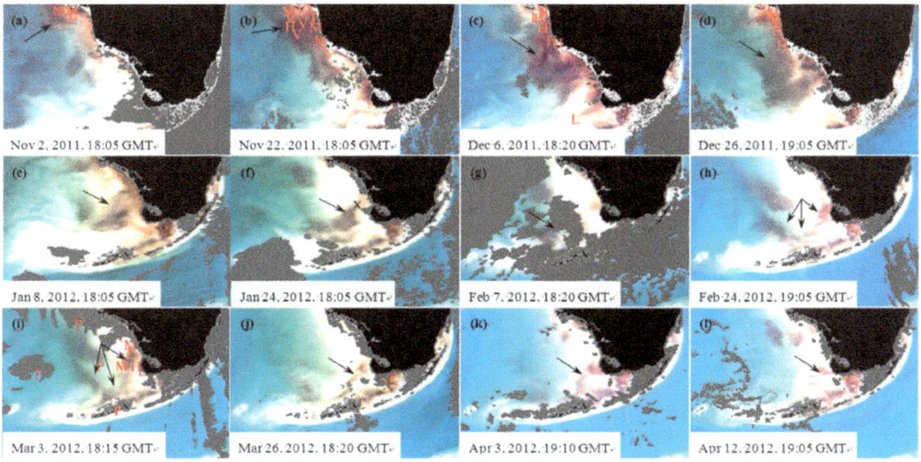

Figure 3. Representative MODIS/Aqua enhanced RGB (ERGB) images showing the development and progression of the 2012 black water event, with positions of the black water annotated with arrows. Remote sensing reflectance ($R_{rs} = nL_w/F_o$ where F_o is the extraterrestrial solar irradiance) data from the station (x) in (f) were extracted and shown in Figure 8. Taxonomy data collected on 1–3 November 2011, 21–23 November 2011, 5–7 December 2011, 25–27 December 2011, and 1–2 March 2012 were overlaid in **(a)**, **(b)**, **(c)**, **(d)**, and **(i)**: N—*K. brevis* not present; P—present < 10^3 cells·L^{-1}; L—low between 10^3 and 10^4 cells·L^{-1}; M—medium between 10^4 and 10^5 cells·L^{-1}; H—high between 10^5 and 10^6 cells·L^{-1}; V—very high > 10^6 cells·L^{-1}.

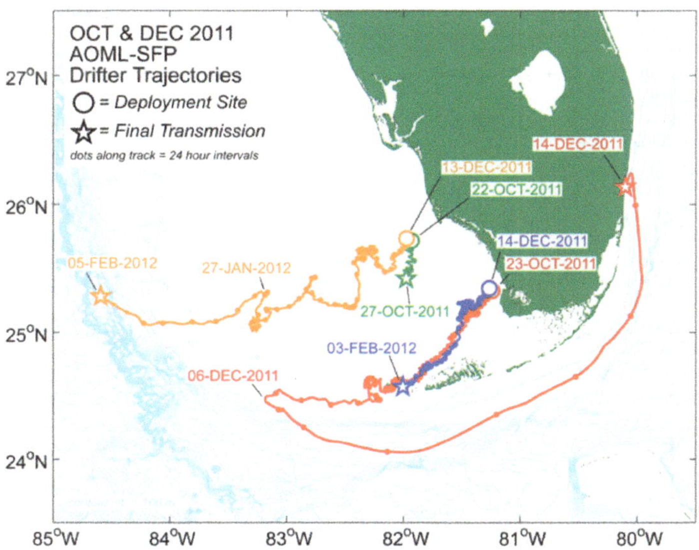

Figure 4. Surface trajectories of drifters released between October and December 2011 during Atlantic Oceanographic and Meteorological Laboratory (AOML)-South Florida Program (SFP) cruises (Figure 1) at Cape Romano and Shark River. The drifters are CODE/Davis style with 1-m surface drogue. Data were transmitted via satellites once every 30 min.

Figure 5 shows surface salinity, chlorophyll and CDOM fluorescence, and turbidity measured by the flow-through system onboard the R/V *Walton Smith*. There was a high chlorophyll patch west of Cape Sable (Figure 5(a)) where CDOM and turbidity were both low. Field measurements also showed high chlorophyll concentrations of 3.6, 4.2, 5.9, and 3.8 mg·m^{-3} for Stations 53, 56, Opt2, and 60 (Figure 1), respectively. Salinity for the high-chlorophyll region was somewhat higher than long-term historical values obtained from the bimonthly routine surveys (http://www.aoml.noaa.gov/phod/sfp/data/ship_obs.php).

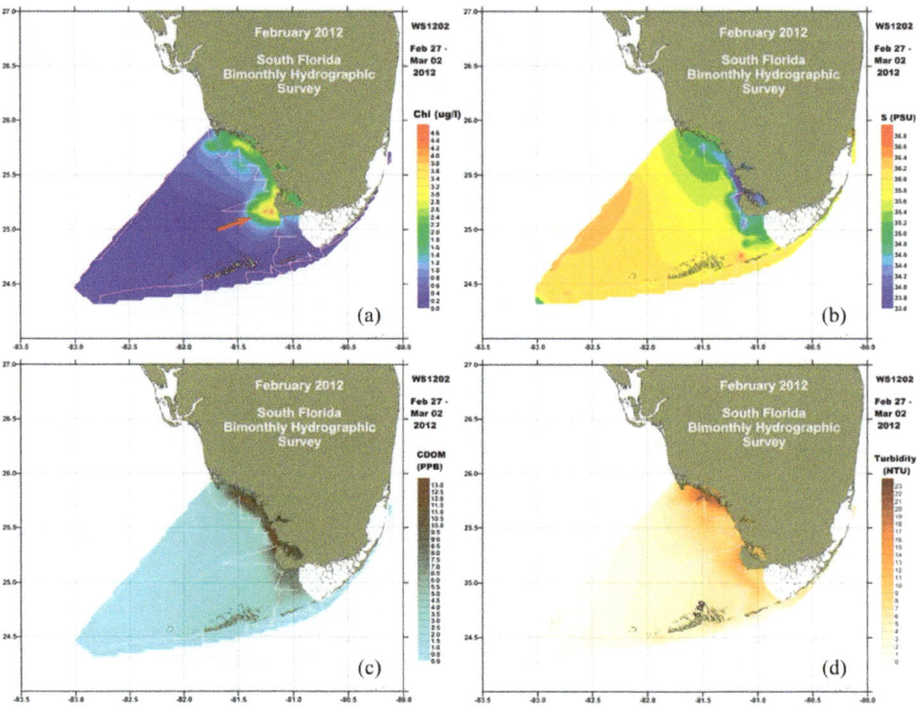

Figure 5. Chlorophyll fluorescence (**a**), salinity (**b**), CDOM fluorescence (**c**), and turbidity (**d**) measured by a flow-through system during a cruise survey from 27 February to 2 March 2012. The white line shows the ship track. The red arrow in (a) shows a patch with high chlorophyll fluorescence. Water samples collected from the patch showed low to high concentrations of *K. brevis* (Figure 3(i)).

Using $nL_w(443)$ as a measure of black water events has followed [10], based on the principle that $nL_w(443)$ will generally decrease with increasing phytoplankton concentration and/or CDOM absorption, as both phytoplankton and CDOM strongly absorb blue light. Figure 6a shows scatter plots of MODIS $nL_w(443)$ against "concurrent" total absorption from phytoplankton pigment ($a_{ph}(443)$, m^{-1}) and CDOM ($a_g(443)$, m^{-1}) from the field measurements. A time window of ±3 h between satellite and field measurements was first tried, with only a few matching pairs found. Then, the time window was relaxed to ±12 h. Results showed a generally decreasing pattern of

$nL_w(443)$ with increasing $a_{ph}(443) + a_g(443)$. For $a_{ph}(443) + a_g(443) > 0.3\ m^{-1}$, $nL_w(443)$ was generally lower than 2 $mW \cdot cm^{-2} \cdot \mu m^{-1} \cdot sr^{-1}$. Yet it is difficult to derive a universal, time-independent $nL_w(443)$ threshold to delineate black water events because each event may contain variable amounts of chlorophyll-a and CDOM, as shown by the different threshold values for the 2002 and 2012 events.

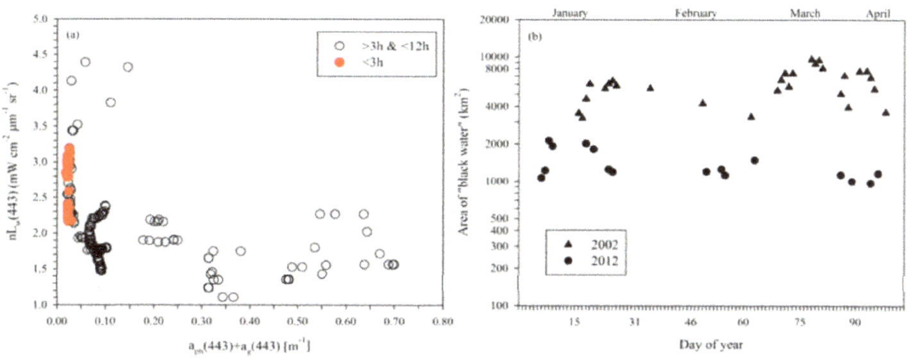

Figure 6. (**a**) Scatter plot of MODIS/Aqua $nL_w(443)$ against phytoplankton plus CDOM absorption coefficients determined on the same day during the 2012 black water event. (**b**) Comparison between the spatial coverage of the 2002 and 2012 black water events.

Even though the threshold values may change with different black water events, it is still useful to put the 2012 black water event in the context of historical events. The 2002 black water has been reported in detail by [8,9]. The event was first reported in early January 2002, and the black water reached Florida Bight in mid-January. It was retained within the Florida Bight by an anti-cyclonic eddy, and then reached the Dry Tortugas in late March, after which the water was entrained into the Loop Current and advected into the Florida Strait. The 2002 black water event ended around 20 April.

Scatter plots for the spatial coverage of the 2002 and 2012 black water events over time are shown in Figure 6(b). The area of black water ranged from 3,229 km^2 to 9,550 km^2 with a mean of 6,063 (±1,782) km^2 for 2002, compared with from 964 $km^2$to 2,123 km^2 with a mean of 1,369 (±379) km^2 for 2012. Although the difference between their temporal coverage was not large, the spatial coverage in 2002 was much larger than in 2012. Generally, in terms of spatial and temporal distributions, the black water event in 2012 was weaker than in 2002.

DISCUSSION

Causes of the 2012 Black Water Event

A clear understanding of factors that cause black water events and their consequences on the FKNMS ecosystems can improve the efficacy of ecosystem based management strategies. Therefore, the potential sources for the 2012 black water were examined in detail. As shown in Figure 3, a *K. brevis* bloom patch entered the Florida Bight in late 2011. Therefore, we hypothesize that the 2012 black water patch in the Florida Bight was formed from the red tide bloom that originated earlier off central West Florida and was advected and concentrated by the coastal ocean circulation. Results from *in situ* water samples collected on 1 and 2 March 2012 revealed that the black water in Florida Bight contained *K. brevis* ranging from 333 to 88,000 cells·L^{-1}(Figure 3(i)), which can cause discolored water [18] and supported our hypothesis. The patch also contained abundant non-toxic *Rhizosoliniaceae* diatom.

In order to examine the possible effects of river discharge on the black water events, monthly mean discharge of Shark River Slough is shown in Figure 7. The monthly discharges during November 2011 and October–November 2001 were higher than the corresponding monthly climatologies. Although the departures were not statistically significant and the events initiated from earlier *K. brevis* red tide events, these higher-than-usual discharges could have contributed to the evolution of the black water events. Indeed, the linkage between river discharge and algal blooms has been widely recognized in the scientific community. As the river discharge carries "new" nutrients to the coastal ocean, which may add fuel to support and maintain the growth of phytoplankton, the role of river discharge for maintaining the black water and red tide events cannot be ignored [18].

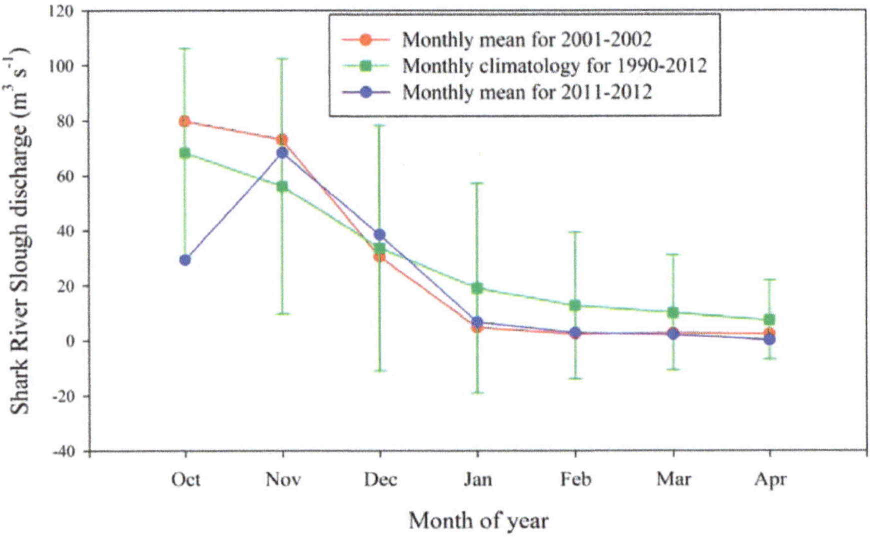

Figure 7. Monthly mean discharges from the Shark River Slough (location shown inFigure 1). Data shown here are the composites of discharge data from S12A, S12B, S12C, and S12D. Data courtesy of South Florida Water Management District.

Although it is generally difficult to differentiate what water constituents are contained in the black water patch by using satellite data alone, Figure 8 shows an exception where the spectral shape of the MODIS R_{rs} could be used to identify the phytoplankton type in the black water patch. The MODIS spectrum was extracted from a location in the black water patch on 24 January 2012 (Figure 3(f)). The spectral curvatures between 469 and 555 nm (*i.e.*, high-low-high-low-high in the R_{rs} spectrum) are unique features for the N-fixing cyanobacterium *Trichodesmium* due to their specific pigment composition [19]. Hu *et al.*[19] used extensive MODIS and *in situ* data to show that such spectral curvatures were only found from *Trichodesmium* blooms and not from other blooms. Therefore, *Trichodesmium* could be an important source of nitrogen required by the black water patch, a phenomenon also noted for sustaining phytoplankton blooms in the Great Barrier Reef lagoon [20]. Also, submarine groundwater discharge in the vicinity of the black water might have provided additional nutrient supplies [7].

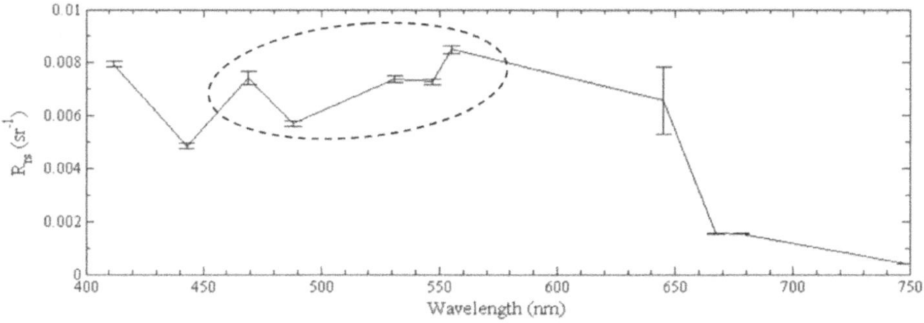

Figure 8. MODIS R_{rs} spectrum from 3 × 3 pixels centered at the station marked in Figure 3(f), measured on 24 January 2012 (GMT 18:05). The vertical bars show the standard deviations. The unique spectral curvatures between 469 and 555 nm (high-low-high-low-high, as outlined by the dashed circle) indicates the presence of N-fixation *Trichodesmium.*

Impacts of Black Water Events on Coral Reef Health

There have been earlier reports about the impacts of black water as mentioned above, such as decline of corals and die-offs of sponge colonies. Here, we give an example of how another black water event affected the benthic habitat. A MODIS fluorescence line height (FLH) image on 19 May 2008 (Figure 9(a)) shows an algal bloom patch at the "Rock Pile" reef (marked with a black cross). This bloom was also associated with a black water event. An *in situ* survey indicated that the bloom was followed by blooms of the chlorophyte *Cladophora liniformis* (Figure 9(b)) on the bottom and over the skeletons of corals, which died in the early 1990's as a result of excessive nutrient loading from Shark River Slough [21]. Figure 9(c)shows the spatial distribution of MODIS K_d_488_lee, which cannot well distinguish the black water patch (as indicated by the high FLH values in the vicinity of the "Rock Pile" reef) from nearby coastal waters because suspended sediments also contribute to K_d_488_lee, making it impractical to delineate black water. However, when combined with bathymetry data, K_d_488_lee could be used to derive the bottom available light (as a percentage of surface light) as exp(−K_d_488_lee × z) where z is the bottom depth. The light available on the bottom in the bloom region was much less than in the nearby non-bloom region (Figure 9(d)). Less light availability has been shown to contribute to the death of corals [22], and the same mechanism could explain the chlorophyte bloom over the skeleton of corals at the "Rock Pile" reef.

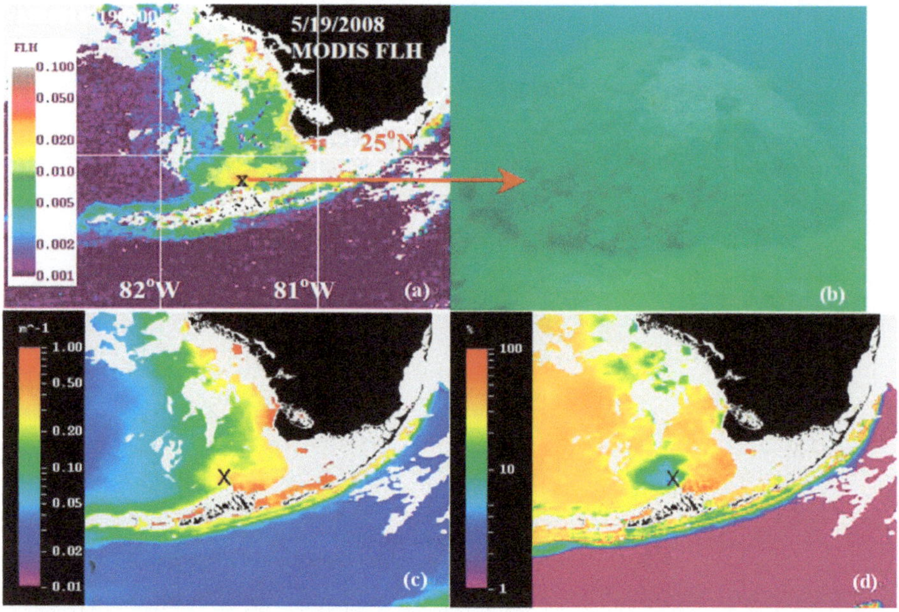

Figure 9. MODIS fluorescence line height (FLH) image (**a**) shows an algal bloom patch at the Rock Pile (marked with a black cross), followed by a bloom of the chlorophyte*Cladophora liniformis* on the bottom and over dead corals (**b**). The corresponding K_d_488_lee map is shown in (**c**). The bottom light availability in percentage, which was calculated from $\exp(-K_d\_488\_lee \times z)$ (z is the bottom depth), is shown in (**d**). The available bottom light in the bloom region is much lower than elsewhere.

Coral reefs are well known to be sensitive to slight increases in nutrient concentrations [20,23] and historical decreases in coral cover in the FKNMS over the past several decades support this precept. Increased flows from Shark River Slough in the early 1990s were linked not only to phytoplankton blooms, but also benthic macroalgal blooms that can overgrow adult corals and inhibit recruitment of juvenile corals (e.g., Figure 9(b)). Figure 10 shows the time series of discharge from Shark River Slough from 1984 to 2011. Between 1991 and 1995, discharges from Shark River Slough increased with parallel increases in nitrogen concentrations, especially ammonium, at Looe Key in the FKNMS [24]. Following this nitrogen enrichment, increased coral disease and die-off reduced coral cover in the FKNMS to a historical low of 6.4% coral cover between 1996 and 1999 [7]. Accordingly, because both the 2002 and 2012 black water events followed decades of increasing environmental stress and coral loss in the FKNMS, impacts of these subsequent events were relatively minor in terms of coral loss. Although there is relatively little living coral remaining in the FKNMS as compared with historical coverage, it is important to recognize the potential connection between water quality and coral mortality and recovery in the FKNMS. Thus, it is important to continue the satellite and field observations of the water environments of these delicate ecosystems. Such observations not only serve as a direct means to monitor the water environments and anomaly events, but also provide critical information to establish baseline data to assess the long-term changes in the water environments of the delicate coral reef ecosystems, which can help resource managers make ecosystem-based management decisions.

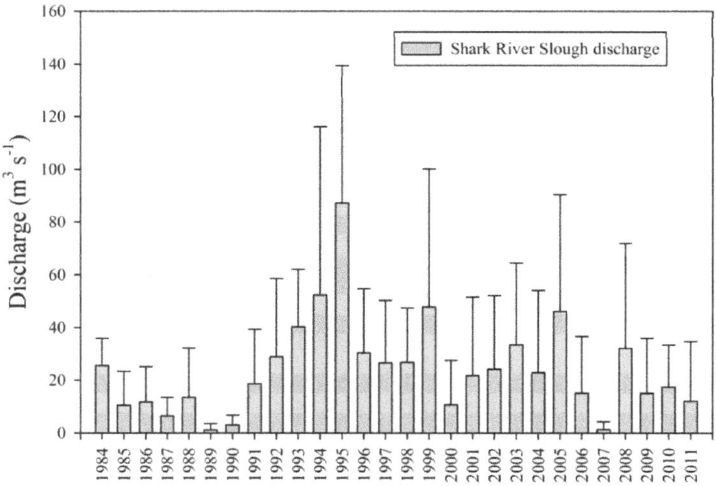

Figure 10. Annual mean discharge of Shark River Slough (summation of stations S12A, S12B, S12C, and S12D) from 1984 to 2011. The standard deviations are also shown. Data courtesy of the South Florida Water Management District.

CONCLUSION

The 2012 black water patch remained in the Florida Bight from early January through mid-April. The patch contained varying concentrations of *K. brevis* that was formed from the *K. brevis* bloom initiated north of Charlotte Harbor in the fall of 2011. Despite similar causes of the event as compared to the 2002 black water event, the black water patch in early 2012 was weaker and much smaller in terms of spatial coverage. The average size of the black water patch is 6,063 (±1,782) km^2 for 2002, compared with 1,369 (±379) km^2 for 2012. However, in general, previous black water patches have negative effects on benthic coral reef communities, such as die-offs of corals and sponge colonies.

The case study here demonstrates the importance of continuous ocean observations by all means, including satellite measurements, *in situ* surveys, and environmental monitoring (e.g., river discharge and nutrient flux). Of these, satellite ocean color observations provide unique information at synoptic time scales to effectively monitor and trace adverse water quality events, from which further insights may be obtained in the understanding of variations in water quality and ecosystem health.

ACKNOWLEDGMENTS

This study was supported by NASA's Gulf of Mexico program and Ocean Biology and

Biogeochemistry program. The SFP of NOAA AOML has been funded by the NOAA/OAR Ship Charter Fund, NOAA's Center for Sponsored Coastal Ocean Research, NOAA's Deepwater Horizon Supplemental Appropriation, and the US Army Corps of Engineers. We thank the NASA Ocean Biology Processing Group for providing MODIS and SeaWiFS data. We also thank Gerardo Toro-Farmer (USF) and Maria Vega-Rodriguez (USF) for collecting water samples and thank Jennifer Cannizzaro (USF) and Jennifer Wolny (Florida Institute of Oceanography) for analyzing some of the water samples. Three anonymous reviewers provided substantial comments that helped improve this manuscript, whose effort is acknowledged here.

REFERENCES

1. Causey, B.D. The role of the Florida Keys National Marine Sanctuary in the South Florida Ecosystem Restoration Initiative. In *The Everglades, Florida Bay, and Coral Reefs of the Florida Keys: An Ecosystem Sourcebook*; Porter, J.W., Porter, K.G., Eds.; CRC Press: Boca Raton, FL, USA, 2002; pp. 883–894.
2. Keller, B.D.; Causey, D.B. Linkages between the Florida Keys National Marine Sanctuary and the South Florida Ecosystem Restoration Initiative. *Ocean Coast. Manage* **2005**, *48*, 869–900.
3. Hu, C.; Muller-Karger, F.E. On the Connectivity and "Black Water" Phenomena near the FKNMS: A Remote Sensing Perspective. In Proceedings of Science, People and Policy in the Florida Keys National Marine Sanctuary Connectivity Colloquium, *Key West, FL, USA, 19–21 August 2004*; Keller, B.D., Wilmot, F.C., Eds.; Department of Commerce National Oceanic and Atmospheric Administration, National Marine Sanctuary Program: Silver Spring, MD, USA, 2007; pp. 47–55.
4. Lapointe, B.E.; Matzie, W.R.; Barile, P.J. Biotic Phase-Shifts in Florida Bay and Fore Reef Communities in the Florida Keys: Linkages with Historical Freshwater Flows and Nitrogen Loading from Everglades Runoff. In *The Everglades, Florida Bay, and Coral Reefs of the Florida Keys: An Ecosystem Sourcebook*; Porter, J.W., Porter, K.G., Eds.; CRC Press: Boca Raton, FL, USA, 2002; pp. 629–648.
5. Lapointe, B.E.; Matzie, W.R.; Barlie, P.J. Anthropogenic nutrient enrichment of seagrass and coral reef communities in the Lower Florida Keys: Discrimination of local *versus* regional nitrogen sources. *J. Exp. Mar. Biol. Ecol* **2004**, *308*, 23–58.
6. Lapointe, B.E.; Matzie, W.R.; Clark, M.W. Phosphorus Inputs and Eutrophication on the Florida Reef Tract. In *Global Coral Reefs: Health, Hazards, and History*; Ginsberg, R.N., Ed.; Rosenstiel School of Marine and Atmospheric Sciences, University of Miami: Miami, FL, USA, 1993; pp. 106–112.
7. Porter, J.W.; Kosmynin, V.; Patterson, K.L.; Porter, K.G.; Jaap, W.C.; Wheaton, J.L.; Hackett, K.; Lybolt, M.; Tosolos, C.P.; Yanev, G.; *et al.* Detection of Coral Reef Change by the Florida Keys Coral Reef Monitoring Project. In *The Everglades, Florida Bay, and Coral Reefs of the Florida Keys: An Ecosystem Sourcebook*; Porter, J.W., Porter, K.G., Eds.; CRC Press: Boca Raton, FL, USA, 2002; pp. 749–769.

8. The South-West Florida Dark-Water Observations Group (SWFDOG). Satellite images track "black water" event off Florida coast. *EoS Trans. AGU* **2002**, *83*, 281–285.

9. Neely, M.B.; Bartels, E.; Cannizzaro, J.; Carder, K.L.; Coble, O.; English, D.; Heil, C.; Hu, C.; Hunt, J.; Ivey, J.; *et al.* Florida's Black Water Event. In Proceedings of the Xth International Conference on Harmful Algae, *Saint Petersburg Beach, FL, USA, 21–25 October 2002*; Steidinger, K.A., Landsberg, J.H., Tomas, C.R., Vargo, G.A., Eds.; Florida Marine Research Institute: Saint Petersburg, FL, USA, 2004; pp. 377–379.

10. Hu, C.; Hackett, K.E.; Callahan, M.K.; Andréfouët, S.; Wheaton, J.L.; Porter, J.W.; Muller-Karger, F.E. The 2002 ocean color anomaly in the Florida Bight: A cause of local coral reef decline? *Geophys. Res. Lett.* **2003**.

11. Hu, C.; Muller-Karger, F.E.; Vargo, G.A.; Neely, M.B.; Johns, E. Linkages between coastal runoff and the Florida Keys ecosystem: A study of a dark plume event. *Geophys. Res. Lett.* **2004**.

12. Hedley, J.D.; Roelfsema, C.M.; Phinn, S.R.; Mumby, P.J. Environmental and sensor limitations in optical remote sensing of coral reefs: Implications for monitoring and sensor design. *Remote Sens.* **2012**.

13. Cannizzaro, J.P; Carder, K.L.; Chen, F.R.; Heil, C.A.; Vargo, G.A. A novel technique for detection of the toxic dinoflagellate, *Karenia brevis*, in the Gulf of Mexico from remotely sensed ocean color data. *Cont. Shelf Res* **2008**, *28*, 137–158.

14. Lee, Z.P.; Carder, K.L.; Arnone, R.A. Deriving inherent optical properties from water color: A multiband quasi-analytical algorithm for optically deep waters. *Appl. Opt* **2002**, *41*, 5755–5772.

15. Zhao, J.; Barnes, B.; Melo, N.; English, D.; Lapointe, B.; Muller-Karger, F.; Schaeffer, B.; Hu, C. Assessment of satellite-derived diffuse attenuation coefficients and euphotic depths in south Florida coastal waters. *Remote Sens. Environ***2013**, *131*, 38–50.

16. Hu, C.; Lee, Z.; Ma, R.; Yu, K.; Li, D.; Shang, S. Moderate Resolution Imaging Spectroradiometer (MODIS) observations of cyanobacteria blooms in Taihu Lake, China. *J. Geophys. Res.* **2010**.

17. Lee, T.N.; Johns, E.; Wilson, D.; Williams, E.; Smith, N. Transport processes linking south florida coastal ecosystems. In *The Everglades, Florida Bay, and Coral Reefs of the Florida Keys, an Ecosystem Source Book*; Porter, J.W., Porter, K.G., Eds.; CRC Press: Boca Raton, FL, USA, 2002; pp. 309–342.

18. Yentsch, C.S.; Lapointe, B.E.; Poulton, N.; Phinney, D.A. Anatomy of a red tide bloom off the southwest coast of Florida. *Harmful Algae* **2008**, *7*, 817–826.

19. Hu, C.; Cannizzaro, J.; Carder, K.L.; Muller-Karger, F.E.; Hardy, R. Remote detection of Trichodesmium blooms in optically complex coastal waters: Examples with MODIS full-spectral data. *Remote Sens. Environ* **2010**, *114*, 2048–2058.

20. Bell, P.R.F. Eutrophication and coral reefs—Some examples in the Great Barrier Reef lagoon. *Water Res* **1992**, *26*, 553–568.

21. Lapointe, B.E.; Bedford, B.J.; Littler, M.M.; Littler, D.S. Shifts in coral overgrowth by sponges and algae. *Coral Reef.* **2007**.

22. Yentsch, C.S.; Yentsch, C.M.; Cullen, J.J.; Lapointe, B.E.; Phinney, D.A.; Woodman, S.F. Sunlight and water transparency: Cornerstones in coral research. *J. Exp. Mar. Biol. Ecol* **2002**, *268*, 171–183.

23. Howarth, R.; Anderson, D.; Cloem, J.; Elfring, C.; Hopkinson, C.; Lapointe, B.; Malone, T.; Marcus, N.; McGlathery, K.; Sharkley, A.; *et al*. Nutrient pollutions of coastal rivers, bays, and seas. *Issue. Ecol* **2000**, *7*, 1–15.

24. Lapointe, B.E.; Bedford, B.; Baumberger, R; Looe Key, FL. Nutrients and Climate Change Pose Threat to Coral Reefs. In *Effects of Nutrient Enrichment in the Nation's Estuaries: A Decade of Change*; Bricker, S., Longstaff, B.J., Dennison, W.C., Jones, A.B., Boicourt, K., Wicks, E.C., Woerner, J.L., Eds.; National Centers for Coastal Ocean Science: Silver Spring, MD, USA, 2007; pp. 104–105.

7
CHAPTER

SATELLITE-DERIVED PHOTIC DEPTH ON THE GREAT BARRIER REEF: SPATIO-TEMPORAL PATTERNS OF WATER CLARITY

Scarla Weeks, P. Jeremy Werdell, Britta Schaffelke, Marites Canto , Zhongping Lee , John G. Wilding and Gene C. Feldman

Keywords: Water clarity, photic depth, Secchi depth, satellite, Great Barrier Reef, spatio-temporal patterns, biophysical processes

INTRODUCTION

The quantification of environmental variables in space and time is essential to understand the ecology of marine organisms and their responses to a changing environment. For benthic organisms, temperature, salinity and the availability of light and nutrients are key parameters controlling their distribution and productivity. Ocean color satellites, such as the NASA Sea-viewing Wide Field-of-view Sensor (SeaWiFS) and the NASA Moderate Imaging Spectroradiometer onboard Aqua (MODIS-Aqua), provide oceanographic data records on spatial and temporal scales unattainable using conventional *in situ* sampling schemes. The current state of operational satellite remote sensing is that algorithms for measuring geophysical parameters, such as chlorophyll concentration or water clarity, are reasonably reliable over deep water, but still limited in optically shallow regions. Considerable progress has, however, been made in the application of polar orbiting satellite data to mapping nearshore optical properties [1–3]. Recent progress in remote sensing applications includes the development of bio-optical algorithms for determining the photic depth ($Z_{\%}$; units of m) from the observed satellite

Citation: Weeks, S, Werdell , P., Schaffelke, B, Canto, M., Lee, Z. Wilding, J.G. Feldman, G.C. Satellite-Derived Photic Depth on the Great Barrier Reef: Spatio-Temporal Patterns of Water Clarity. Remote Sens. 2012, 4, 3781-3795, doi:10.3390/rs4123781.

radiances. $Z_{\%}$ is a measure of water clarity [4] with, for example, $Z_{1\%}$ reflecting the depth where only 1% of the surface irradiance (PAR, photosynthetic available radiation; units of $\mu E \cdot cm^{2} \cdot s^{-1}$) remains. The value of 1% of surface PAR has been conveniently considered as an ecologically important threshold, the "euphotic depth", or lower limit of the euphotic zone, or "compensation depth", a depth below which most photo-autotrophic organisms cannot achieve positive net daily production [5,6]. The euphotic depth relates inversely to light attenuation and, thus, provides a direct, intuitive measure of the clarity or transparency of the overlying water column [5].

Detecting changes to the transparency of the water column is critical for understanding the responses of benthic organisms to light availability, especially in shallow coastal and shelf regions, such as the Great Barrier Reef (GBR), Australia [7,8]. Degradation in the condition of coastal and inshore ecosystems due to poor water quality is a key issue for the GBR [9], as terrestrial runoff of excess nutrients and sediments affects turbidity and sedimentation regimes and intermittently increases water column productivity, in turn leading to the deterioration of inshore coral reefs at local scales [10]. Water clarity is considered a key measure of water quality on the GBR, with Secchi depth (Z_{SD}; units of m) being the most widely used measure. Measurements of optical properties of the water are usually carried out during dedicated bio-optical research cruises and, therefore, are limited in number, while Z_{SD} is a common and easily obtained parameter in monitoring, making it suitable for satellite data evaluation [11]. A water quality trigger level of $Z_{SD} < 10$ m has been formulated for the GBR [12], below which macroalgal cover rapidly increases and coral richness declines [8]. However, the links between reef health and water quality at GBR-wide scales remain the subject of debate due to the limited availability of large-scale data. For large ecosystems, such as the GBR, efficient ocean-color algorithms that can produce accurate synoptic satellite imagery and estimates of water-column optical properties, such as water clarity, are essential for understanding the links between oceanographic processes and the biological response.

The circulation in the GBR region is complex, with variability in the physical processes significantly influencing the patterns of water clarity. The GBR ecosystem extends along the continental shelf of northeastern Australia between 9°S and 24°S and covers ~345,000 km², including approximately 2,900 coral reefs. The direction of water currents on the inner GBR shelf is primarily northward, driven by the predominant southeast wind regime. Along the outer shelf, the South Equatorial Current bifurcates to form the northward-flowing Hiri Current and the southward-flowing East Australian Current (EAC) [13]. The location of the bifurcation varies seasonally between 14°S, during austral summer, when the EAC is strongest, and 20°S during austral winter. Drifter studies have suggested another bifurcation at around 19°S, and within the GBR lagoon, a strong and consistent northwest flow north of 18°S and a weaker and more variable flow south of 18°S, with limited water exchange between the two latitudinal regions [14]. In the southern GBR, the deep (80–150 m) Capricorn Channel separates the inner and the outer shelf into two distinct reef systems, while a stable cyclonic eddy, the Capricorn Eddy, typically forms in the lee of the shelf bathymetry [15]. Estimates of water residence times for the GBR lagoon differ between recent studies, ranging from weeks [14,16,17] to several months [13,18]. Analyses of satellite imagery of riverine flood plumes suggest water residence times of several weeks in the coastal and inshore GBR [19,20].

The primary objectives of this study were: (1) to develop a photic depth product for the GBR; (2) to detect changes in GBR water clarity in space and time based on the optimal, available algorithm for light attenuation through the water column; and (3) to determine the dominant modes of variation and identify the physical drivers that influence the spatio-temporal patterns of water clarity across the GBR ecosystem.

DATA AND METHODS

Since sufficient GBR *in situ* optical data were not readily available for validation purposes, regional measurements of Z_{SD} collected by the Australian Institute of Marine Science (AIMS) and the Queensland Department of Primary Industries and Fisheries (QDPI) were used to serve as a first order ground-truth for the satellite data products, with over 5,000 records acquired. These data were somewhat biased towards the inshore region (Figure 1), due to the focus of the regional research and monitoring programs, with Z_{SD} often quite low (<5 m). Such values are not unusual in GBR inshore waters, which experience intermittent high turbidity due to resuspension and flood inputs. Higher Z_{SD} values were generally found further offshore. The timeframes of the data (1994–1999; 1992–2010) largely pre-dated the MODIS-Aqua satellite data (2002–2010), hence SeaWiFS data (1997–2010) were also used to validate and improve the photic depth product for the GBR ecosystem.

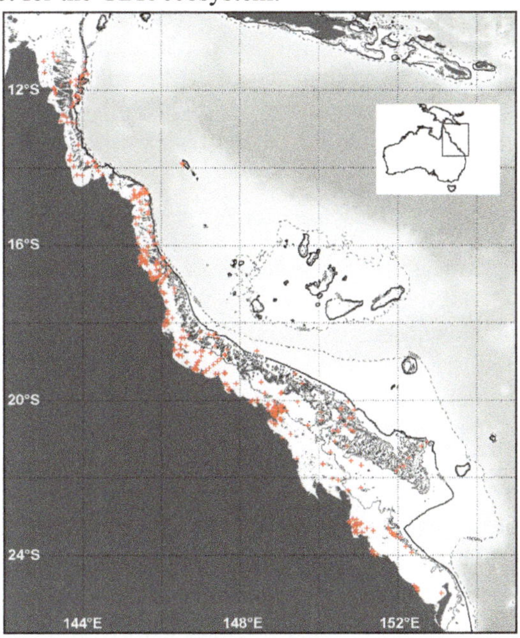

Figure 1. Map of Secchi depth measurement locations (red) in the Great Barrier Reef from 1994 to 2010.

Two operational photic depth algorithms have been implemented into the standard NASA ocean color satellite data processing environment, both of which are available via the SeaWiFS Data Analysis System (SeaDAS;http://oceancolor.gsfc.nasa.gov/seadas): an empirical algorithm based on Case-1 assumptions [21] and a semi-analytical algorithm based on the inherent optical properties (IOPs) of the water column [4]. The IOP approach, where photic depth is described as a function of spectral absorption and backscattering coefficients, was selected for this study as the water-clarity algorithm to determine light penetration depth in eutrophic, Case-2 coastal waters [6]. This algorithm has been tested and validated for oceanic and coastal waters in the Arabian Sea, Monterey Bay, Gulf of Mexico and Yellow Sea [4], but not in the GBR.

In this study, IOPs from the quasi-analytical algorithm (QAA; [2]) were used to determine the depth where 10% of the surface light (PAR) level was still available, $Z_{10\%}$, which was in turn regionally tuned to Z_{SD} using "matchups" with *in situ*Z_{SD} data. The satellite-to-*in situ* matchups for the AIMS and QDPI *in situ* Z_{SD} dataset were acquired from the NASA Ocean Biology Processing Group (OBPG) [22]. Matchups were generated for both SeaWiFS and MODIS-Aqua. These matchups were generated using the methods and quality assurance procedures described in Bailey and Werdell [23], which included: (1) locating satellite observations made within ± 3-hours of the time of *in situ* data collection; (2) calculating the filtered median of all unflagged pixels within a 5 × 5 pixel box centered on the *in situ* target; and (3) excluding any matchups where the coefficient-of-variation of the unflagged satellite pixels exceeded 0.15. To the best of our knowledge, we excluded matchup stations in optically shallow water, where light reflected off of the sea floor contributes to the water-leaving signal observed by the satellite.

A regression of the *in situ* Z_{SD} values against the matching satellite estimates of $Z_{10\%}$ was used to adjust the satellite-derived $Z_{10\%}$ to Z_{SD}. A Type II linear regression (RMA) of log-transformed satellite and *in situ* data was used to estimate Z_{SD} for the GBR according to:

$$Z_{SD} = 10\text{^}[\{\log_{10}(Z_{10\%}) - a_0\}/a_1] \quad (1)$$

where a_0 and a_1 are 0.518 and 0.811 for SeaWiFS ($N = 235$; $r^2 = 0.78$; RMSE = 0.148) and 0.529 and 0.816 for MODIS-Aqua ($N = 71$; $r^2 = 0.83$; RMSE = 0.096). This method for regional tuning is ultimately insensitive to our choice of a specific $Z_{\%}$, as the assignment of percent light level in Lee *et al.*[4] changes only the absolute magnitude of the depth estimates, not their variability. For example, had we used $Z_{37\%}$ in Equation (1), a_0 and a_1 would be different than reported above, but their derived Z_{SD} would not. Finally, this GBR-validated Z_{SD} algorithm was implemented into the SeaDAS environment and applied to the full regional time series of MODIS-Aqua data (2002–2012). Daily, monthly and decadal means and anomalies were generated at 1-km^2 resolution and analyses undertaken. For the purpose of this study, the analyses focused on the central to southern GBR (S-GBR, 18°S–24°S).

To represent the temporal evolution of water clarity on the S-GBR over the last decade (July 2002 to June 2012), latitude-time Hovmöller diagrams were generated for both the S-GBR continental shelf and inshore lagoon regions. Bathymetry masks were created to limit the spatial domains to 0–200 m and 0–35 m for the shelf and inshore regions, respectively, excluding topographic reef features. Monthly mean Z_{SD} values were averaged along each ~0.01° line of latitude (1-km resolution) for each of the selected masked areas. Pixels coinciding with missing values (<3%) and over topographic reef features were ignored in the computation.

To characterize the variability of the satellite-derived Z_{SD} on the S-GBR, an Empirical Orthogonal Function (EOF) Analysis was performed on the ten-year monthly Z_{SD} dataset. An EOF Analysis is a statistical technique used to decompose the spatial and temporal variability of time series datasets into a few independent dominant modes, each representing a certain percentage of the total variance of the dataset. In this study, we employed the singular value decomposition (SVD) method [24] to calculate the spatial eigenfunctions and temporal eigenvectors. The former describes the spatial pattern associated with the mode, while the latter provides the corresponding temporal variation, which can be interpreted in relation to physical processes. The EOF assumes that there is no gap in the data time series. Hence, prior to doing the analysis, missing data over small gaps (<3% in total) were filled for each monthly dataset, using a kernel interpolation [25]. The bathymetry mask was used to limit the spatial domain to the shelf area (0–200 m). The long-term (10 years) temporal mean was removed from the time series at each pixel location prior to analysis to reduce the dominant seasonal signal [24].

RESULTS AND DISCUSSION

The results of the satellite-to-*in situ* "matchups" for the AIMS/QDPI Z_{SD} dataset were far better than expected, given that there had been no regional refinement of the IOP photic depth algorithm. The results showed primarily a simple bias offset between the *in situ* and satellite retrievals, with similar slopes and intercepts for SeaWiFS and MODIS-Aqua (Equation (1)). Using a linear regression of log-transformed data to generate the GBR-validated Z_{SD} algorithm eliminated negative values returned for Z_{SD} when the regression was applied in normal space. Figure 2 shows the MODIS-Aqua GBR satellite-to-*in situ* matchups tuned using log-transformed data in the regression.

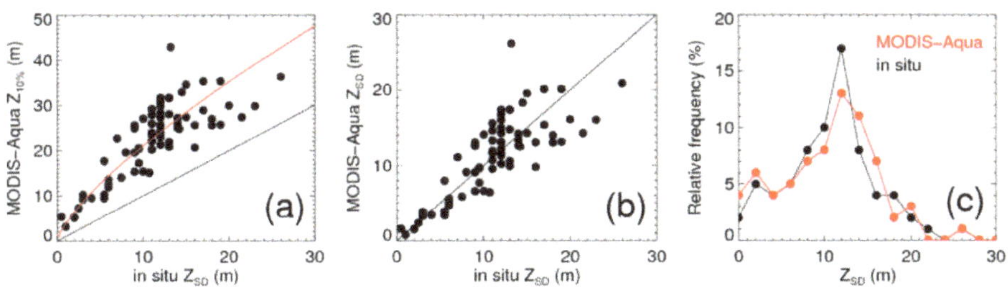

Figure 2. MODIS-Aqua GBR satellite-to-*in situ* matchups using the GBR-validated algorithm. Panel (**a**) shows *in situ* Z_{SD} *versus* MODIS-Aqua $Z_{10\%}$ with the best fit (Equation (1)) shown in red. Panels (**b**) and (**c**) compare *in situ* Z_{SD} with MODIS-Aqua Z_{SD} from the GBR-validated algorithm. A Type II linear regression analysis of the results presented in (b) yields N = 71, r^2 = 0.63, RMSE = 3.45 and slope = 1.03, and intercept =−0.25).

The GBR-validated Z_{SD} time series of monthly and decadal means generated from MODIS-Aqua data showed considerable seasonality in water clarity, with distinct patterns of spatial distribution. On the GBR shelf, inshore waters were generally more turbid than further offshore, with waters most turbid during March and most transparent during September (Figure 3). This distinct seasonal variation is due to variability in the ocean-atmosphere dynamics, with more turbid waters in the austral summer (Figure 3(a)), the "wet season" on the GBR, when the prevailing north-westerlies and the Coriolis effect result in primarily offshore surface flow of sediment-laden waters from river discharge and increased vertical mixing. During austral winter, or the "dry season" on the GBR, the prevailing southeasterly trade winds dominate, limiting cross-shelf surface flow. Furthermore, seasonal strengthening of the southward-flowing EAC in early spring (September–October), when the opposing trade winds begin to subside [26–28], results in strong intrusions of clear, oligotrophic EAC waters through the Palm and Magnetic Passages in the central GBR [29], with southward flow along the mid-shelf channel (Figure 3(b)). At the southern end of the GBR, the Capricorn Eddy forms most strongly during spring when the EAC strengthens, with resultant intrusions of clear EAC waters into the Capricorn and Curtis Channels [15], further enhancing separation of the inner and outer shelf regions (Figure 3(b)).

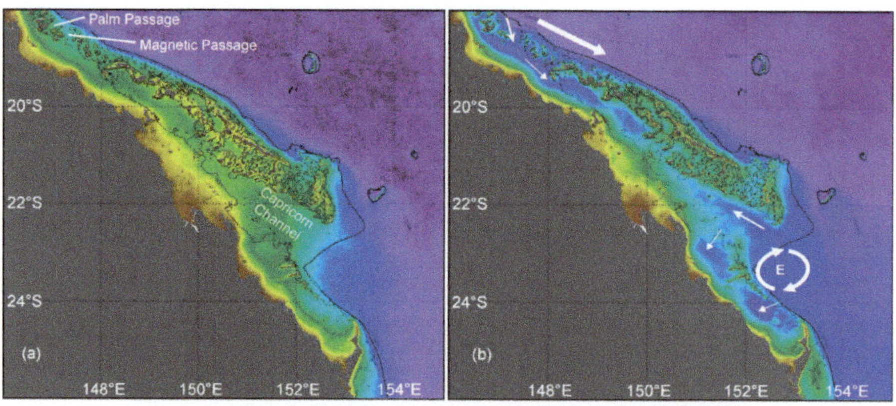

Figure 3. Long-term (2002–2012) monthly mean Secchi depth (m) on the central-southern Great Barrier Reef showing waters most turbid (**a**) during March and most transparent (**b**) during September. The solid black line denotes the 200 m isobath. The large white arrow represents the southward flow of the East Australian Current, with smaller white arrows showing the directions of oceanic intrusions onto the shelf. The circular arrows show the location of the Capricorn Eddy (E) that forms in the lee of the shelf bathymetry.

The Z_{SD} anomalies clearly showed areas and times where water clarity was anomalously poor, following periods of increased cyclonic or storm activity. For example, intense positive anomalies followed excessive storm activity in March 2009, with the cross-shelf extent and turbid water track of Tropical Cyclone (TC) Hamish, a powerful category 5 storm, clearly visible along the continental shelf edge of the

S-GBR, even extending offshore as far as the 1,000 m isobath (Figure 4(a)). Similarly, the newly implemented GBR Z_{SD} product allowed near-realtime tracking of the intrusion of anomalous conditions following TC Yasi (category 5) that impacted the central GBR in February 2011 (Figure 4(b)). In contrast, the strongest negative anomalies occurred in September 2008 (Figure 4(c)), when intrusions of clear oceanic waters appeared particularly intense, both into the central GBR channel and from the south, into the Capricorn and Curtis Channels.

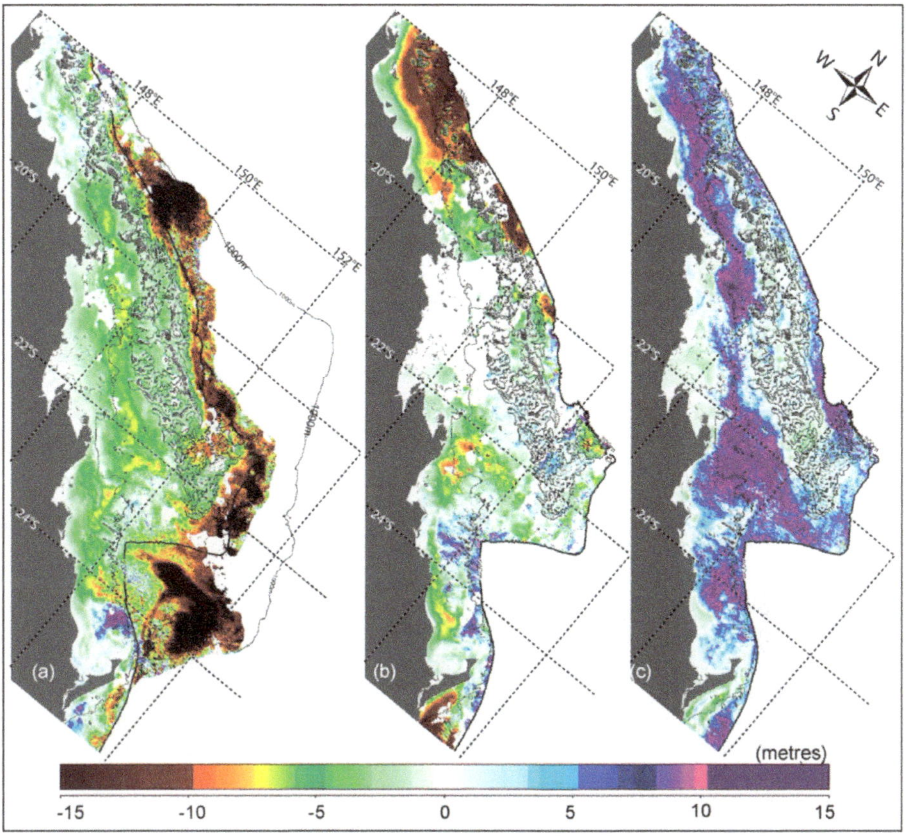

Figure 4. Secchi depth anomalies (m) on the central-southern Great Barrier Reef (panel orientations rotated by 40°). The dark and light black lines denote the 200 m and 1,000 m isobaths, respectively. Panel (**a**) shows intense positive anomalies and the turbid water track of Tropical Cyclone (TC) Hamish along the shelf edge from 14 to 17 March 2009. Panel (**b**) shows intrusions of turbid waters on 4–6 February 2011 following TC Yasi; and panel (**c**) shows strong negative anomalies from 12 to 17 September 2008 following intrusions of clear oceanic waters onto the shelf.

The Hovmöller diagrams illustrate the variation of water clarity on the S-GBR in space (latitude) and time (July 2002 to June 2012), both across the entire continental shelf (Figure 5) and along the inshore region (Figure 6). A distinct seasonal signal is apparent on the S-GBR shelf (Figure 5), with water generally more turbid in late summer (February to April) and clearest in late winter/early spring (August to October). Spatially, mean water clarity across the shelf is higher in latitudinal bands coincident with intrusions of oceanic waters in the central (between 18.25°S and 19.25°S) and southern GBR (between 22.25°S and 23.25°S). However, most notable is the decline in water clarity in recent years, with a significant decrease in mean shelf Z_{SD} of 2.00 m (from 13.9 m to 11.9 m) over the last decade, after removal of the seasonal signal ($F = 10.58$; $p < 0.0015$; $n = 120$). A progressive increase in space-time for which the average cross-shelf Z_{SD} values remained below the critical water quality level for the GBR of <10 m Z_{SD}[8] is also apparent. Poor water clarity was especially pronounced in April to May 2012 (Figure 5), when mean cross-shelf Z_{SD} values of <8 m were experienced in the S-GBR, declining at locations to as low as 6 m in May, likely due to strong vertical mixing and intrusions onto the shelf of anomalously turbid waters following sustained, strong winds in the greater study region (Australian Bureau of Meteorology,http:// www.bom.gov.au/).

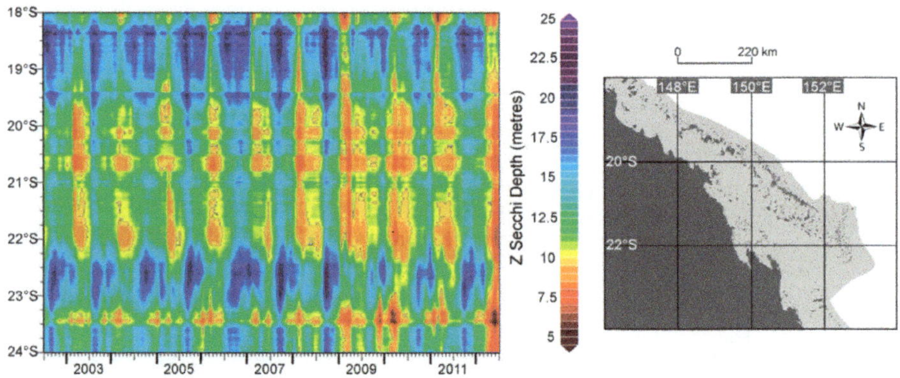

Figure 5. Hovmöller diagram illustrating the variation of water clarity on the central-southern Great Barrier Reef in space (18°S–24°S) and time (July 2002–June 2012). Areas with mean cross-shelf Secchi depth <10 m are highlighted by grey contour. The right-hand panel shows the bathymetry mask used to limit the spatial domain to 0–200 m, excluding reef features.

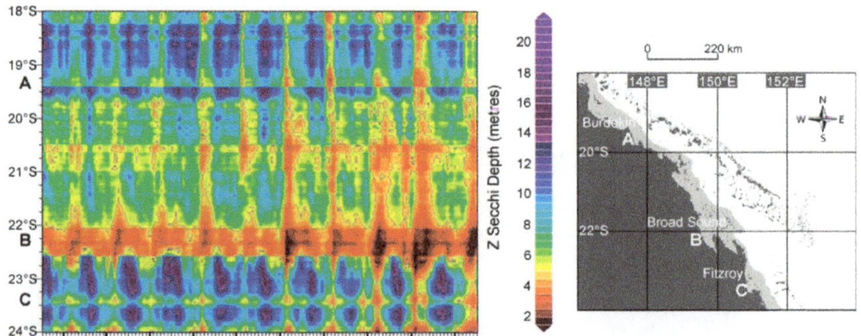

Figure 6. Hovmöller diagram illustrating the variation of water clarity on the inner shelf of the

central-southern Great Barrier Reef in space (18°S–24°S) and time (July 2002–June 2012). Areas with mean Secchi depth <5 m are highlighted by grey contour. (**Right**) panel shows the bathymetry mask used to limit the spatial domain to 0–35 m, excluding reef features. The locations of (A) the Burdekin River, (B) Broad Sound, and (C) the Fitzroy River are annotated (**Left**).

On the inner shelf (0–35 m), a distinct seasonal signal is also apparent (Figure 6), along with the turbid water signatures of the seasonal riverine inputs from the Burdekin (~19.4°S) and Fitzroy Rivers (~23.5°S), the two largest rivers in Queensland, which had above-median flows for the Burdekin from 2007–2012 and above-median flows for the Fitzroy in 2008 and 2010–2012, respectively. Also clearly visible is Broad Sound (~22.1°S–22.4°S), an embayment where high tidal energy causes strong vertical mixing of the shallow waters, leading to low water clarity all-year-round. The decline in water clarity in recent years is even more pronounced in this inshore area where a significant decrease in mean inner shelf Z_{SD} of 2.1 m (from 8.3 m to 6.2 m) occurred over the last decade, after removal of the seasonal signal ($F = 10.58$; $p < 0.0015$; $n = 120$). From summer 2007/08 onwards, the areas and time periods markedly increased where Z_{SD} values remained below 5 m (Figure 6), which is half of the Z_{SD} trigger level for GBR waters [8,12]. These negative trends in water clarity coincided with the five years during which the GBR experienced strong La Niña conditions in 2007/08, 2008/09, 2010/11 and 2011/12 (Australian Bureau of Meteorology). In northeastern Australia, La Niña events are typically associated with an earlier onset of and a more vigorous summer monsoon, characterized by enhanced tropical cyclone activity and wetter summers [30–33]. The associated increased river discharge and vertical mixing is likely to be the main reason for the measured decrease in water clarity, especially of the shallow inner shelf waters.

The EOF Analysis decomposed the spatial and temporal variability of the Z_{SD} time series on the S-GBR shelf over the last decade into a few independent dominant modes, each representing a certain percentage of the total variance of the dataset. Only the first two dominant modes are presented here, explaining 51.3% and 10.7% of the total variance, respectively. Mode 1 (Figure 7), which accounts for over half of the total variance of the dataset, indicates a strong seasonal signal in photic depth variability on the S-GBR, with considerable interannual variation. The spatial pattern (Figure 7(a), normalized from −1.0 to +1.0) shows a consistent phase (positive in Figure 7(a)) across the entire shelf, with the most positive values (purple) in the mid-shelf region. The temporal eigenvectors for Mode 1 (Figure 7(b)) provide the temporal variation associated with the spatial pattern (Figure 7(a)) over the 10-year period. More positive eigenvalues indicate months when the shelf waters were clearer, whereas more negative eigenvalues represent months when water clarity was poorer. The greatest variance explained by Mode 1 occurred at locations where values are most positive in the spatial map (Figure 7(a)), with the clearest waters here during months when the eigenvalues (Figure 7(b)) were most positive, or conversely, the most turbid waters here during months when the eigenvalues were most negative. A pronounced pattern is apparent in the temporal variation, with more turbid waters in the first half of the year, generally from March to May, and clearer waters in the latter half, generally in September and October. Interannual variation is apparent across the time series with a change to reduced water clarity in more recent years.

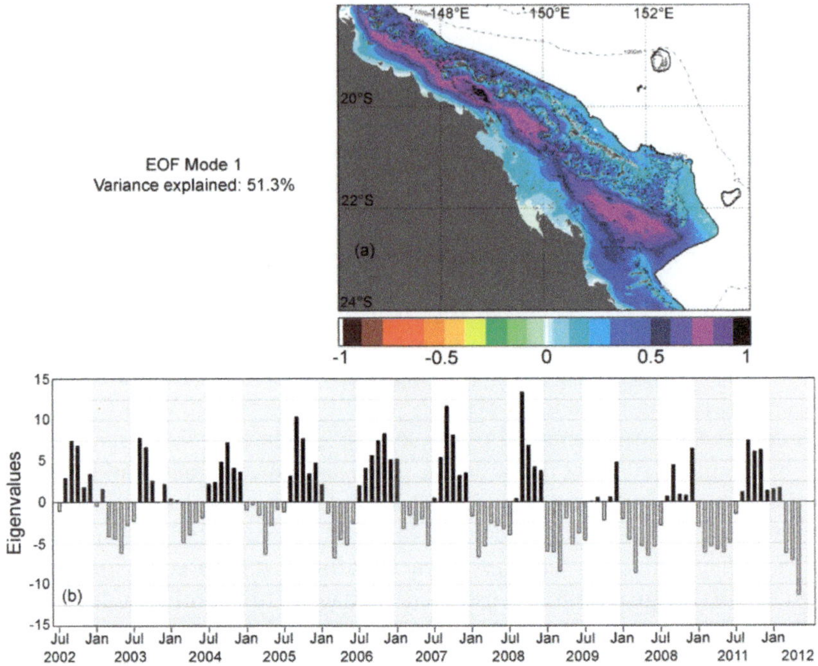

Figure 7. Mode 1 of the Empirical Orthogonal Function Analysis explaining 51.3% of the total variance of water clarity on the central-southern Great Barrier Reef from July 2002 to June 2012. The spatial pattern (panel (**a**), normalized from −1.0 to +1.0) shows a positive phase across the entire shelf with the most positive values (purple) in the mid-shelf region. Panel (**b**) provides the temporal variation associated with the spatial pattern over the 10-year period.

However, most notable in Figure 7 is that the greatest variation in water clarity occurred along the mid-shelf, not in the inshore region directly exposed to seasonal river outflow, inferring a strong influence of oceanic intrusions during certain months. The combined spatial and temporal patterns show the clearest waters intruding through the Palm and Magnetic Passages and along the mid-shelf channel, and similarly, in the Capricorn Channel in the south. Furthermore, water clarity is generally greatest during early spring (Figure 7(b)), when the EAC strengthens and the southeast trade winds relax. This springtime acceleration of the EAC is known to lead to intrusions in the central GBR [27,29] and further to the south, the Capricorn Eddy forms most strongly and regularly in spring, with associated intrusions up the Capricorn Channel [15,28]. These intrusions onto the mid-shelf would somewhat separate the water bodies of the inner-shelf from the outer shelf waters. During the early months of the year, when the GBR is under the influence of the summer monsoon and associated increased river outflows, northwesterly winds dominate, enhancing cross-shelf flow of more sediment-laden waters. The influence of the strong La Niña conditions experienced over the recent five years is apparent in the interannual variation (Figure 7(b)), where enhanced tropical cyclone activity, wetter summers, stronger northwesterlies and increased vertical mixing have impacted the entire shelf waters for many consecutive months.

The dominance of Mode 1 (51.3%) further reinforces the strong influence of the varying physical dynamics on the spatio-temporal patterns of water clarity on the S-GBR.

Mode 2 of the EOF Analysis (Figure 8) explains 10.7% of the total variance of Z_{SD} on the S-GBR shelf over the last decade. The spatial pattern (Figure 8(a), normalized from −1.0 to +1.0) shows generally positive values inshore and along the outer reefs, contrasting with negative values along the shelf edge, the mid-shelf and in the south. The associated temporal eigenvectors for Mode 2 (Figure 8(b)) show generally positive eigenvalues in autumn and winter, the "dry season" on the GBR, with negative eigenvalues in spring and summer during the "wet season". Hence, during spring/summer for example, the negative eigenvalues combined with positive loadings in the spatial map show areas of poor water clarity and, where combined with negative spatial loadings, show areas of increased water clarity during those same months.

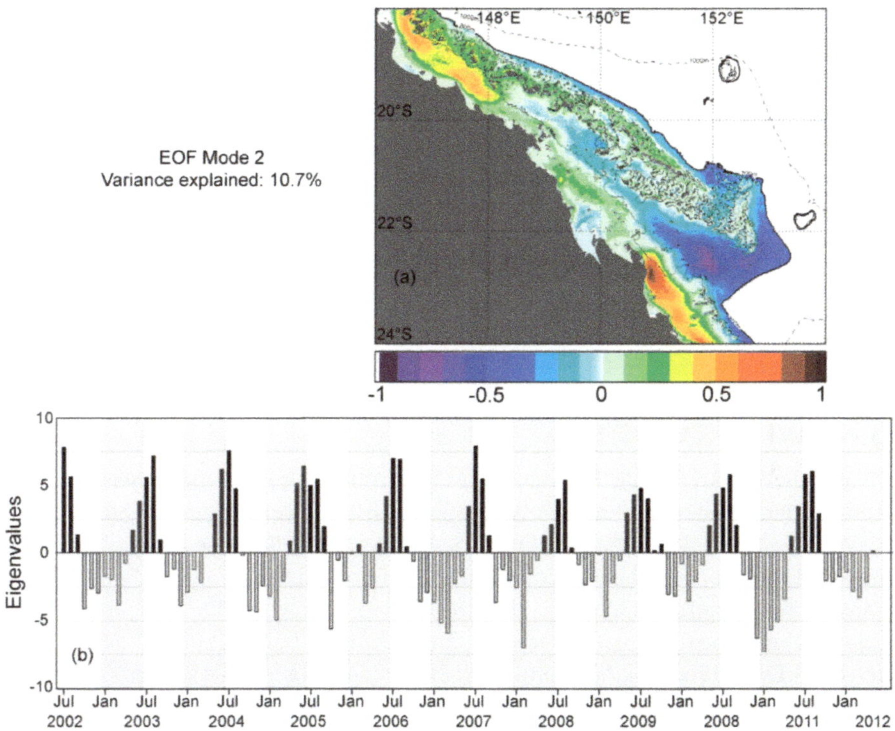

Figure 8. Mode 2 of the Empirical Orthogonal Function Analysis explaining 10.7% of the total variance of water clarity on the central-southern Great Barrier Reef (GBR) from July 2002 to June 2012. The spatial pattern (panel (**a**), normalized from −1.0 to +1.0) shows strong positive values intruding on the central GBR, and opposing positive/negative values in the south, relating most turbid waters coincident with strong East Australia Current dynamics. Panel (**b**) provides the temporal variation associated with the spatial pattern over the 10-year period.

In the central GBR, the greatest variance appeared on the mid-shelf, primarily due to increased vertical mixing and intrusions onto the shelf of more turbid waters during summer, such as occurred when TC Yasi approached the central GBR in January 2011, as opposed to clearer waters in this area during winter. However, the greatest variance explained by Mode 2 occurred in the south, largely representing the offshore and spatial extent of river outflow. Here, the opposing positive/negative values in the spatial map show that the most turbid waters coincided with a strong EAC in summer when clear oceanic waters flowed along the shelf-edge, skirted the Swains reefs in the south and intruded up the Capricorn Channel, containing any large river outflow inshore. This is especially notable in the long wet summer of 2010/11, when southeast Queensland experienced extreme flooding. The 2010–2011 wet season was characterized by extreme events in the GBR region, driven by a very strong La Niña, with the intense and prolonged rainfall resulting in the Fitzroy River having its largest flow in the instrumental record (approximately 38 million ML) [34]. However, of interest is that while the variance explained by Mode 2 appears related to storms and cyclonic events in summer, and consequent increased river discharge, variability in water clarity in the S-GBR was generally not most pronounced directly inshore where Z_{SD} remained persistently low.

CONCLUDING REMARKS

The Z_{SD} product generated from the implementation of the GBR-validated Z_{SD} algorithm allowed us to produce a long-term time series of water clarity data since 2002. Potential future applications of these data are, e.g., (i) to define baseline ranges of water clarity along and across the GBR to identify and quantify future changes in water clarity as a result of land use practice changes or other environmental changes such as climate change; and (ii) to analyze spatial and temporal patterns of change in reef communities in response to changes in water clarity.

ACKNOWLEDGMENTS

We thank the NASA Ocean Biology Processing Group for assistance with the large volume of data processing and the development of the web-based validation search/display utility. Special thanks to B. Franz, S. Bailey, and C. Proctor for implementation of the Secchi depth algorithm into the SeaDAS environment and also to Bruce Monger for kindly providing his EOF code. The Australian Institute of Marine Science (www.aims.gov.au) and the Queensland Department of Primary Industries and Fisheries (now Queensland Department of Agriculture, Fisheries and Forestry, www.daff.qld.gov.au) are gratefully acknowledged for provision of *in situ* Secchi depth data. This research was funded by the Australian Research Council (Project LP100100342) and the Great Barrier Reef Foundation. Z.P. Lee was partially supported by the NASA Energy and Water Cycle Program (#NNX09AV97G).

REFERENCES

1. Lee, Z. *Remote Sensing of Inherent Optical Properties: Fundamentals, Tests of Algorithms, and Applications*; Report of an IOCCG Working Group on Ocean-Colour Algorithms; International Ocean-Colour Coordinating Group: Dartmouth, NS, Canada, 2006.

2. Lee, Z.; Carder, K.L.; Arnone, R.A. Deriving inherent optical properties from water color: A multiband quasi-analytical algorithm for optically deep waters. *Appl. Opt* **2002**, *41*, 5755–5772.
3. Qin, Y.; Brando, V.E.; Dekker, A.G.; Blondeau-Patissier, D. Validity of SeaDAS water constituents retrieval algorithms in Australian tropical coastal waters. *Geophys. Res. Lett* **2007**, *34*, L21603.
4. Lee, Z.P.; Weidemann, A.; Kindle, J.; Arnone, R.; Carder, K.L.; Davis, C. Euphotic zone depth: Its derivation and implication to ocean-color remote sensing. *J. Geophys. Res.* **2007**.
5. Falkowski, P.G.; Raven, J.A. *Aquatic Photosynthesis*; Blackwell Science: Malden, MA, USA, 1997; p. 375.
6. Kirk, J.T.O. *Light and Photosynthesis in Aquatic Ecosystems*, 2nd ed.; Cambridge University Press: Cambridge, UK, 1994; p. 509.
7. Fabricius, K.E.; De'ath, G.; Humphrey, C.; Zagorskis, I.; Schaffelke, B. Intra-annual variation in turbidity in response to terrestrial runoff on near-shore coral reefs of the Great Barrier Reef. *Estuar. Coast. Shelf Sci* **2012**.
8. De'ath, G.; Fabricius, K. Water quality as a regional driver of coral biodiversity and macroalgae on the Great Barrier Reef. *Ecol. Appl* **2010**, *20*, 840–850.
9. Brodie, J.E.; Kroon, F.J.; Schaffelke, B.; Wolanski, E.C.; Lewis, S.E.; Devlin, M.J.; Bohnet, I.C.; Bainbridge, Z.T.; Waterhouse, J.; Davis, A.M. Terrestrial pollutant runoff to the Great Barrier Reef: An update of issues, priorities and management responses. *Mar. Pollut. Bull* **2012**, *65*, 81–100.
10. Fabricius, K.E. Effects of terrestrial runoff on the ecology of corals and coral reefs: Review and synthesis. *Mar. Pollut. Bull* **2005**, *50*, 125–146.
11. Kratzer, S.; Hakansson, B.; Sahlin, C. Assessing secchi and photic zone depth in the Baltic Sea from satellite data.*AMBIO* **2003**, *32*, 577–585.
12. Great Barrier Reef Marine Park Authority. *Water Quality Guidelines for the Great Barrier Reef Marine Park*; Great Barrier Reef Marine Park Authority: Townsville, QLD, Australia, 2009.
13. Brinkman, R.; Wolanski, E.; Deleersnijder, E.; McAllister, F.; Skirving, W. Oceanic inflow from the Coral Sea into the Great Barrier Reef. *Estuar. Coast. Shelf Sci* **2002**, *54*, 655–668.
14. Choukroun, S.; Ridd, P.V.; Brinkman, R.; McKinna, L.I.W. On the surface circulation in the western Coral Sea and residence times in the Great Barrier Reef. *J. Geophys. Res.* **2010**.
15. Weeks, S.; Bakun, A.; Steinberg, C.; Brinkman, R.; Hoegh-Guldberg, O. The Capricorn Eddy: A prominent driver of the ecology and future of the southern Great Barrier Reef. *Coral Reef* **2010**, *29*, 975–985.
16. Hancock, G.J.; Webster, I.T.; Stieglitz, T.C. Horizontal mixing of Great Barrier Reef waters: Offshore diffusivity determined from radium isotope distribution. *J. Geophys. Res.-Oceans* **2006**.
17. Wang, Y.; Ridd, P.V.; Heron, M.L.; Stieglitz, T.C.; Orpin, A.R. Flushing time of

solutes and pollutants in the central Great Barrier Reef lagoon, Australia. *Mar. Freshwater Res* **2007**, *58*, 778–791.

18. Luick, J.L.; Mason, L.; Hardy, T.; Furnas, M.J. Circulation in the Great Barrier Reef Lagoon using numerical tracers and *in situ* data. *Cont. Shelf Res* **2007**, *27*, 757–778.
19. Devlin, M.; Schaffelke, B. Spatial extent of riverine flood plumes and exposure of marine ecosystems in the Tully coastal region, Great Barrier Reef. *Mar. Freshwater Res* **2009**, *60*, 1109–1122.
20. Schroeder, T.; Devlin, M.J.; Brando, V.E.; Dekker, A.G.; Brodie, J.E.; Clementson, L.A.; McKinna, L. Inter-annual variability of wet season freshwater plume extent into the Great Barrier Reef lagoon based on satellite coastal ocean colour observations. *Mar. Pollut. Bull* **2012**, *65*, 210–223.
21. Morel, A.; Huot, Y.; Gentili, B.; Werdell, P.J.; Hooker, S.B.; Franz, B.A. Examining the consistency of products derived from various ocean color sensors in open ocean (Case 1) waters in the perspective of a multi-sensor approach. *Remote Sens. Environ* **2007**, *111*, 69–88.
22. McClain, C.R. A decade of satellite ocean color observations. *Ann. Rev. Mar. Sci* **2009**, *1*, 19–42.
23. Bailey, S.W.; Werdell, P.J. A multi-sensor approach for the on-orbit validation of ocean color satellite data products.*Remote Sens. Environ* **2006**, *102*, 12–23.
24. Yoder, J.A.; Schollaert, S.E.; O'Reilly, J.E. Climatological phytoplankton chlorophyll and sea surface temperature patterns in continental shelf and slope waters off the northeast US coast. *Limnol. Oceanogr* **2002**, *47*, 672–682.
25. Mühlenstädt, T.; Kuhnt, S. Kernel interpolation. *Comput. Stat. Data Anal* **2011**, *55*, 2962–2974.
26. Andrews, J.C.; Furnas, M.J. Subsurface intrusions of Coral Sea water into the central Great Barrier Reef. Structures and shelf-scale dynamics. *Cont. Shelf Res* **1986**, *6*, 491–514.
27. Burrage, D.M.; Black, K.P.; Ness, K.F. Long-term current predictions in the central GBR. *Cont. Shelf Res* **1994**, *14*, 803–829.
28. Ridgway, K.R.; Godfrey, J.S. Seasonal cycle of the East Australian current. *J. Geophys. Res.* **1997**.
29. Berkelmans, R.; Weeks, S.J.; Steinberg, C.R. Upwelling linked to warm summers and bleaching on the Great Barrier Reef. *Limnol. Oceanogr* **2010**, *55*, 2634–2644.
30. Evans, J.L.; Allan, R.J. El-Nino Southern Oscillation modification to the structure of the monsoon and tropical cyclone activity in the Australasian region. *Int. J. Climatol* **1992**, *12*, 611–623.
31. Lough, J.M. Climate Variability Change on the Great Barrier Reef. In *Oceanographic Processes of Coral Reefs: Physical and Biological Links in the Great Barrier Reef*; Wolanski, E., Ed.; CRC Press: Bocan Raton, FL, USA, 2001; pp. 269–300.
32. Hung, C.W.; Yanai, M. Factors contributing to the onset of the Australian summer

monsoon. *Quar. J. Roy. Meteorol. Soc***2004**, *130*, 739–758.

33. Lough, J.M. Climate and Climate Change on the Great Barrier Reef. In *Climate Change and the Great Barrier Reef: A Vulnerability Assessment*; Johnson, J.E., Marshall, P.A., Great Barrier Reef Marine Park Authority, Eds.; Great Barrier Reef Marine Park Authority: Townsville, QLD, Australia, 2007; p. 818.
34. Devlin, M.; Brodie, J.; Wenger, A.; da Silva, E.; Alvarez-Romero, J.G.; Waterhouse, J.; McKenzie, L. Extreme Weather Conditions in the Great Barrier Reef: Drivers of Change? Proceedings of the 12th International Coral Reef Symposium, Cairns, QLD, Australia, 9–13 July 2012.

8
CHAPTER

RS-GIS BASED ASSESSMENT OF RIVER DYNAMICS

Archana Sarkar, R. D. Garg and Nayan Sharma

Keywords: Remote Sensing, GIS, NDWI, Brahmaputra River, Bank Erosion/Deposition

INTRODUCTION

The river Brahmaputra has been the lifeline of northeastern India since ages. This mighty river runs for 2880 kms through China, India and Bangladesh. Any alluvial river of such magnitude has problems of sediment erosion-deposition attached with it; the Brahmaputra is no exception. The problems of flood, erosion and drainage congestion in the Brahmaputra basin are gigantic. The Brahmaputra River is characterized by its exceedingly large flow, enormous volume of sediment load, continuous changes in channel morphology, rapid bed aggradations and bank line recession and erosion. The river has braided channel in most of its course in the alluvial plains of Assam. The lateral changes in channels cause severe erosion along the banks leading to a considerable loss of good fertile land each year. Bank oscillation also causes shifting of outfalls of its tributaries bringing newer areas under waters. Thousands of hectares of agricultural land is suffering from severe erosion continuously in the Brahmaputra basin covering parts of states like Assam, Arunachal Pradesh, Meghalaya, Nagaland and Manipur.

Stream-bank erosion and its effects on channel evolution are essential geomorphic

Citation: A. Sarkar, R. Garg and N. Sharma, "RS-GIS Based Assessment of River Dynamics, "Journal of Water Resource and Protection, Vol. 4 No. 2, 2012, pp. 63-72, doi: 10.4236/jwarp.2012.42008.

research problems with relevance to many scientific and engineering fields. Remote sensing satellite data having ability to provide comprehensive, synoptic view of fairly large area at regular interval with quick turn-around time integrated with Geographical Information System (GIS) techniques makes it appropriate and ideal for studying and monitoring river erosion and its bank line shifting. Various studies in this regard have been carried out for some major rivers all over the world [1-6]. Several investigators have used remotely sensed data for ascertaining channel changes of Brahmaputra River and its tributaries. NRSA has done the river migration study of the Brahmaputra using airborne scanner survey and has carried out repetitive survey to monitor changes in landuse, river channels and banks to provide a base for estimating the response of the rivers to flood events [7]. Bardhan [8] studied the channel behavior of the Barak River using satellite imagery and other data to identify the river stretches, if any, which remained reasonably stable during the period 1910-1988. SAC (Space Application Centre), Ahmedabad and Brahmaputra Board jointly took up a study to access the extent of river erosion in Majuli island in order to identify and delineate the areas of the island which have undergone changes along the bank line due to dynamic behavior of the river [9]. Naik [10] studied the erosion at Kaziranga National Park using remote sensing data. Goswami [11] carried out a study on river channel changes of the Subansiri (northern tributary of Brahmaputra River) in Assam, India. Mani [12] studied the erosion in Majuli Island using remote sensing data. Bhakal [13] have quantified the extent of bank erosion in Brahmaputra River near Agyathuri in Assam, India over a period of thirty years (1973-2003) using remote sensing data integrated with GIS. Kotoky [14] studied selected reach of Brahamputra (from Panidihing Reserve Forest to Holoukonda Bil of the Brahmaputra River) with two sets of Survey of India toposheets (1914 and 1975) and a set of IRS satellite images (1998, IRS-1B, LISS II B/W geocoded data. Sarma [15] studied the nature of bank line migration of Burhi Dihing River (southern tributary of Brahmaputra River) course during a period of time from 1934 to 2004. Das and Saraf [16] made a study in respect to a trend in river course changes of Brahmaputra River and influence of various surrounding geotectonic features for varying period between 1970-2002 for different sections of the river using Landsat MSS, TM and ETM images. However, a comprehensive study of the bank erosion and channel migration of the entire Brahmaputra in India with most recent satellite data has not yet been reported in the literature. The present paper briefly describes a study of the Brahmaputra river—its entire course in Assam from upstream of Dibrugarh up to the town Dhubri near Bangladesh border for a stretch of around 620 kms for a period of 18 years (1990-2008) using an integrated approach of Remote Sensing and GIS. The satellite data has provided the information on the channel configuration of the river system on repetitive basis revealing much needed data on the changes in river morphology, erosion/deposition pattern and its influence on the land, stable and unstable reaches of the river banks, changes in the main channel of the Brahmaputra River, etc. It is envisaged that the conjunctive use of the information generated out of this study with other ground based data will contribute substantially in more meaningful approach towards planning and execution of means and measures to combat recurring floods and erosion.

STUDY AREA

The Brahmaputra River, termed a moving ocean, is an antecedent snowfed river which flows across the rising young Himalayan Range. Geologically, the Brahmaputra is the youngest of the major rivers of the world. It originates at an altitude of 5300 m about 63 km south-east of the Mansarovar lake in Tibet. The river is known as Tsangpo in Tibet. Flowing eastward for 1625 km over the Tibetan plateau, the Tsangpo enters a deep narrow gorge at Pe (3500 m) and continues southward across the east-west trending ranges of the Himalayas, viz. the Greater Himalayas, Middle Himalayas and sub-Himalayas. After crossing the Indo-China border near Pasighat the river is called as the Siang or the Dihang. Two major rivers namely the Dibang and the Lohit join the Dihang at a short distance upstream of Kobo to form the river Brahmaputra. The river flows westward through Assam for about 700 km distance from Dhola until downstream of the town Dhubri, where it abruptly turns south and enters Bangladesh. The gradient of the Brahmaputra River is as steep as 4.3 to 16.8 m/km in the gorge section upstream of Pasighat, but near Guwahati it is as flat as 0.1 m/km. In the course of its 2880 km journey, the Brahmaputra receives as many as twenty two major tributaries in Tibet, thirty three in India and three in Bangladesh. The northern and southern tributaries differ considerably in their hydro-geomorphologic characteristics owing to different geological, physiographic and climatic conditions. Due to the colliding Eurasian (Chinese) and Indian tectonic plates, the Brahmaputra valley and its adjoining hill ranges are seismically very unstable.

For the present study, a reach of 620 km on the main stem of Brahmaputra River, i.e., its entire course in Assam from upstream of Dibrugarh up to the town Dhubri near Bangladesh border has been considered. The geographical bounds of the area are 25°45' to 28°N latitude and 90° to 96°30'E longitude. **Figure 1** shows the study reach of the river.

MATERIAL AND METHOD

Data

The basic data used in this study are digital satellite images of Indian Remote Sensing (IRS) LISS-I (Path/Row 13/49; 14/47,48,49; 15/47,48; 16/47,48) and LISS-III (Path/Row 110/51,52; 111/51,52; 112/51,52; 113/51,52; 114/53) sensor, comprising of scenes for the years 1990 and 2008. The other collateral data used in the present study are Survey of India toposheets nos 78/J, K, N, O; 82/L, P; 83/B, E, F, I, J, M at 1:250,000 scale and Landsat ETM images (Path/Row 135/41, 42; 136/42; 137/42; 138/42;), which are freely downloadable from the website www.glcf.umiacs.umd.edu/data/landsat. ERDAS Imagine 9.3 image processing software has been used for processing the satellite images. ArcMap 9.3 GIS software has been used for analysis of the river bank data.

Methodology

Satellite images for years 1990 and 2008 have been processed to obtain the required information about the study area. The following section describes the preparation of satellite images for information extraction.

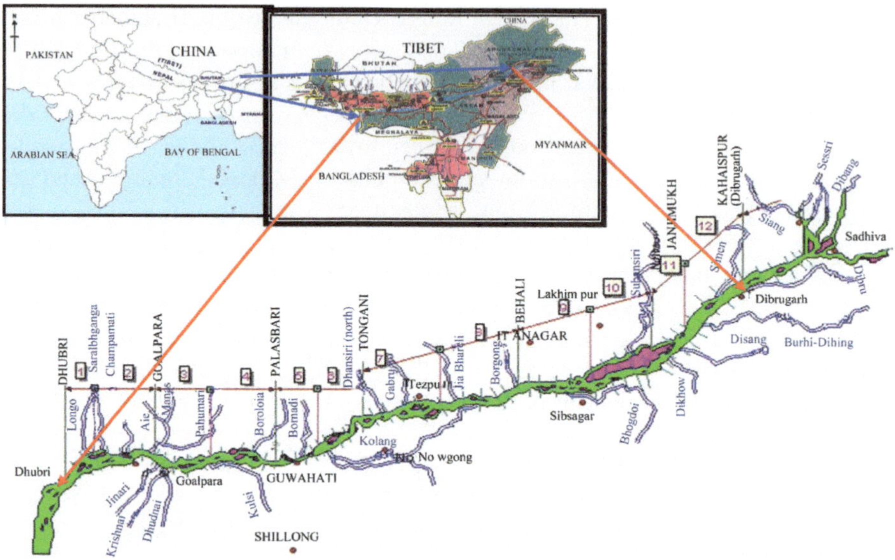

Figure 1. Study area.

Geo-Referencing and Digital Processing of Satellite Data

The satellite images procured from the NDC (NRSC Data Center), the official data supplier for IRS satellite images in India, do not have the global coordinate and projection systems. Therefore, the data need to be georeferenced before commencing the analysis. Since the topographic maps procured from Survey of India belong to year 1965, there are increased chances of error, if the toposheet data is taken as a reference for geo-referencing the satellite images. Therefore, Landsat ETM images have been taken as reference. Toposheets have been used for assistance in identification of features. The Landsat images are precisely ortho-rectified with the following parameters:

Projection Type: Universal Transverse Mercator (UTM)

Spheroid Name: WGS 84 Datum Name: WGS 84 Zone: 46-47 Landsat ETM images for the study area have been downloaded and a mosaic has been prepared. Then individual image of IRS has been registered with respect to the Landsat image. For each image an average of 18 GCPs (ground control points) have been taken and georeferencing has been carried out using second-order polynomial with nearest neighborhood resampling keeping grid size as 23 m. RMS (root mean square) error has been kept below 0.8 pixel. Survey of India toposheets have been utilized for accurately identifying the GCPs on IRS and Landsat images. The geo-referenced images of the same year have been mosaiced together. Brahmaputra River is covered with clouds for most part of the year, therefore the satellite images of one year couldn't be procured for same date or month. There is a time gap of about 2 to 4 months in the satellite images of one year. Therefore, radiometry of the images differ a lot. Making a mosaic without balancing the radiometry will not give good results. Therefore, radiometric normalization has been performed before mosaicing by histogram equalization and matching. This has given very good results, providing uniform brightness levels for similar features for the entire river basin. Also, feather option has been used for mosaicing to get the seamless boundaries between different images of same year. The mosaics for the years 1990 and 2008 have been prepared in this manner.

Preparation of NDWI Image

NDWI (Normalized Difference Water Index) shows the moisture content in soil and vegetation. NDWI image for IRS LISS-III data has been prepared using the following formula [17]

$$NDWI = \frac{SWIR - NIR}{SWIR + NIR}$$

Since the IRS LISS-I doesn't have SWIR band, therefore its NDWI image can be created using the following formula [18]

$$NDWI = \frac{Green - NIR}{Green + NIR}$$

NDWI is useful to assess water content in a normalized way. This index increases with vegetation water content or from dry soil to free water. The SWIR reflectance indicates changes in soil and vegetation water content, while the NIR reflectance is affected by internal structure of leaf and soil but not by water content directly. The combination of the NIR with the SWIR removes variations induced by internal structure of leaf and soil, thus improving the accuracy in retrieving the vegetation and soil water content [19].

Delineation/Digitization of River Bank Line

The entire river from Dhubri to upper Assam beyond Dibrugarh has been divided into 120 strips at an equal spacing of approximately 5 km, reference cross sections have been drawn at the boundary of each strip. Each ten cross sections have been grouped as a reach with numbering from downstream to upstream of the river (of equal base length). Base line of latitude 25.966°N and longitude 90°E has been taken as permanent reference line, so that they maintain their identity when the morphology is changed. **Table 1** shows some important locations within various reaches of the river.

The river bank-line has been identified and delineated for all the satellite image mosaics of 1990 and 2008. NDWI image proved very useful in identifying the river bank line. While the shallow water channels have been considered the part of river, old and new soil/sand deposits at the river banks posed some ambiguity of interpretation as the river. A few of these soil/sand patches are at considerable distance from active water channel, but have dark tone on satellite image, indicating higher moisture. On the other hand, several soil patches very close to the active water channel bear bright signature, an indication of low moisture. NDWI has been used in delineating the soil/sand areas which have higher moisture content. It has been observed that the areas with recent soil deposits have higher moisture in comparison to other areas adjacent to river bank, due to the river related activities associated with them. Either the river was flowing through that area in the recent past, or that area was submerged in water, when there was high flow in the river. These areas have been marked as a part of river. Thus the use of NDWI image helped in delineating/demarcating the river bank line.

The identified river bank lines for the left (south) and right (north) banks of the river, have been digitized using ArcMap software. Two river bank lines have been prepared for the years 1990 and 2008. The length of arcs of both the left and right banks for all the above years have been calculated using GIS software. Erosion and deposition area has been estimated through area estimation using GIS software tools for polygon areas with the shifting bank-lines in study period.

RESULTS AND DISCUSSION

Mosaic images of year 1990 and 2008 with digitised bank lines and reference cross-sections drawn are presented in Figures 2(a) and (b). **Table 2** accounts for the reach wise erosion and deposition area for the period 1990-

Table 1. Identification of reaches in respect of the location in the vicinity

Reach Number	Locations in Vicinity
1	Dhubri
2	Goalpara
3	Palasbari
4	Guwahati
5	Morigaon (Near Mangaldai)
6	Morigaon (Near Dhing)
7	Tezpur
8	U/s of Tezpur (Near Gohpur)
9	Majuli (Near Bessamora)
10	U/s of Majuli (Near Sibsagar)
11	Dibrugarh
12	U/s of Dibrugarh

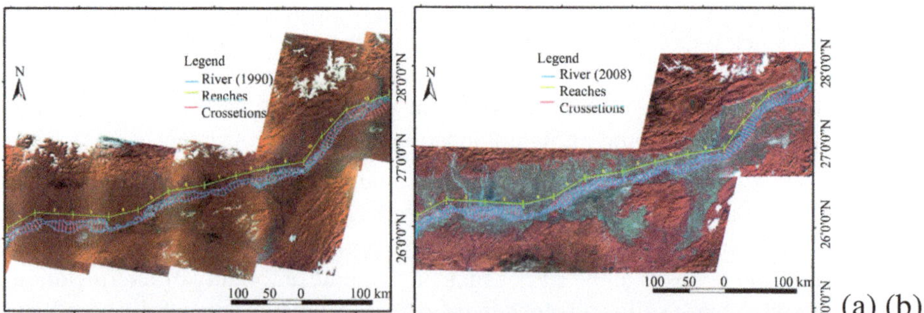

Figure 2. Satellite image of entire Brahmaputra river from U/s Dibrugarh to Dhubri (a) 1990; (b) 2008.

It is seen from **Table 2** that there has been significant erosion in the river during a period of 18 years as compared to deposition. Erosion is more pronounced in the southern bank and overall rate of erosion and deposition work out to be 62.7 km^2/year and 12.6 km^2/year respectively. The erosion rate has been calculated after excluding forest area of southern bank in Reach 12.

Figure 3 depicts the river bank lines in the year 1990 and 2008. The integration of river bank lines of year 1990 and 2008 revealed that the Brahmaputra River has shifted its bank line drastically causing severe damage to the good agriculture as well as habitat area on its both sides. The erosion sites are located throughout its reach between U/s of Dibrugarh to Dhubri within India. **Figure 4** depicts the erosion and deposition in study reach of the river for a period of 18 years. Figure 5 gives a graphical plot of reach wise total erosion/deposition in both banks of the river. It is evident from this graph that three to four major geological control points are present in the Brahamputra in Assam flood plains which are holding the river. These geological control points are in the vicinity of Jogighopa near Goalpara, Pandu near Guwahati, Tezpur and Bessemora near Majuli. Each control point usually has well defined hydrographic profile. Within geologically controlled reaches, intermittent fanning out and fanning in are exhibited which are being temporally severed in the geological increasing time scale.

Table 2. Erosion/deposition area along Brahmaputra for the period 1990 to 2008

Reach No.	South (Left) Bank		North (Right) Bank	
	Total Erosion km^2	Total Deposition km^2	Total Erosion km^2	Total Deposition km^2
1 (Dhubri)	195.005	4.410	124.461	0.698
2 (Goalpara)	18.411	3.484	79.046	3.502
3 (Palasbari)	23.663	9.208	51.970	7.425
4 (Guwahati)	6.831	0.963	4.618	10.546
5 (Morigaon)	99.799	1.490	35.781	5.877
6 (Morigaon)	11.253	8.204	29.057	10.828
7 (Tezpur)	16.628	7.766	38.758	6.566
8 (U/s Tezpur)	26.098	16.616	32.831	1.053
9 (Majuli)	32.788	9.068	25.562	13.851
10 (U/s Majuli)	43.088	0.375	64.273	0.735
11 (Dibrugarh)	47.525	3.069	37.896	1.330
12 (U/s Dibrugarh)	399.163	3.830	20.376	83.455
TOTAL	920.251	68.483	544.631	145.866

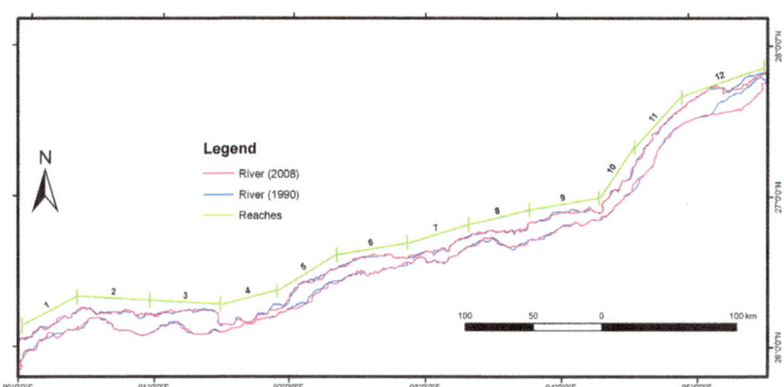

Figure 3. River bank lines in discrete years.

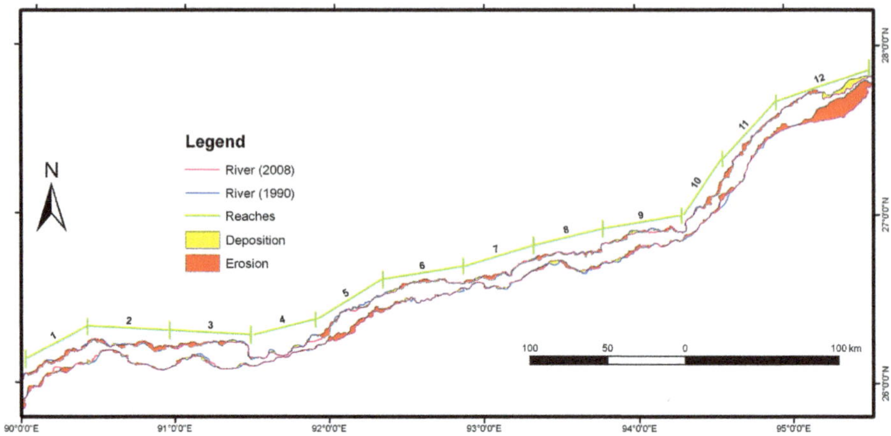

Figure 4. Erosion & deposition (1990-2008).

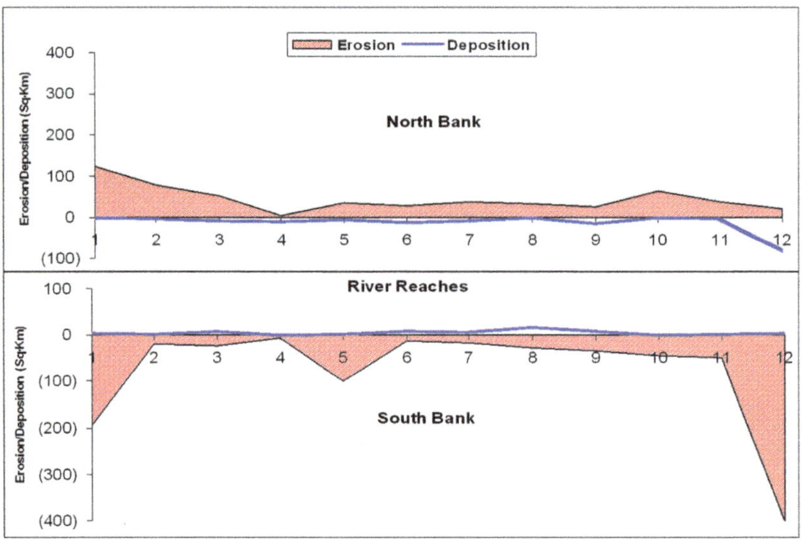

Figure 5. Total Erosion & Deposition over eighteen years (1990-2008) in various reaches of the Brahmaputra River.

Figure 6 depicts erosion and deposition in all the twelve reaches of the river. In reach-1 (in the vicinity of Dhubri) the river takes a southwesterly turn which is mainly due to the existence of a highland in the northern part of the turn which has prevented further westward flow of the river. This reach exhibits huge erosion in both the banks with more north-westward expansion. In reach- 2 and reach-3 near Goalpara and Palasbari respectively, the Brahmaputra River flows along the northern edge of the Shillong Plateau and through numerous inselbergs (granitic and gneissic hills) belonging to an exposed part of the Shillong Massif [16]. In these two reaches, right bank (northern) shows considerable shifting by following a preferred course closer to the northern edge of the Shillong Plateau with narrowing down to incised channel at Pandu near Guwahati. This change has probably occurred due to the developed lowlands along the fault affected plateau edge causing the river to slowly respond to that tectonic change. Reach-4 is in the vicinity of Guwahati and it is seen that in downstream of Guwahati, river tends to move towards north side whereas, in upstream it tends towards south side keeping the control point at Guwahati invariant. Erosion is also minimum in both the banks compared with other reaches.

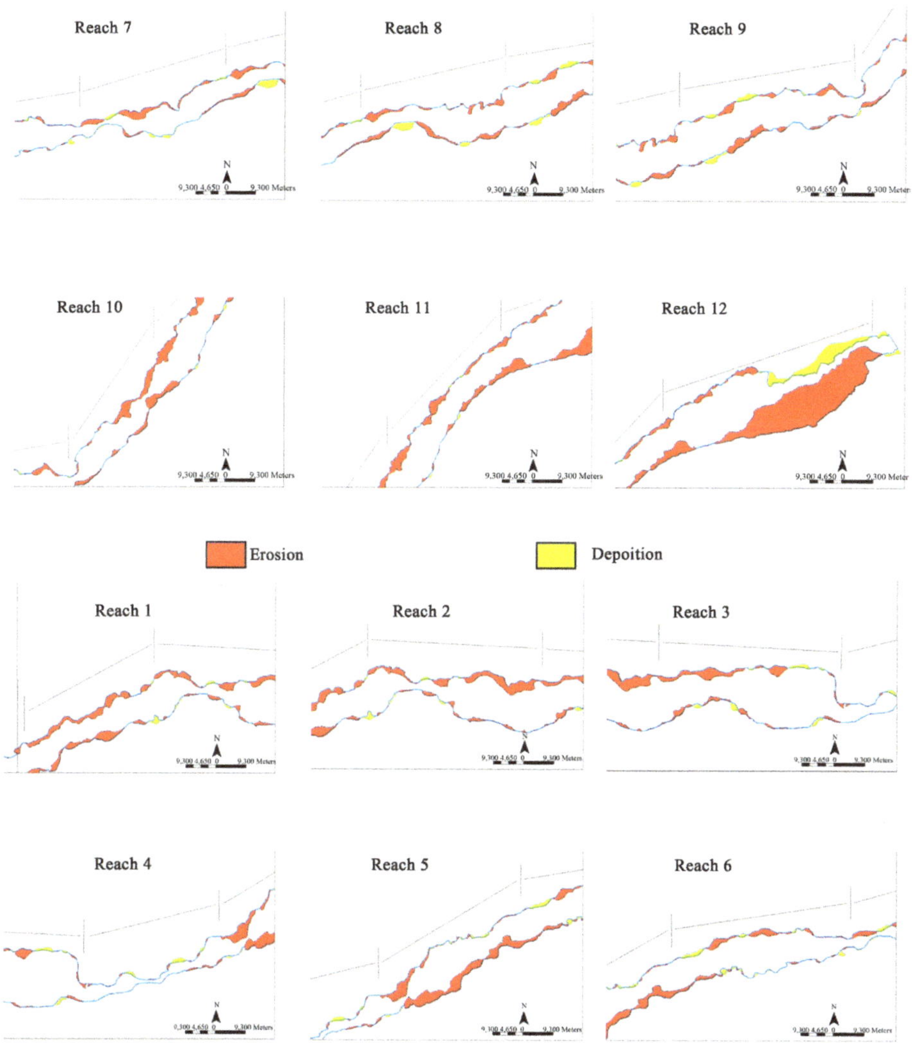

Figure 6. Erosion/deposition in various reaches of the brahmaputra river over a period of 18 years (1990-2008).

In reach-5 and reach-6 near Mangaldai and Dhing respectively, Brahmaputra River takes a huge northward curvature and the images depict huge erosion in the south bank (left bank). It is inferred that the Main Frontal Thrust (MFT) has developed in the Himalayan foothills, north of these two reaches which might have caused vertical tectonic loading of sediments of the basin resulting in the development of a northward slope and very strong northward migration. Reach-7 and reach-8 in the vicinity of Tezpur and upstream of Tezpur (near Gohpur) respecttively are north of the Mikhir Hills Massif (MHM) and the river basin becomes very narrow in these two reaches. Das & Saraf [16] have reported that the Brahmaputra River first shifted northward in front of MHM and later the river started shifting southward mainly due to the southward migration of Siwalik fold belts of the Himalayan region which is probably forcing the river to shift back in these two reaches. It is evident from the higher erosion in the north bank over last eighteen years.

Reach-9 is in the vicinity of Majuli Island and it can be seen from the river bank lines for this reach that the Brahmaputra River which was flowing in a south-westward direction upstream of this reach takes a sudden westward turn at Neamati, south of Majuli Island with higher erosion of the south bank during 1990-2008. The sudden westward turn of the river at Neamati is probably influenced by the occurrence of relatively higher landform to the south of the river and by the subsurface east-west trending Jorhat Fault zone, located along the southern bank of the river in this reach. Sarma and Phukan [20] suggested existence of the Brahmaputra River along a northern course around the present Majuli Island in the late seventeenth century but it avulsed its course southward. Reach- 10 in the vicinity of Sibsagar (U/s of Majuli) shows higher erosion in the north bank compared to the south bank during 1990-2008. This reach indicates a tendency of a northward expansion of the river course that should be due to the development of a northward slope of the basin.

The next reach, i.e., reach-11 in the vicinity of Dibrugarh shows a southward expansion with higher erosion of the south bank compared to the north bank. Although the main river channel has expanded in the southern direction, into the Dibrugarh town, the northward expansion is of much higher order. This is probably due to the formation of the northward dipping Main Frontal Thrust (MFT) in the Himalayan foothills very close to the Brahmputra River causing vertical tectonic loading of sediments resulting in the development of northward sloping land. Avulsion activities are seen during this period in the upstream of Dibrugarh (reach-12) where Lohit River changed its path and avulsed southward. This reach shows a very high amount of erosion in the south bank during 1990-2008 and it is marked by numerous abandoned channels, a meandering scar and water saturated zones.

Table 3 presents prioritization of various reaches in both banks of the river with respect to the land area lost. In the south bank, reach-5 in the vicinity of Morigaon (U/s of Guwahati) has undergone maximum erosion followed by reach-10 which is upstream of the Majuli Island. In north bank, reach-1 in the vicinity of Dhubri followed by reach-3 in the vicinity of Palasbari (D/s of Guwahati) exhibit maximum erosion. These prioritized reaches require in depth study for erosion management in the vicinity.

Table 3. Prioritization with respect to Land Area Lost

South Bank		North Bank	
Prioritize Reaches	**Land Area Lost (in decreasing order) km^2**	**Prioritize Reaches**	**Land Area Lost (in decreasing order) km^2**
5 (*Morigaon*)	**131.75**	**1 (*Dhubri*)**	**104.70**
10 (U/S Majuli)	125.72	3 (Palasbari)	97.49
11 (Dibrugarh)	85.31	2 (Goalpara)	81.16
8 (U/S Tezpur)	80.48	11 (Dibrugarh)	51.99
7 (Tezpur)	68.20	9 (Majuli)	26.70
9 (Majuli)	49.17	4 (Guwahati))	17.97
6 (Morigaon)	48.00	6 (Morigaon)	16.29
4 (Guwahati)	14.24	10 (U/S Majuli)	15.22
1 (Dhubri)	8.40	7 (Tezpur)	9.48
3 (Palasbari)	7.63	8 (U/S Tezpur)	5.12
2 (Goalpara)	4.65	5 (Morigaon)	4.01

CONCLUSIONS

The present work using remote sensing and GIS based approach with on multi-date satellite data has revealed sharp changes in fluvial land form in recent years resulting in considerable inhabited land loss. The braided Brahmaputra River within the study reaches exhibits differential rate of erosion and deposition during eighteen years (1990-2008). It is observed that in general the river has eroded both the banks throughout its course except at a few sites where its banks are well defined as the river is constricted due to presence of rocks. River adjustment processes that affected fluvial system of the Brahmaputra include the channel control points, surmised to be the major co-actors, working in unison with other forcing functions like channel degradation and aggradation, lateral river migration, widening or narrowing, avulsion, changes in the quantity and character of the sediment load at spatial and temporal scale, intensely powerful monsoon regime, recurring earthquakes and adverse impact of anthropogenic factors. Many reaches along the Brahmaputra River have been perceived as suffering from high erosion that endanger nearby settlements and infrastructure. The reaches have been prioritized with respect to the land area lost in eighteen years. There arises the need for in-depth study of interaction of geo-tectonic activities conjunctively with fluvial regime in the region to understand the complex physical processes completely for suggesting more practical result oriented river management interventions.

This study has proved the utility and application of satellite remote sensing which allows a retrospective, synoptic viewing of large regions and so provides the opportunity for a spatially and temporally detailed assessment of changes in river channel erosion/deposition. This study has further demonstrated how the use of GIS has been expedient in organization of geo-spatial databases and facilitation of channel position mapping and measurement. Conjunctive use of the information compiled and generated through this study with other collateral data on climate, flood control measures and in situ observations will serve towards taking up long term flood/erosion protection measures in the Brahmaputra basin.

REFERENCES

1. N. Surian, "Channel Changes Due to River Regulation: The Case of the Piave River, Italy," Earth Surface Processes and Landforms, Vol. 24, No. 12, 1999, pp. 1135- 1151.doi:10.1002/(SICI)1096-9837(199911)24:12<1135::AID-ESP40>3.0.CO;2-F
2. X. Yang, M. C. J. Damen and R. A. van Zuidam, "Satellite Remote Sensing and GIS for the Analysis of Channel Migration Changes in the Active Yellow River Delta, China," International Journal of Applied Earth Observation and Geoinformation, Vol. 1, No. 2, 1999, pp. 146- 157. doi:10.1016/S0303-2434(99)85007-7
3. M. Rinaldi, "Recent Channel Adjustments in Alluvial Rivers of Tuscany, Central Italy," Earth Surface Processes and Landforms, Vol. 28, No. 6, 2003, pp. 587-608. doi:10.1002/esp.464
4. C. Fuller, A. R. G. Large and D. J. Milan, "Quantifying Channel Development and Sediment Transfer Following Chute-Off in a Wandering Gravel-Bed River," Geomorphology, Vol. 54, No. 3-4, 2003, pp. 307-323. doi:10.1016/S0169-555X(02)00374-4
5. L. Q. Li, X. X. Lu and Z. Chen, "River Channel Change during the Last 50 Years in the Middle Yangtze River: An Example of the Jianli Reach," Geomorphology, Vol. 85, No. 3-4, 2007, pp. 185-196. doi:10.1016/j.geomorph.2006.03.035
6. M. Kummu, X. X. Lub, A. Rasphonec, J. Sarkkulad and J. Koponen, "Riverbank Changes along the Mekong River: Remote Sensing Detection in the Vientiane-Nong Khai Area," Quaternary International, Vol. 186, No. 1, 2008, pp. 100-112. doi:10.1016/j.quaint.2007.10.015
7. NRSA, "Brahmaputra Flood Mapping and River Migration Studies-Airborne Scanner Survey," National Remote Sensing Agency, Hyderabad, 1980.
8. M. Bardhan, "Channel Stability of Barak River and Its Tributaries between Manipur-Assam and Assam-Bangladesh Borders as Seen from Satellite Imagery," Proceedings National Symposium on Remote Sensing Applications for resource Management with Special Emphasis on N.E. Region, Guwahati, 25-27 November 1993, pp. 481-485.
9. SAC and Brahmaputra Board, "Report on Bank Erosion on Majuli Island, Assam: A Study Based on Multi Temporal Satellite Data," Space Application Centre, Ahmedabad and Brahmaputra Board, Guwahati, 1996.
10. S. D. Naik, S. K. Chakravorty, T. Bora and I. Hussain, "Erosion at Kaziranga National Park, Assam, a Study Based on Multitemporal Satellite Data," Project Report. Space Application Centre (ISRO) Ahmedabad and Brahmaputra Board, Guwahati, 1999, p. 70.

11. U. Goswami, J. N. Sarma and A. D. Patgiri, "River Channel Changes of Subansiri in Assam, India," Geomorphology, Vol. 30, No. 3, 1999, pp. 227-244. doi:10.1016/S0169-555X(99)00032-X
12. P. Mani, R. Kumar and C. Chatterjee, "Erosion Study of a Part of Majuli River-Island Using Remote Sensing Data," Journal of Indian Society of Remote Sensing, Vol. 31, No. 1, 2003, pp. 11-18. doi:10.1007/BF03030747
13. L. Bhakal, B. Dubey and A. K. Sarma, "Estimation of Bank Erosion in the River Brahmaputra near Agyathuri by Using Geographic Information System," Journal of Indian Society of Remote Sensing, Vol. 33, No. 1, 2005, pp. 81-84. doi:10.1007/BF02989994
14. P. Kotoky, D. Bezbaruah, J. Baruah and J. N. Sarma, "Nature of Bank Erosion along the Brahmaputra River Channel, Assam, India," Current Science, Vol. 88, No. 4, 2005, pp. 634-640.
15. J. N. Sarma, D. Borah and U. Goswami, "Change of River Channel and Bank Erosion of the Burhi Dihing River (Assam), Assessed Using Remote Sensing Data and GIS," Journal of Indian Society of Remote Sensing, Vol. 35, No. 1, 2007, pp. 94-100.doi:10.1007/BF02991837
16. J. D. Das and A. K. Saraf, "Remote Sensing in the Mapping of the Brahmaputra/Jamuna River Channel Patterns and Its Relation to Various Landforms and Tectonic Environment," International Journal of Remote Sensing, Vol. 28, No. 16, 2007, pp. 3619-3631.doi:10.1080/01431160601009664 [Paper reference:3]
17. B. C. Gao, "NDWI—A Normalized Difference Water Index for Remote Sensing of Vegetation Liquid Water from Space," Remote Sensing of Environment, Vol. 58, No. 3, 1996, pp. 257-266. doi:10.1016/S0034-4257(96)00067-3
18. S. K. Jain, A. K. Saraf, A. Goswami and T. Ahmad, "Flood Inundation Mapping Using NOAA AVHRR Data," Water Resources Management, Vol. 20, No. 6, 2006, pp. 949- 959.doi:10.1007/s11269-006-9016-4
19. P. Ceccato, S. Flasse, S. Tarantola, S. Jacquemond and J. M. Gregoire, "Detecting Vegetation Water Content Using Reflectance in the Optical Domain," Remote Sensing of Environment, Vol. 77, 2001, pp. 22-33. doi:10.1016/S0034-4257(01)00191-2
20. J. N. Sarma and M. K. Phukan, "Origin and Some Geomorphological Changes of Majuli Island of the Brahmaputra River in Assam, India," Geomorphology, Vol. 60, No. 1-2, 2004, pp. 1-19. doi:10.1016/j.geomorph.2003.07.013

9
CHAPTER

INTER-COMPARISON BETWEEN VIIRS AND MODIS RADIANCES AND OCEAN COLOR DATA PRODUCTS

Rong-Rong Li, Mark David Lewis , Richard W. Gould Jr. , Adam Lawson, Ruhul Amin, Sonia C. Gallegos and Sherwin Ladner

Keywords: MODIS, VIIRS, remote sensing, water-leaving radiance

INTRODUCTION

The Visible Infrared Imaging Radiometer Suite (VIIRS) [1] currently onboard the Suomi National Polar-orbiting Partnership (NPP) Spacecraft was successfully launched into space in October of 2011 (http://npp.gsfc.nasa.gov/viirs.html). VIIRS is now routinely generating Environment Data Records (EDRs) of the land surface, ocean color and temperature, and atmosphere. VIIRS enables the scientific community to continue the comprehensive Earth observation recorded by the Moderate Resolution Imaging Spectrometer (MODIS) [2,3] onboard the NASA Aqua Spacecraft.

The VIIRS EDRs are derived from the measured radiances, which are part of the VIIRS Sensor Data Records (SDRs) [4]. Errors in VIIRS SDRs will propagate to VIIRS EDRs. In order to monitor the performance and stability [5] of the VIIRS reflective solar bands (RSB) in the visible and near-IR spectral region, radiometric inter-comparison between VIIRS and MODIS has been made using an extended simultaneous nadir overpass (SNO) approach [6,7,8]. For the Suomi and Aqua spacecrafts, SNOs occur most frequently at high latitude polar regions. They occasionally occur at low latitude

Citation: Li, R.-R.; Lewis, M.D.; Gould, R.W., Jr.; Lawson, A.; Amin, R.; Gallegos, S.C.; Ladner, S. Inter-Comparison between VIIRS and MODIS Radiances and Ocean Color Data Products. Remote Sens. 2015, 7, 2193-2207. doi:10.3390/rs70202193.

regions. Part of the rationale for this MODIS and VIIRS data consistency investigation is to understand how the data consistency or inconsistency might impact scene fusion. Thus, our goal here is to assess the data consistency at off-nadir viewing angles.

In spite of the relatively good agreement between VIIRS and MODIS radiance data in the visible and near-IR spectral regions, the retrieved VIIRS ocean color data products, as reported in several recent publications [9,10,11], do not agree well. For example, Turpie [9] has shown that the area-averaged 8-day composite chlorophyll concentrations retrieved from VIIRS data with the standard NOAA and NASA algorithms have large differences over global deep ocean waters in certain seasons in 2012. Pahlevan [11] has also investigated the consistency between VIIRS and MODIS SDRs and EDRs over various oceanic/open waters for the time period of March through October 2012. In order to avoid possible uncertainties in atmospheric conditions at large scan angles, their cross-comparisons are limited to measurements made in the near-nadir viewing angles, *i.e.*, <±10 degrees from the nadir direction. The opportunities for SNO observations with both VIIRS and MODIS over a given geographic region, such as the Chesapeake Bay in the eastern coastal area of the U.S., are quite limited due to the intrinsic orbital properties of the two satellites. However, because of the large swath widths for both the VIIRS and MODIS instruments, scene pairs with non-nadir sensor zenith angles and nearly simultaneous MODIS and VIIRS observations are available.

In this paper, we report inter-sensor comparisons between VIIRS and MODIS data acquired over the Chesapeake Bay and nearby areas not only with SNO measurements, but also with relatively large differences in sensor zenith angles. We compare total radiance (Lt) at the top of the atmosphere (TOA), and the ocean color (OC) data products, such as normalized water-leaving radiance (nLw) and chlorophyll concentration, derived from both the VIIRS and MODIS data using the Automatic Processing System (APS), which was developed at the Naval Research Laboratory, Stennis Space Center (NRL/SSC) to produce operational bio-optical products.

INSTRUMENT CHARACTERISTICS

The VIIRS instrument is a scanner with a swath width on the ground of 3040 km. It collects scientific data from an altitude of approximately 830 km in an ascending polar orbit, with 22 narrow bands located in the 0.4–12.5 μm ranges [1]. It achieves global coverage every day and has a repeat cycle of 16 days. The seven VIIRS Visible/NearInfrared (VisNIR) bands at a "moderate" spatial resolution of 750 m in the wavelength interval between 0.4 and 0.9 micron, referred to as M1–M7 and listed in Table 1, have important applications for global remote sensing of oceans.

One MODIS instrument [2,3] is currently onboard the Aqua Spacecraft at an altitude of 705 km in an ascending polar orbit. It is also a scanner but with a narrower swath width on the ground of 2330 km. It achieves global coverage in two days and has a repeat cycle of 16 days. It has a dedicated set of ocean color bands, which are also listed in Table 1. From this table, it is seen that a number of MODIS and VIIRS ocean color bands are very similar, except for minor differences in band center position and full width at half maximum (FWHM). In the present study, we did not take into account the small wavelength differences between the two sensors. Also, we did not perform a comprehensive uncertainty analysis to evaluate all possible sources of uncertainties.

Table 1. The VIIRS VisNIR band names, positions, full widths at half maximum (FWHMs), and characteristics of the corresponding MODIS ocean color bands

VIIRS			MODIS		
Band	**Wavelength (µm)**	**FWHM (µm)**	**Band**	**Wavelength (µm)**	**FWHM (µm)**
M1	0.410	0.020	B8	0.412	0.015
M2	0.443	0.018	B9	0.443	0.010
M3	0.486	0.020	B10	0.488	0.010
M4	0.551	0.020	B12	0.547	0.010
M5	0.671	0.020	B13	0.667	0.010
M6	0.745	0.015	B15	0.748	0.010
M7	0.862	0.039	B16	0.869	0.015

DATA ANALYSIS AND SAMPLE RESULTS

We made quantitative inter-sensor comparisons between VIIRS and MODIS data acquired over the turbid Chesapeake Bay areas and the nearby clear ocean areas not only with SNO measurements, but also with relatively large differences in the sensors' view angles. Our data analysis and sample results are presented below.

SNO Measurements

For the Suomi and Aqua satellite orbits, SNOs occur most frequently at high latitude polar regions [5]. Through analysis of orbital tracks of Suomi and Aqua, we identified a number of overlapping VIIRS and MODIS scenes over the Chesapeake Bay area. After screening out cloud-contaminated scenes, we were able to identify fairly clear pairs of VIIRS and MODIS images. Figure 1 shows MODIS and VIIRS images over the Chesapeake Bay area acquired on 10 January 2013. Figure 1a,b are MODIS and VIIRS images, respectively. The time difference between the two sensors passing over the Chesapeake Bay area was only about seven minutes. The solar zenith angles of the two sensors are both approximately 63 degrees. The view zenith angles at Chesapeake Bay areas are less than 20 degrees for both MODIS and VIIRS sensors.

Eight cloud-free water areas near the Chesapeake Bay, as marked in Figure 1, were randomly selected for inter-sensor comparison purposes. The radiances of the seven ocean color channels (see Table 1) and the corresponding retrieved ocean color products, such as normalized water-leaving radiances (nLw) and chlorophyll concentrations, are compared. In order to ensure the two sensors covered the same areas, we selected rectangular regions over homogeneous water areas with the same latitudes and longitudes at the four corners of each area. This reduced uncertainties associated with the spatial resolution of the two sensors. The yellow star (*) marked in the image is the location where the ground measurements from a HyperPro spectrometer were taken on the same day. The data acquisition times for MODIS-aqua, VIIRS, and the HyperPro spectrometer are 13:10, 13:03, and 11:10 EST, respectively. MODIS and VIIRS are multispectral sensors with 1000 m and 750 m spatial resolutions, while ground measurements from the HyperPro spectrometer are spatial point measurements and spectrally cover the contiguous wavelength range from 0.35 to 0.80 μm at a spectral resolution of 0.01 μm.

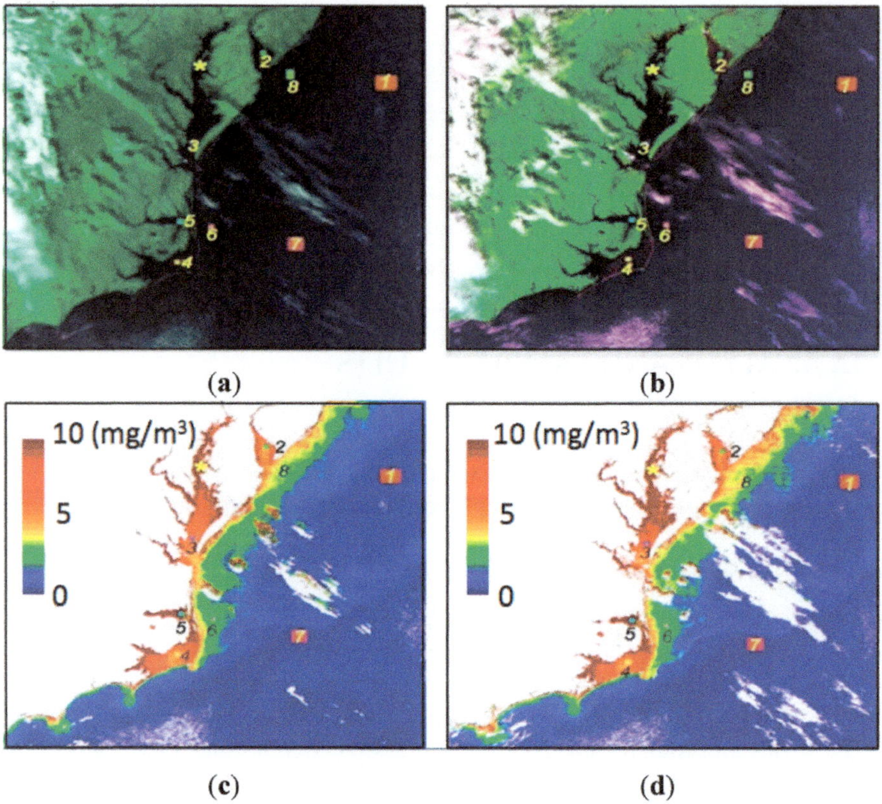

Figure 1. Satellites image over the Chesapeake Bay, acquired on 10 January 2013: (**a**) Aqua-MODIS RGB image; (**b**) VIIRS RGB image; (**c**) MODIS chlorophyll concentration (mg/m^3); and (**d**) VIIRS chlorophyll concentration (mg/m^3). Eight regions of interest (ROIs) were randomly chosen for inter-sensor comparisons on the seven ocean color channels. The yellow star (*) in the image is the location where the HyperPro *in situ* measurement is made.

Vicarious calibration [12] is applied during atmospheric correction for ocean color applications. It is a process for establishing gain adjustments to improve and fine-tune sensor measurement. During atmospheric correction, nLw values are derived from the Lt values. To compute the vicariously calibrated gains, sensor scenes over *in situ* data locations are first processed with standard gains (gains equal to 1) for all spectral bands. During this process, atmospheric scattering and absorption factors, such as Rayleigh and aerosol scattering radiances and atmospheric gas absorption coefficients, are generated at each wavelength band. After the satellite-derived nLw values have been computed, the various scattering and absorption factors are still held in memory. To perform the vicarious calibration, the satellite-derived nLw values are then replaced by *in situ* nLw values. All the atmospheric correction scattering and absorption factors are then added to the *in situ* nLw values to estimate the Lt values that would yield these measured nLw values. Thus, these spectral vicarious total radiances (vLt) represent the adjusted Lt value needed to retrieve the *in situ* nLw measurements after atmospheric correction.

When vLt values are generated from multiple dates of imagery over *in situ* data locations, a multi-date statistical sample is established. The ratio of the vLt/Lt over this multi-day data set provides a vicariously calibrated gain factor for each wavelength that, when multiplied by the Lt value, provides a "tuned" Lt value that will yield more accurate nLw values after atmospheric correction. All pixels in the same scene (*i.e.*, away from the *in situ* data locations), or scenes at other times can be processed using the vicariously calibrated gain factors. NASA uses the Marine Optical Buoy (MOBY) [13], moored near Lanai, Hawaii, to provide *in situ* in-water measurements used in vicarious calibration of ocean color sensors. The resulting vicariously calibrated gains then become the accepted gains to use for the sensor. The MOBY and VIIRS vicariously calibrated gains used to process data in this study were generated using MOBY *in situ* data. Figure 2a shows Lt measurements over Region 2. The dotted blue line is for the MODIS data, and the dashed red line for VIIRS data. The shapes of the spectral wavelength dependences, *i.e.*, the decreases in radiances with increasing wavelengths, for both the MODIS and VIIRS data have similar spectral trends to the Rayleigh and aerosol scattering properties. The regression analysis of the scatter plot of the MODIS and VIIRS Lt measurements in Figure 2b has a high value of coefficient of determination (R^2), where R is the correlation coefficient. The slope from the regression analysis is 1.08. The line in Figure 2b is a least-squares fit to the data, which contains the Lt measurements from seven channels of both sensors. If a data point in Figure 2b deviates from this line, we can conclude that there are differences in the radiometry for the corresponding MODIS or VIIRS channel. The use of the Figure 2b scatter plot to view data, in principle, allows for quick identification of possible band discrepancies between corresponding sensor bands. The 8% deviation from 1.0 in the slope could be attributed to sensor calibration, differences in derived Rayleigh scattering, and/or differences in derived aerosol scattering between the MODIS and VIIRS data.

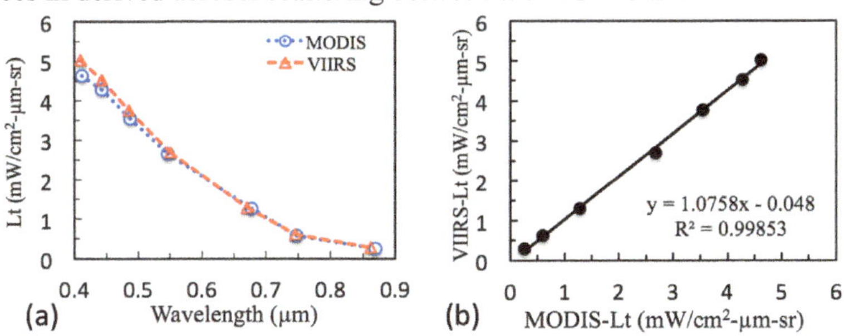

Figure 2. (**a**) Top-of-Atmosphere multi-channel total radiances (Lt) in unit of mW/(cm²-µm-sr) of MODIS and VIIRS over Region 2, and (**b**) scatter plot of MODIS *versus* VIIRS Lt plus the best fit least-square regression line.

The Automatic Processing System (APS) developed at NRL/SSC [14] was used to generate ocean color data products, such as water-leaving radiance and chlorophyll concentration, from both MODIS and VIIRS data. APS is based on, and is consistent with, the NASA SeaWiFS Data Analysis Software (SeaDAS). Figure 3a shows the nLw values retrieved from MODIS and VIIRS data over Region 2. After atmospheric corrections, the retrieved nLw curves of the two sensors coincide well. Figure 3b is the scatter plot of VIIRS *versus* MODIS normalized water-leaving radiances, which are shown in Figure 3a. A regression analysis of the seven channels in Figure 3b gives a slope of 0.96 with a very high R^2 value. This demonstrates good agreement between derived VIIRS and MODIS nLw across all wavelengths using APS atmospheric correction.

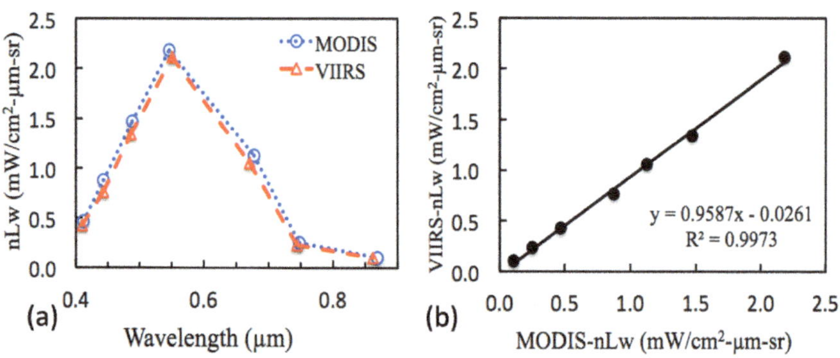

Figure 3. (**a**) Multi-channel normalized water-leaving radiances (nLw) of MODIS and VIIRS over Region 2 retrieved by APS, and (**b**) scatter plot of VIIRS *versus* MODIS nLw with the best-fit least-squares regression line.

To further demonstrate the agreements between VIIRS and MODIS, in Figure 4a we show the scatter plot of VIIRS *versus* MODIS Lt measurements for seven ocean color channels at the eight selected surface areas, as marked in Figure 1. Regression analysis, shown in Figure 4a, has a slope of 1.07 and an offset of 0.05. This indicates that the VIIRS Lt is about 7% higher than that of MODIS, considering all wavelengths and at all 8 locations. Figure 4b shows the scatter plot of the MODIS and VIIRS APS-derived nLw estimates for all channels and at all 8 locations. The regression analysis in Figure 4b shows that all the data points are tightly clustered around a line with a slope of 0.94 and a small offset. The differences between APS-retrieved VIIRS and MODIS nLw estimates are approximately 10%, with VIIRS lower than MODIS. This is a good agreement in view of the fact that roughly 80%–90% of atmospheric scattered path radiances in Figure 4a are removed from the Lt measurement when computing the nLw during the atmospheric correction process.

In order to evaluate the retrieval results on a channel-by-channel basis, we did the inter-comparison for each channel. Figure 5 shows the scatter plots of nLw for each wavelength over the eight locations, as marked in Figure 1. The least square fitting lines and R^2 values are calculated at each wavelength, and are shown in each plot on Figure 5. These R^2 values provide information on how well the data are matched between sensors for each channel. The results show that, for the channels at 0.443 μm or longer wavelengths, the VIIRS-nLw and MODIS-nLw agreed well with high R^2 values. For the 0.410-μm channel, the retrieved nLw values were not clustered as tightly as the other channels, but still had R^2 values greater than 0.9. The inconsistency can be attributed to minor errors in sensors' calibrations and atmospheric corrections.

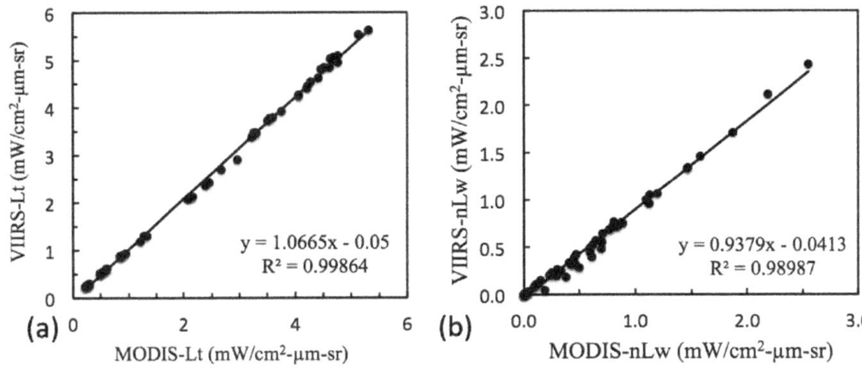

Figure 4. (**a**) Scatter plot of VIIRS *versus* MODIS Lt measurements for the eight locations marked in Figure 1, and (**b**) Scatter plot of VIIRS *versus* MODIS APS-derived nLw estimates for the eight locations with the best-fit least-squares regression line.

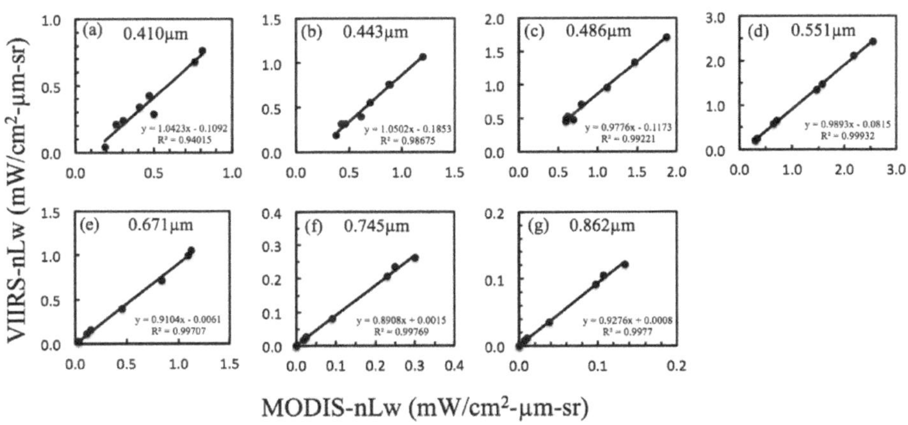

Figure 5. Scatter plots of VIIRS *versus* MODIS APS derived nLw estimates for the eight locations marked in Figure 1 and with each of the seven ocean color channels plotted separately.

Table 2. Statistical results on the comparisons of nLw between MODIS and VIIRS for data collected over Chesapeake Bay areas on 10 January 2013

Parameters	0.410 μm	0.443 μm	0.486 μm	0.551 μm	**0.671 μm**	**0.745 μm**	**0.862 μm**	**Overall**
N	8	8	8	8	8	8	8	56
Regression	1.042x − 0.109	1.05x − 0.185	0.978x − 0.117	0.989x − 0.082	0.910x − 0.006	0.891x + 0.002	0.928x + 0.001	0.97x − 0.071
R^2	0.940	0.9867	0.9922	0.9993	0.9971	0.998	0.998	0.987
PD (%)	25.37	25.70	16.53	12.87	14.81	5.87	0.94	14.58
APD (%)	25.37	25.70	16.53	12.87	14.89	8.66	7.74	15.97

The fitting results for each wavelength and for data acquired on 10 January 2013 are summarized in Table 2. Here, the average Percent Differences is denoted as PD, and the average Absolute Percent Differences denoted as APD of N total number of locations. PD determines the bias between the quantities being compared, while APD estimates the average uncertainty [15]. The values of PD are calculated by averaging the percent differences between the matchups of the individual observations. Percentage difference of the jth matchup, PDj, is calculated by:

$$PD_j = 100 * \frac{\left(\sum_{i=0}^{N-1} \frac{(M_{j,i} - V_{j,i})}{M_{j,i}}\right)}{N} \qquad (1)$$

where i is the index for the samples, j for the sensor channels, N the total number of locations in the comparisons, and $V_{j,i}$ and $M_{j,i}$ stand for the VIIRS and MODIS data at the jth channel and the ith sample, respectively. Similarly, APD_j values are calculated by the following:

$$APD_j = 100 * \frac{\left(\sum_{i=0}^{N-1} \frac{|M_{j,i} - V_{j,i}|}{M_{j,i}}\right)}{N} \qquad (2)$$

To verify the validity of the VIIRS and MODIS nLw estimates, we also compared them with *in situ* measured nLw data collected with the HyperPro hyperspectral radiometer in the Chesapeake Bay area, which is marked with a yellow star (*) in Figure 1 images. The HyperPro spectral measurements were made on the same day, approximately two hours prior to the VIIRS and MODIS overpasses. Figure 6 shows the nLw as a function of wavelength from VIIRS, Aqua MODIS, and HyperPro. The MODIS data are represented by the circles, VIIRS by the triangles, and HyperPro by the plus. Since MODIS and VIIRS are multi-channel sensors, they do not have the spectral resolution that HyperPro does. In general, the nLw values from the three sensors are comparable although those from the satellite data are lower than those from the ground measurements, especially in short wavelength regions. Multiple factors, such as the spatial resolution of the satellite sensors (MODIS spatial resolution is 1 km and VIIRS is 750 m) *versus* point measurements from the HyperPro and possible errors in atmospheric correction of the satellite data, are likely responsible for the discrepancies.

It should be pointed out that both VIIRS and MODIS lack narrow channels to capture the weak water-leaving radiance peak centered near 0.70 μm. The two instruments also lack narrow channels to capture the major water-leaving radiance peak centered near 0.58 μm. Therefore, both the multi-channel VIIRS and MODIS instruments are unable to capture the full spectral information contained in the HyperPro-measured normalized water-leaving radiance spectrum over a turbid water area.

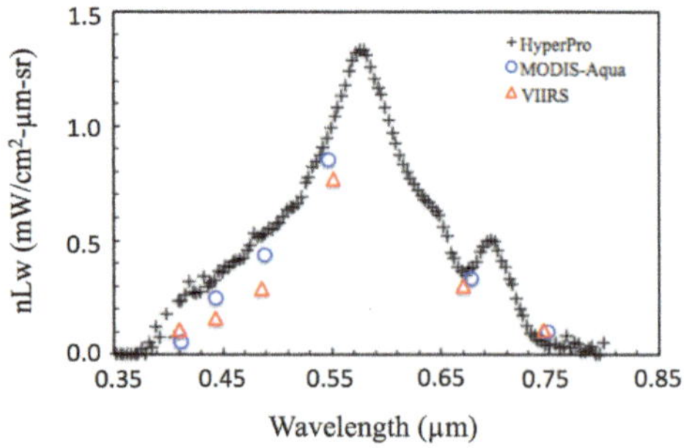

Figure 6. A comparison of normalized water-leaving radiances (nLw) retrieved on 10 January 2013 from VIIRS data, MODIS data, and a field-measured spectrum over a turbid water area in the Chesapeake Bay.

The retrieved chlorophyll concentrations (chl) from APS for MODIS and VIIRS data are shown in Figure 1c,d. The chlorophyll variation patterns in both images are quite similar. Figure 7 shows the scatter plot of VIIRS chlorophyll concentration *versus* MODIS chlorophyll concentration for the selected regions marked in the Figure 1 images. A regression analysis of these data points produced a line with a slope of 1.09 and offset of 0.07 with R^2 value of 0.985.

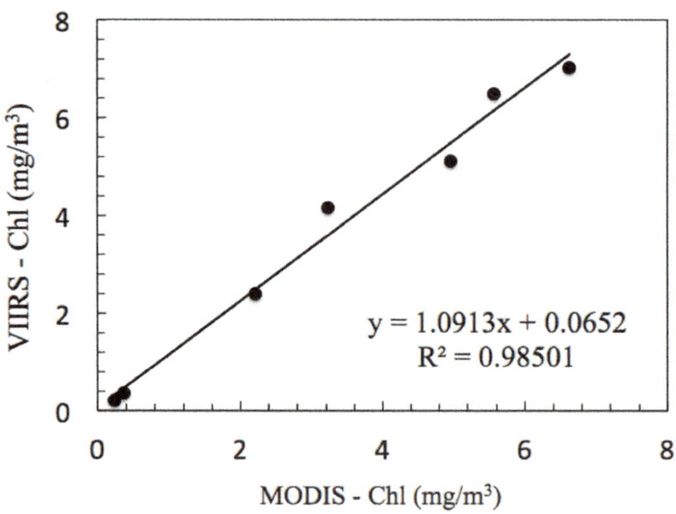

Figure 7. Scatter plot of VIIRS chlorophyll concentration *versus* MODIS chlorophyll concentration from coincident scenes on 10 January 2013, for the selected regions marked in Figure 1 images with the best-fit least-squares regression line.

Measurements with Large Differences in View Zenith Angles

As described previously, due to the intrinsic Suomi VIIRS and Aqua MODIS satellite orbital properties, it is very difficult to find coincident cloud-free SNO VIIRS and MODIS scenes over the Chesapeake Bay area. However, because of large swath widths of both sensors, coincident cloud-free scenes covering the Chesapeake Bay area with large differences in the sensor viewing angles can be found. Figure 8 shows a pair of VIIRS and MODIS images acquired on 9 March 2013. The Chesapeake Bay is located near nadir for VIIRS (Figure 8a), but close to the far right edge of the MODIS image (Figure 8b), as shown in yellow rectangular box. The sensor view angles over the Chesapeake Bay are approximately 25 degrees for VIIRS and 53 degrees for MODIS.

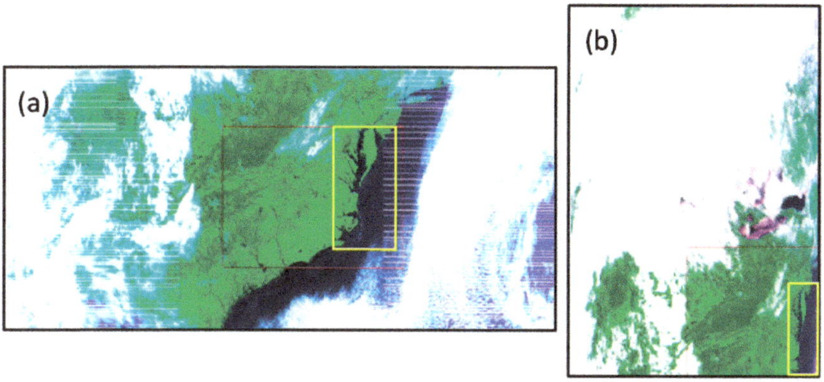

Figure 8. (**a**) VIIRS image and (**b**) MODIS image acquired over the Chesapeake Bay on 9 March 2013.

Figure 9a shows portions of the VIIRS image (after geo-registration) covering the Chesapeake Bay area. The parallel white lines in this image do not contain real VIIRS data but represent data gaps due to the VIIRS read out circuit design for data at higher sensor zenith angles. Figure 9b is for the corresponding MODIS image. Figure 9c is the APS-derived chlorophyll concentration image from the VIIRS data, while Figure 9d is the chlorophyll concentration image obtained from MODIS data.

For inter-sensor comparisons, eight areas similar to the locations shown in Figure 1 are marked in Figure 9c,d. Figure 10a shows the Lt measurements over Region 1. The dotted line is for the multi-channel MODIS data, and the dashed line for the VIIRS data. Due to view angle differences between MODIS and VIIRS, the MODIS scene had significantly more Rayleigh scattered radiance (Lr) than the VIIRS scene. This is confirmed in Figure 10b, in which the APS-calculated Lr for MODIS and VIIRS based on solar and view angles are shown. Figure 10c shows the nLw estimates retrieved from MODIS and VIIRS data over the region. Although the Lt radiances of MODIS and VIIRS (Figure 10a) are quite different, the retrieved nLw are well matched (Figure 10c). This is due, in part, to the fact that a larger Lr value was subtracted out from the MODIS Lt in the atmospheric correction process. Figure 10d is the scatter plot of VIIRS *versus* MODIS nLw values, as shown in Figure 10c. A regression analysis of the seven channel data points in Figure 10d gives a slope of 0.896 and offset 0.0484 with a high R^2value. The plots in Figure 10, in particular Figure 10c, have demonstrated that nLw can be recovered reasonably well regardless of the sensor view angles.

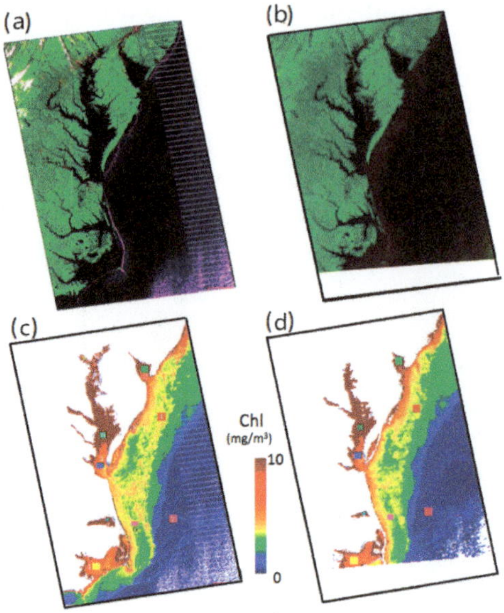

Figure 9. Satellite images over the Chesapeake Bay acquired on 9 March 2013: (**a**) VIIRS RGB image; (**b**) MODIS RGB image; (**c**) VIIRS Chlorophyll concentration; and (**d**) MODIS Chlorophyll concentration.

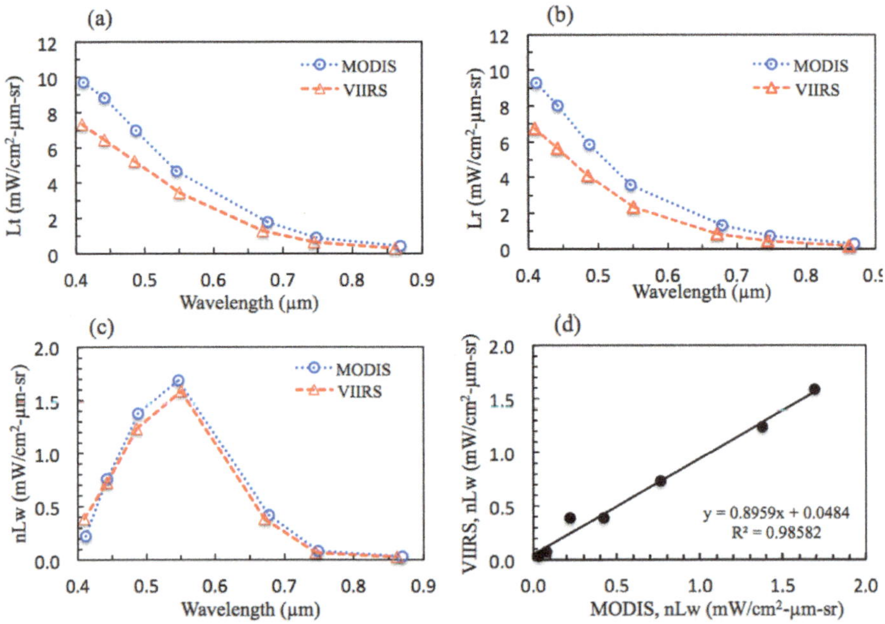

Figure 10. (**a**) Top-of-Atmosphere multi-channel radiances (Lt) of MODIS and VIIRS over Re-

gion 1; (**b**) APS-calculated Rayleigh path radiances (Lr) of MODIS and VIIRS; (**c**) Spectra of normalized water-leaving radiances (nLw) of MODIS and VIIRS retrieved with APS; and (**d**) scatter plot of VIIRS *versus* MODIS nLw estimates.

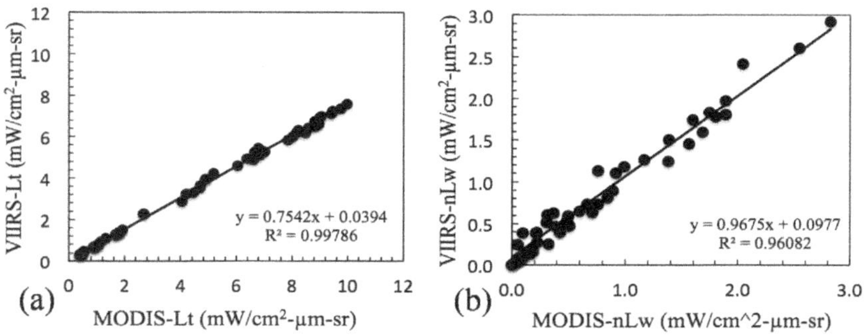

Figure 11. Data from MODIS and VIIRS imagery collected on 9 March 2013: (**a**) scatter plot of VIIRS *versus* MODIS Lt for the eight locations marked in Figure 9 with best-fit least-squares regression line, and (**b**) scatter plot of VIIRS *versus* MODIS nLw with best fit least-squares regression line.

To increase the number of VIIRS and MODIS data for ocean color comparisons, we show in Figure 11a the scatter plot of VIIRS *versus* MODIS Lt for all eight selected regions, as marked in Figure 9 images, and with all seven channels listed inTable 1. A regression analysis using all the data points in Figure 11a gives a slope of about 0.754. MODIS-measured Lt values are about 25% higher than those of VIIRS. This is due to the larger view angles of the MODIS measurements (at the rightmost edge of Figure 8b) with longer path radiances and stronger Rayleigh scattering. Figure 11b is a scatter plot similar to Figure 11a but for the APS-derived nLw from the Figure 11a VIIRS and MODIS Lt. A regression analysis shows that all the data points in Figure 11b are clustered around a line with a slope of approximately 0.97 and a R^2 value of 0.96.

The detailed comparisons for each channel over the eight locations are shown in Figure 12. Each scatter plot represents nLw over all the locations, as marked in Figure 9. Regression analysis results are presented in each of the plots. For channels at 0.488 micron or longer wavelengths, the VIIRS-nLw and MODIS-nLw agreed well with high R^2 values. This demonstrates that nLw can be retrieved reasonably well from MODIS and VIIRS radiances acquired with large differences in satellite sensors' view zenith angles, particularly for channels centered at 0.488 micron or longer wavelengths.

Table 3. Statistical results on the comparisons on nLw between MODIS and VIIRS for data collected over Chesapeake Bay areas on 9 March 2013

Parameters	**0.410 μm**	**0.443 μm**	**0.486 μm**	**0.551 μm**	**0.671 μm**	**0.745 μm**	**0.862 μm**	**Overall**
N	5	8	8	8	8	8	8	53
Regression	0.939x + 0.258	0.753x + 0.294	0.930x + 0.121	1.060x − 0.044	1.028x − 0.0086	1.123x − 0.024	1.1795x − 0.0113	1.002x + 0.0836
R^2	0.778	0.816	0.911	0.97	0.99	0.977	0.977	0.917
PD (%)	−43.06	−20.80	−5.89	−0.98	0.22	4.91	1.09	−9.215
APD (%)	34.45	16.65	12.99	9.30	8.36	10.59	9.79	14.589

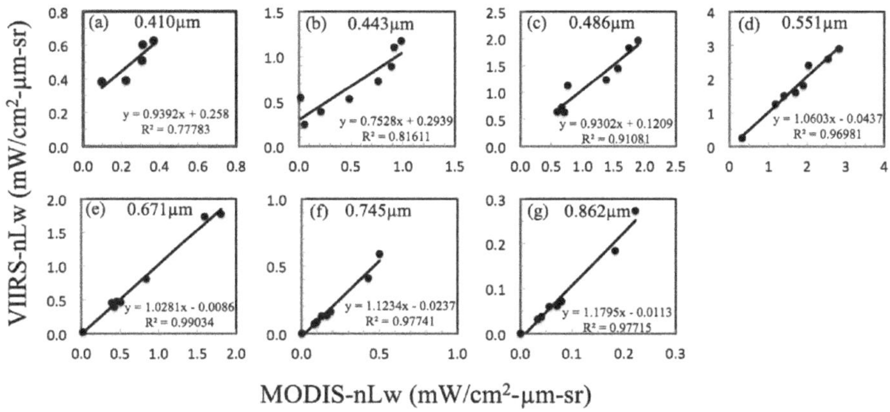

Figure 12. Scatter plots of VIIRS *versus* MODIS APS derived nLw estimates from 9 March 2013 for the eight locations marked in Figure 9 and with each of the seven ocean color channels plotted separately.

The regression analysis results on the comparisons for each wavelength for data acquired on 9 March 2013 are listed in Table 3. There are lower R^2 values for retrieved nLw estimates for the channels at 0.410 μm and 0.443 μm between MODIS and VIIRS, which can be attributed to errors in atmospheric correction procedures. For channels at 0.486 μm or longer wavelengths, there are higher R^2 values and smaller nLw differences.

CONCLUSIONS

We report on inter-sensor comparisons between VIIRS and MODIS data acquired over the Chesapeake Bay and nearby areas not only with SNO measurements, but also with relatively large differences in sensors' zenith angles. We compare radiances at the top of the atmosphere and normalized water-leaving radiances and ocean color data products derived with the Automatic Processing System from both VIIRS and MODIS data. We observed that, although there are large differences between the measured radiances at the top of the atmosphere of the two sensors when there are significant differences in the sensor zenith angles, the retrieved ocean color products, such as normalized water-leaving radiances, matched approximately 15% (mean difference, averaged across all wavelengths). The results show that the variation in satellite view zenith angles is not a main factor affecting the retrieval of ocean color from different satellite sensors (*i.e.*, the atmospheric correction routine adequately removes the view-angle dependence). Thus, VIIRS and MODIS provide consistent ocean color data that can generally be reliably merged to create fused composite imagery, even when viewing angles differ considerably.

ACKNOWLEDGMENTS

This research is supported by the U.S. Naval Research Laboratory and the Office of Naval Research. We thank Michael Ondrusek (NOAA) for graciously providing HyperPro data for the Chesapeake Bay.

AUTHOR CONTRIBUTIONS

Rong-Rong Li had the original idea for the study and carried out the analyses. Mark David Lewis generated MODIS and VIIRS data products using APS. Richard W. Gould, Jr. acquired relevant in situ data. Adam Lawson and Sherwin Ladner were involved APS software and technique. Ruhul Amin and Sonia Gallegos contributed to the study. Rong-Rong Li drafted the manuscript, which was revised by Mark David Lewis and Richard W. Gould, Jr. All authors read and approved the final manuscript.

REFERENCES

1. Murphy, R.E.; Guenther, B.; Ip, J.; Jackson, J.; Olenijczak, D.; Iisager, B.; Hutchison, K. Update on the algorithmic basis and predicted performance of selected VIIRS environmental data records. In Proceedings of IEEE International Conference on Geoscience and Remote Sensing Symposium, IGARSS 2006, Denver, CO, USA, 31 July–4 August 2006.
2. Salomonson, V.V.; Barnes, W.L.; Maymon, P.W.; Montgomery, H.E.; Ostrow, H. MODIS: Advanced facility instrument for studies of the earth as a system. *IEEE Trans. Geosci. Remote Sens.* **1989**, *27*, 145–153.
3. King, M.D.; Menzel, W.P.; Kaufman, Y.J.; Tanre, D.; Gao, B.-C.; Platnick, S.; Ackerman, S.A.; Remer, L.A.; Pincus, R.; Hubanks, P.A. Cloud and aerosol properties, precipitable water, and profiles of temperature and humidity from MODIS. *IEEE Trans. Geosci. Remote Sens.* **2003**, *41*, 442–458.
4. Cao, C.; De Luccia, F.J.; Xiong, X.; Wolfe, R.; Weng, F. Early on-orbit performance of the Visible Infrared Imaging Radiometer Suite onboard the Suomi National Polar-Orbiting Partnership (S-NPP) satellite. *IEEE Trans. Geosci. Remote Sens.* **2014**, *52*, 1142–1156.
5. Wu, A.; Xiong, X. NPP VIIRS and Aqua MODIS RSB comparison using observations from simultaneous nadir overpasses (SNO). *Proc. SPIE* **2012**.
6. Cao, C.; Weinreb, M.; Xu, H. Predicting simultaneous nadir overpasses among polar-orbiting meteorological satellites for the intersatellite calibration of radiometers. *J. Atmos. Oceanic Technol.* **2004**, *4*, 537–542.
7. Nagle, F.W.; Holz, R.E. Computationally efficient methods of collocating satellite, aircraft, and ground observations. *J. Atmos. Oceanic Technol.* **2009**, *26*, 1585–1595.
8. Uprety, S.; Cao, C.; Xiong, X.; Blonski, S.; Wu, A.; Shao, X. Radiometric intercomparison between Suomi-NPP VIIRS and aqua MODIS reflective solar bands using simultaneous nadir overpass in the low latitudes. *J. Atmos. Oceanic Technol.* **2013**, *30*, 2720–2736.
9. Turpie, K.R.; Balch, B.; Bowler, B.; Franz, B.A.; Frouin, R.; Gregg, W.; McClain, C.R.; Rousseaux, C.; Siegel, D.; Wang, M.; *et al.* Quality Assessment of the Visible Infrared Imaging Radiometer Suite (VIIRS) Ocean Color Environmental Data Records (EDR). Available online: http://npp.gsfc.nasa.gov/DEW_NPP_reports/VIIRS%20Ocean%20Color%20Products%20Report%20(Mar-25-2013).

pdf (accessed on 25 March 2013).

10. Davis, C.O.; Tufillaro, N.; Nahorniak, J.; Jones, B.; Arnone, R. Evaluating VIIRS ocean color products for west coast and Hawaiian waters. *Proc. SPIE* **2013**.
11. Pahlevan, N.; Lee, Z.; Arnone, R.; Lawson, A. Investigating the Consistency between VIIRS and MODIS over the Oceans: The Sensor/Environmental Data Records. Available online: https://ams.confex.com/ams/93Annual/webprogram/Paper224049.html (accessed on 10 January 2013).
12. Franz, B.A.; Bailey, S.W.; Werdell, P.J.; McClain, C.R. Sensor-independent approach to the vicarious calibration of satellite ocean color radiometry. *Appl. Optics* **2007**, *46*, 5068–5082.
13. Clark, D.K.; Feinholz, M.; Yarbrough, M.; Johnson, B.C.; Brown, S.W.; Kim, Y.S.; Barnes, R.A. Overview of the radiometric calibration of MOBY. *Proc. SPIE* **2002**, *4486*, 64–76.
14. Martinolich, P. The Automated Processing System. Available online: http://www7331.nrlssc.navy.mil/docs/aps_v5.2/pdf/APS-v5.3.0-Users-Guide.pdf (accessed on 27 December 2013).
15. Hlaing, S.; Harmel, T.; Gilerson, A.; Foster, R.; Weidemann, A.; Arnone, R.; Wang, M.; Ahmed, S. Evaluation of the VIIRS ocean color monitoring performance in coastal regions. *Remote Sens. Environ.* **2013**, *139*, 398–414.

10
CHAPTER

INFLUENCE OF THE TROPICAL ATLANTIC OCEAN'S SEA SURFACE TEMPERATURE

Ana Cleide Bezerra Amorim, Rosane Rodrigues Chaves, Cláudio Moisés Santos e Silva

Keywords: Ocean-Atmosphere, Coastline, Heat Balance, Moisture Fluxes, Flux Divergence

INTRODUCTION

The fluxes of heat, momentum and moisture are very important for the development of atmospheric and oceanic circulations; any significant variation in these fluxes directly affects the circulation and precipitation systems in a given region [1] . The pattern of Sea Surface Temperature Anomaly (SSTA) in the North and South Atlantic, which reflect the occurrence of precipitation anomalies in the Northeast of Brazil (NEB), shows the importance of coupling in the ocean-atmosphere interface. In this sense, the variability of Sea Surface Temperature (SST) causes changes in the pattern of atmospheric circulation and therefore, the behavior of the large scale weather systems responsible for the precipitation over Northern NEB [2] [3] .

The NEB presents different types of climate, the annual precipitation range from 300 mm in semi-arid region to 2000 mm on the coast. The dynamic of the meteorological systems operating in the Northeast were reviewed by [4] . In Northern NEB (NNEB), the higher values of precipitation are due to the Atlantic Intertropical Convergence Zone

Citation: Amorim, A. , Chaves, R. and Silva, C. (2014) Influence of the Tropical Atlantic Ocean's Sea Surface Temperature. Atmospheric and Climate Sciences, 4, 874-883. doi:10.4236/acs.2014.45077.

(ITCZ), which has its climatologically position further South around 4°S in March and April, and further North around 10°N - 14°N in August-September [5] . In the eastern (ENEB), region of interest in this study, the precipitation occurrence is modulated by easterly wave disturbance (EWD) [6] -[10] with maximum activity between May and July

Besides EWD, frontal systems moving from middle latitudes are also observed, especially in the Southern part of the NEB [11] , as well as high level cyclonic vortices [12] .

The interannual variability of NEB precipitation has been the focus of studies for the last decades. Thus, dry or rainy years have been related to inter annual variability associated with the El-Niño-Southern oscillation (ENSO) phenomena. In addition, the SST inter-hemispheric Gradient of Atlantic (GRAD) can influence the precipitation over the NEB, including intensifying or inhibiting the ENSO effects [13] . The simultaneous occurrence of these two phenomena has been investigated [2] [13] -[18] . In some cases, the influence of the Atlantic (mainly the South Atlantic Basin) in the NNEB precipitation is more relevant than the SST of Pacific [16] . The variability in the South Atlantic Ocean (SAO) has been under discussion, since no relation with the North Atlantic Oscillation (NAO) and only a weak relation with ENSO have been reported by some authors [19].

Most of the studies highlight the NNEB in relation to the adjacent oceans analyzing the ITCZ, weather system that determines the rainy season of this region. However, there is an increasing interest in understanding the dynamic of precipitation occurrence in the ENEB region [10] - [20] [21] , since it is in this region that most of the NEB population is concentrated and in where the most extreme precipitation events occur in the region [22]. In this study the main objective is to analyze the precipitation in ENEB from April to June and the processes of ocean-atmosphere interaction, mainly the episodes of GRAD_S.

One of the motivations to study the Atlantic GRAD_S is seen in [23] , which identified through analysis variance (ANOVA) significant evidence concerning the different phases of the tropical Atlantic, in relation of the precipitation in ENEB at a significance level of 5%, mainly in GRAD_S that was associated with precipitation excess, unlike other phases that showed no significant variability. However, in the study of [23] and the adjusted ANOVA model was used specific temporal rainfall series in ENEB. In this study we propose to verify the spatial distribution of precipitation across the Eastern region that also showed significant volumes during episodes of GRAD_S, as well as verifying if atmospheric and oceanic conditions favored the excess of precipitation diagnosed by [23] .

DATA AND METHODOLOGY

Data

Monthly precipitation data were used from 1981 to 2010, from the Global Precipitation Climatology Project (GPCP) of the World Climate Research Programme (WCRP), which aim to quantify the distribution of global precipitation [24] . The GPCP uses information from polar orbit satellites, geostationary and observed data. Finally, the dataset is disposed in a regular grid with a spacing of 0.5° × 0.5°. The period from 1981 to 2010 was chosen for representing the most updated climatological series for precipitation.

The fluxes fields at surface were obtained from the reanalysis [25] of National Center for Environmental Prediction (NCEP). We used the monthly averages, which are provided in a regular grid of 2.5° × 2.5°, for the period of 1982-2012.

The fields of vertically integrated moisture fluxes are from reanalyzes of the global European Centre for Medium-Range Weather Forecasts Interim Re-Analysis (ERA-Interim) model [26] , also for the period from 1982 to 2012, disposed in a grid of 1.5° × 1.5° latitude by longitude.

The SST data were obtained from NCEP [27] for areas located North (5°N - 20°N, 60°W - 30°W) and South (0°S - 20°S, 30°W - 10°E) of the Tropical Atlantic (http://www.cpc.ncep.noaa.gov/data/indices/). Figure 1 show the areas of study, which will be highlighted along the results.

The ENEB shows around 60% of annual precipitation concentrated in the period from April to June, [28] . In Figure 2, it is possible to observe the monthly time series of ENEB precipitation of GCPC, for the period from 1981 to 2011. The GCPC data are in accordance with [28] , with 56.9% of the annual rainfall in the region occurs in April-May-June (AMJ) quarter.

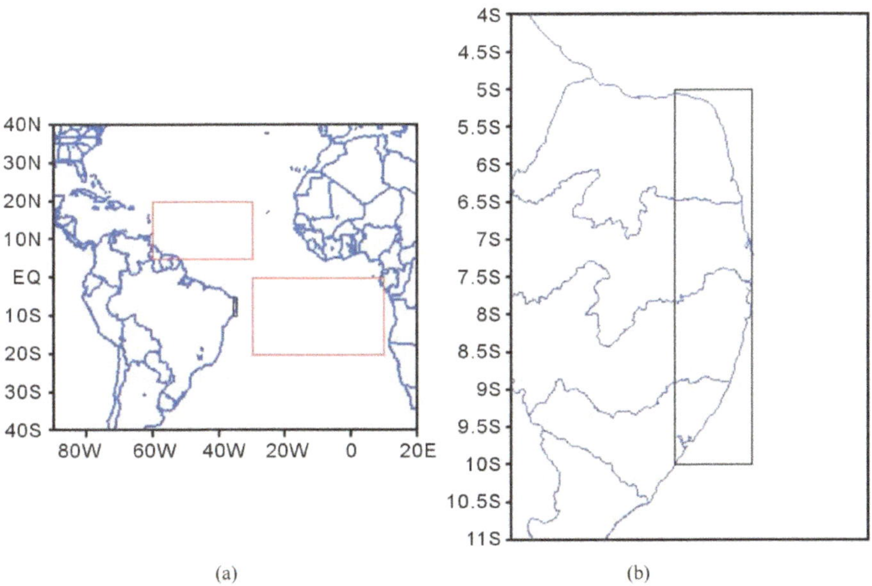

Figure 1. Areas of study. On the left (a) it shows the Atlantic region, and on the right (b), the Eastern coast of the NEB.

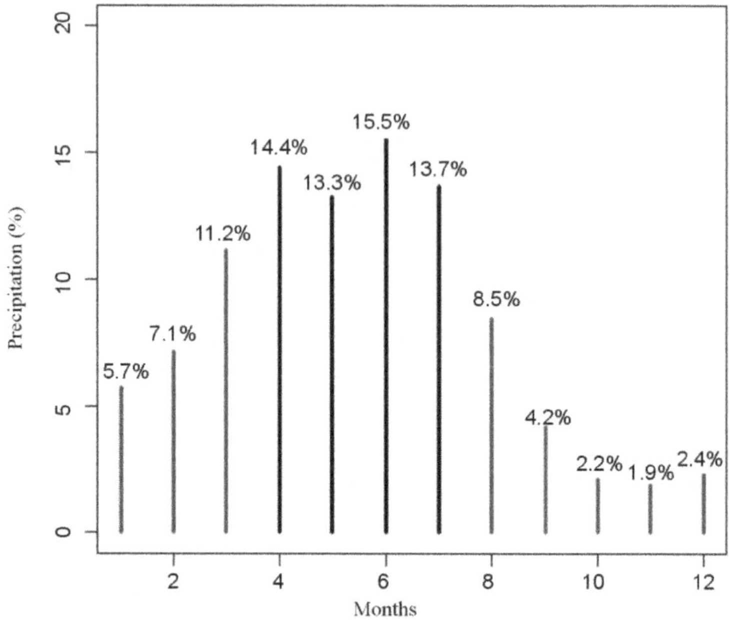

Figure 2. Percentage of monthly precipitation in relation to the annual total for ENEB, with data from GCPC.

Methodology

Selection of Anomalous Years of the Atlantic Ocean

In order to identify the anomalous years in the Atlantic Ocean we used the same criterion applied by [13] . They identified years of GRAD_N Atlantic those that showed simultaneously positive anomalies of SST in the North Atlantic and negative anomalies in the South Atlantic during the months from February to April (FMA). For GRAD_S, the anomalies of SST for the south are positive and negative to the North. Besides the FMA months, therefore GRAD manifests from February, but establishes in MAM (March to May), [13] . The months May and June are also analyzed regarding the presence of GRAD_S, due being rainy months of ENEB.

From this criterion, GRAD_S episodes were 1984, 1986, 1989, 1991, 1994, 2008 and 2009, while the GRAD_N episodes were 1983, 1992, 1997, 2004 and 2005. From these, the composites of AMJ periods of precipitation anomalies were calculated, as well as the total heat flux at the surface, the flux of latent and sensitive heat, and the divergence for both episodes, GRAD_S and GRAD_N.

Heat Balance on the Surface

The heat flux at surface represents how much energy coming from the Sun (shortwave) was partitioned among the sensible and latent flux, and the part that returned to space (long waves), then the balance was calculated by the Equation (1) as suggest by [19] :

$$Ht = QS + QN + LE + SH \tag{1}$$

where: QS is the short wave incident radiation; QN is the long wave radiation emanating from the surface; LE is the latent heat flux by the surface due to the evaporation and SH is the sensible heat flux. The units are $W{\cdot}m^{-2}$.

The Moisture Fluxes was evaluated between the surface and the 700 hPa layer.

$$Q_t = \frac{1}{g}\int_{p_0}^{p_1} qVdp$$

$$Q_\lambda = \frac{1}{g}\int_{p_0}^{p_1} qudp; Q_\varphi = \frac{1}{g}\int_{p_0}^{p_1} qvdp \Rightarrow Q_t = Q_\lambda i + Q_\varphi j \tag{2}$$

where: Q_t—total moisture flux; g—the acceleration of gravity ($\approx$ 9.8 m/s^2); q—specific humidity; p—pressure; V is the magnitude of the wind; p_1 and p_0 represent the pressure at 700 hPa and the surface, respectively [29] .

The moisture flux divergence ($kg{\cdot}m^{-2}s^{-1}$) will be evaluated by the Equation:

$$\nabla.Q_t = \frac{\partial Q_\lambda}{\partial x} + \frac{\partial Q_\varphi}{\partial y} \tag{3}$$

The moisture flux was calculated by multiplying the average value of a by the area, that is in $kg{\cdot}s^{-1}$

$$\oint \vec{Q}.\vec{n}dL = \iint \nabla.\vec{Q}_t dA = \langle \nabla.\vec{Q} \rangle dA = \langle \nabla.\vec{Q} \rangle A \tag{4}$$

where: $\vec{n}$ is the normal vector to the vertical faces; dL is the lateral boundary length, described by the projection of the vertical faces of the area, in the horizontal plane; A is the area of 5° × 1.2° latitude and longitude (5°S - 10°S; 36°W - 34.8°W); $\langle\rangle$ is the average in area A.

RESULT AND DISCUSSION

Anomalies of Precipitation

During the years of Tropical Atlantic GRAD_S, the ENEB showed excess of precipitation in relation to the climatology, except for the years 1991 and 2008, which showed rainfall deficit in the North and South coast, respectively. The year 1991 presented above the normal rainfall patterns for AMJ, only in the coast of Pernambuco, south of Sergipe and northern and eastern Bahia. In 2008, the precipitation excesses were more evident compared to the ones observed in 1991. However, most part of the states of Bahia, Alagoas and Sergipe, east of Pernambuco and Paraíba also registered deficit of precipitation. The variability in precipitation over the NEB is influenced by the Atlantic and Pacific Oceans, studies (e.g., [30]) showed this influence is mainly in GRAD area, although [17] had shown that simultaneous event of ENSO and significant SSTAs in the Atlantic contributed to increase or decrease the anomalies of precipitation. The SST anomalies of the same sign in ATS and eastern Pacific decreased the intensity of precipitation in NEB. Since 1991 was the year of strong El-Niño [31] , and its can explain the negative anomalies of precipitation in this episode (Figure not shown) in most part of NEB. During the 1991 episode, the water vapor given by the ocean near the east coast of Brazil was lower compared to the other years studied, not supplying the atmosphere with water vapor and therefore inhibiting the precipitation in some areas of the region.

Figure 3 shows indeed rainfall excess during episodes of GRAD_S, as well as FMA in northern NEB. De- spite ITCZ remaining in climatologically position further south during the episodes, contributing for the total rainfall at the beginning of ENEB rainy season (more precisely in the months of April and May), the easterly wave disturbances are also responsible for the quantity of rain until August, when ITCZ is already in a further North position. The EWD that possibly reached the Eastern region were strengthened to get closer to the coast. The oceanic region next to the east coast of the NEB on average, show favorable conditions for the occurrence of above the normal rainfall patterns for the period.

Heat Balance on the Surface

The heat balance on the surface involves incident shortwave radiation, long wave radiation emitted, latent and sensible heat, the sum of these energies precisely represents exchanges of energy in the ocean-atmosphere interface [19] . However, in this study, it will highlight the sum of the energies and latent energy.

During the GRAD_S an abnormal heat is generally observed in most part of the southern tropical Atlantic basin. This behavior represents more thermal energy being transferred to the atmosphere as well as more water vapor being removed from the ocean. Once the wind field shows favorable conditions, this humidity may feed the systems that cause precipitation in the region in this period, such as the wave easterly disturbances, which are more usual during the AMJ in ENEB.

In the oceanic region next to the east of the NEB, a negative balance of heat on the surface can be observed, except for the year of 2008. These conditions allow us to state that the oceanic surface layer next to the coast remained, on average, warmer than normal during the trimester, nourishing of moisture EWD that reached the coast during the selected years.

Some regions with surplus of heat on the surface are observed between 5°S - 10°S and 0°W - 25°W over the Atlantic are highlighted when the episodes individually verified (Figure not shown), these small areas can be associated with the easterly waves, the persistence of events during the trimester may reflect to abnormality on these regions.

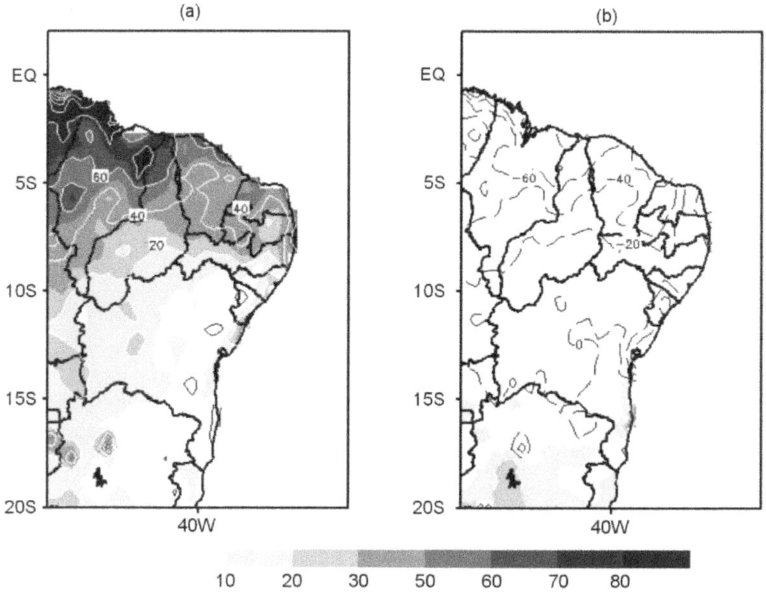

Figure 3. Standard precipitation anomalies in a) GRAD_S and b) GRAD_N episodes in NEB. Negative deviations (positive) of precipitation are represented by dashed lines (grayscale).

Figure 4, a negative balance of heat layer is observed between the equator and 10°N, this region of strong latent heat release is a reflection of the presence of the ITCZ. During the AMJ period, this system is displaced to the Hemisphere of Summer depending on the weather. However, in years of GRAD_S, Figure 4(a) (GRAD_N, Figure 4(b)), studies have reported that the ITCZ remains longer in its position further south (north) influencing the rainy season of the North and Northeast region of Brazil. However, the remaining further south influences the Northern sector of ENEB at the beginning of rainy season, more precisely in April. In the other months (May and June) precipitation has occurred, mainly by EWD and remnants of frontal systems.

The anomalies of latent heat (Figure 5) next to the ENEB also agree with balance of power debated in Figure 4; coastal strip showed anomalies of latent energy below (above) the normal during the rainy seasons of year GRAD_S (GRAD_N) of Tropical Atlantic. Otherwise, over the continental region, areas with positive (negative) anomalies of latent energy, showing that the atmosphere lost more (less) energy than normal through the release of latent heat of condensation of water vapor as verified in the analysis of precipitation occurrence above (below) normal over the continent during episodes of GRAD_S (GRAD_N), Figure 3(a) (Figure 3(b)). In 1991 and 2008, negative anomalies of latent heat were observed, however they were seen in the North coast and far from ENEB coast, respectively.

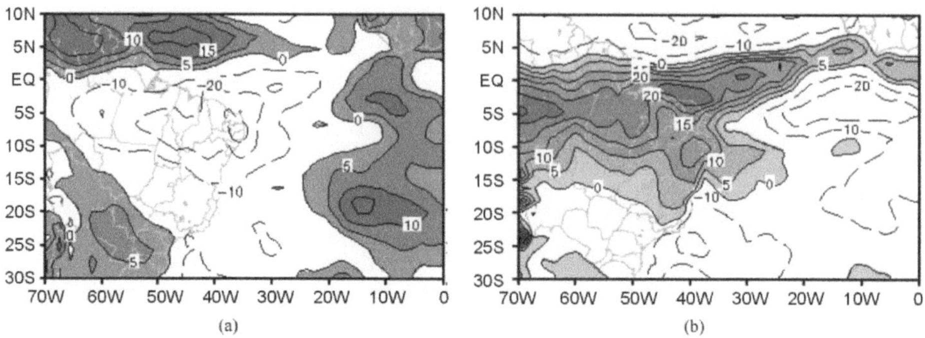

Figure 4. Total heat flux anomalies for the years of a) GRAD_S and b) GRAD_N of Tropical Atlantic in W·m^{-2}. Negative deviation (positive) is with dashed lines and grayscale Gray (continuous line).

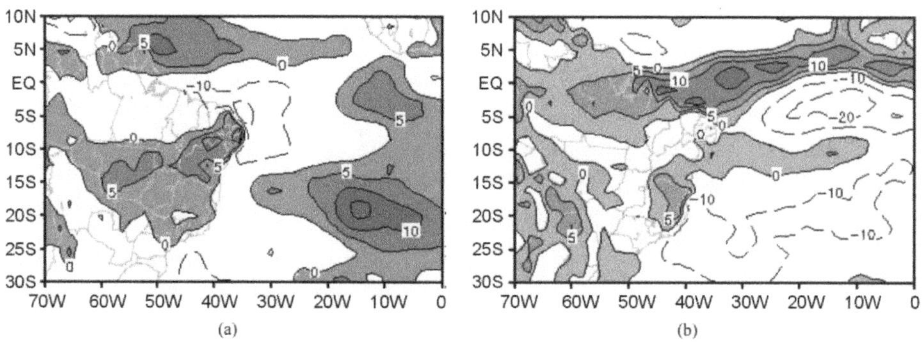

Figure 5. Anomaly of latent heat flux in the years of a) GRAD_S and b) GRAD_N of Tropical Atlantic in the unity W·m^{-2}. Negative deviations (positive) are with dashed lines and grayscale (continuous line).

Moisture Flux Vertically Integrated

The layer between the surface and 700 hPa has the most part of humidity from the air in the atmosphere, diminishing considerably with height. Figure 6 shows water vapor anomalies and respective anomalous flux at lower levels of the atmosphere. The prevalence of abnormalities in ENEB flux is from Northeast/North (southeast/south) during the GRAD_S (GRAD_N). This behavior is reflected in the region of Hadley circulation confluence displaced to the south (North) of the Equator, favoring (disfavoring) appropriate conditions for precipitation over the region. Similar results were observed by [32] for 1985 and for the FMA period. In addition, the pattern GRAD_S (GRAD_N) provides more (less) water vapor to the atmosphere, positive (negative) anomalies in all NEB are observed.

Divergence Vertically Integrated of Moisture Flux

The convergence at low levels, for continuous waves, generates upward vertical movements, favoring the development of clouds and, consequently, the given precipitation due to the availability of moisture. However, the GRAD_S and GRAD_N episodes enhance or inhibit these dynamic effects in the region, making the rainy periods more or less intense [32]

The Southern portion of ENEB in both patterns of the tropical Atlantic is observed the divergence in the first layers of the atmosphere. The further North portion of the region experiences changes as in the intensity as the location of divergence, it is observed that it is intense on the coast of Bahia in both patterns, however in GRAD_ N the values are more intense. During the GRAD_S, the region that shows convergence of moisture flux covers the entire range of 9°S toward the equator over the NEB. When the pattern is contrary, the region of convergence decreases slightly further North, mainly on the coast.

Despite the North region and the east part of the NEB provide moisture convergence (Figure 6) in both patterns of SST of the Atlantic (GRAD_S and GRAD_N), the occurrence of precipitation depends on the availability of water vapor (Figure 5) in the atmosphere, and this behavior was observed during the GRAD_S (GRAD_ N), where positive anomalies (negative) of water vapor were observed in most part of NEB.

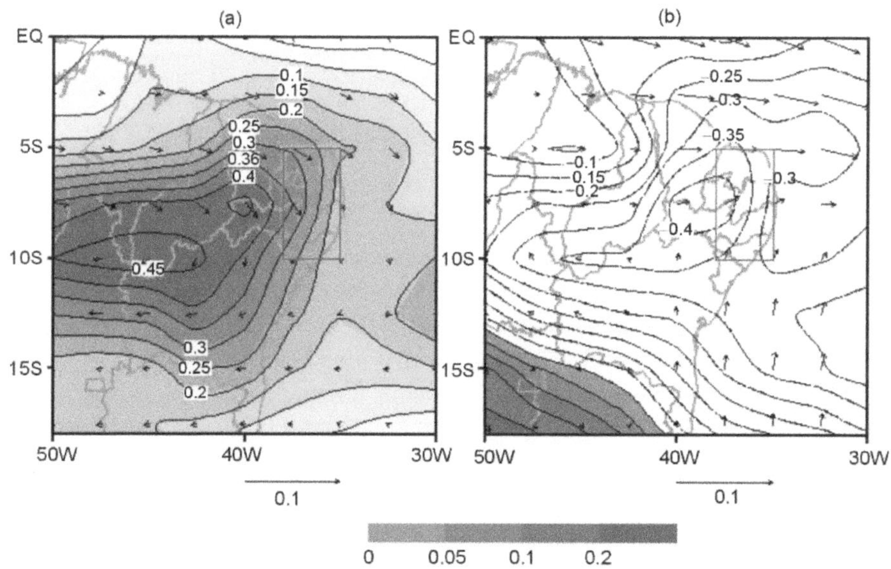

Figure 6. Flux anomalies vertically integrated between the surface and 700 hPa of a) GRAD_S and b) GRAD_N episodes.

Finally, Figure 7 provides the calculated moisture flux, which was multiplied by the divergence average value by the ENEB area. The convergence values explain the precipitation excesses that on average are observed in the coast of ENEB, mainly during episodes of GRAD_S (-162.6×10^{-9}kg·s^{-1}), showing importation of relative humidity above 50% of the value shown in GRAD_N (-58.2×10^{-9} kg·s^{-1}). The water vapor found in the first layers of the atmosphere in ENEB is imported from the oceanic region as rule. Despite the wave adjustment not to be carried out in this study, the results were satisfactory to explain the precipitation anomalies above normal during the GRAD_S in the region ENEB, during the rainy season.

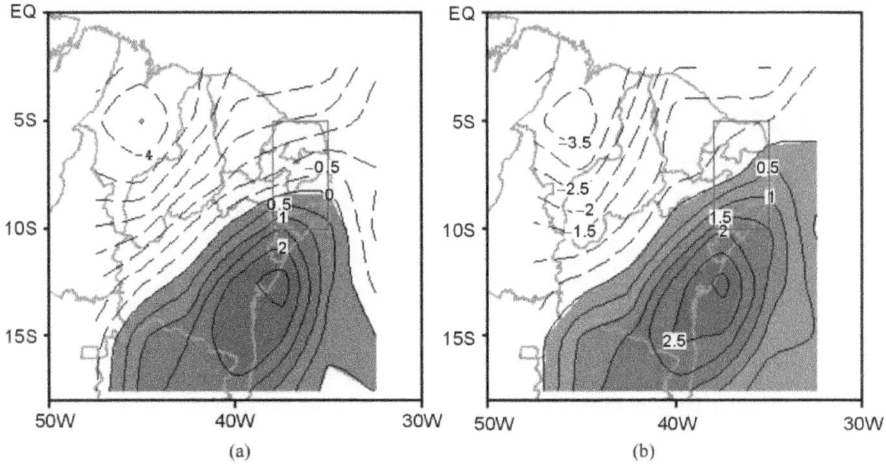

Figure 7. Moisture flux divergence vertically integrated (10^9 kg·s^{-1}) during episodes of a) GRAD_S and b) GRAD_N. The positive (negative) values show divergence (convergence) of moisture and they are highlighted in shades of gray (in dashed line).

CONCLUSIONS

The results showed the importance of ocean-atmosphere interaction in the South Atlantic for precipitation in ENEB. It can be seen that small variations in SST cause strong influences on the flux between the ocean and atmosphere causing impact on weather systems that operate in the region during rainy season.

During episodes where the South Atlantic Ocean was warmer, oceanic and atmospheric favorable conditions were observed, favorable to change in the distribution and volume of precipitation, and inclusive next to the east coast, which supposedly intensified the weather systems that reached the coast.

In this observational study, it can be concluded that the ocean-atmosphere interaction present in the Tropical South Atlantic favors the deviations in rainfall in ENEB. The weather phenomena that operates remotely amplifying or inhibiting the rain also helps the rainy season; however, it appears that the contribution is greater than the spatial distribution of rainfall intensity. In 1991 and 2008 episodes (El Niño and La Niña, respectively), years in which the Pacific Ocean showed the SST anomalies, above normal precipitation were seen on the east coast of NEB; however only the spatial distribution of these precipitation anomalies was affected.

We can also observe that the moisture convergence over the ENEB region in both Atlantic SST patterns (GRAD_S and GRAD_N). However, the occurrence of precipitation depends on the availability of water vapor, situation that depends on the conditions of the oceans, the main source of moisture for the NEB region.

ACKNOWLEDGEMENTS

Ana Cleide Bezerra Amorim, Rosane Rodrigues Chaves, Cláudio Moisés Santos e Silva (CAPES) by doctoral scholarship of the of Planos de Reestruturação e Expansão das Universidades Federais (Reuni), the first author, A. C. B. Amorim.

REFERENCES

1. Doyle, J.D. (2002) Coupled Atmosphere—Ocean Wave Simulations under High Wind Conditions. Monthly Weather Review, 130, 3087-3099. http://dx.doi.org/10.1175/1520-0493(2002)130<3087:CAOWSU>2.0.CO;2
2. Moura, A.D. and Shukla, J. (1981) On the Dynamics of Droughts in Northeast Brazil: Observations, Theory and Numerical Experiments with a General Circulation Model. Journal of the Atmospherical Sciences, 38, 2653-2675. http://dx.doi.org/10.1175/1520-0469(1981)038<2653:OTDODI>2.0.CO;2
3. Souza, E.B. and Nobre, P. (1998) Uma revisão do Padrão Dipolo no Atlântico Tropical. Revista Brasileira de Meteorologia, 13, 31-44.
4. Molion, L.C.B. and Bernardo, S.O. (2002) Uma revisão da dinâmica das chuvas no nordeste brasileiro. Revista Brasileira de Meteorologia, 17, 1-10.
5. Souza, P.E. and Cavalcanti, I.F.A (2009) Atmospheric Centres of Action Associated with the Atlantic ITCZ Position. International Journal of Climatology, 29, 2091-2105.http://dx.doi.org/10.1002/joc.1823
6. Ramos, R.P.L. (1975) Precipitation Characteristics in the Northeast Brazil Dry Region. Journal of Geophysical Research, 80, 1665-1678.http://dx.doi.org/10.1029/JC080i012p01665
7. Yamazaki, Y. and Rao, V.B. (1977) Tropical Cloudiness over the South Atlantic Ocean. Journal of the Meteorological Society of Japan, 55, 205-207.
8. Chou, S.C. (1990) Análise de distúrbios ondulatórios de leste sobre o oceano atlântico equatorial sul. Dissertation, Instituto Nacional de Pesquisas Espaciais, São José dos Campos.
9. Torres, R.R. and Ferreira, N.J. (2011) Case Studies of Easterly Wave Disturbances over Northeast Brazil Using the Eta Model. Weather and Forecast, 26, 225-235. http://dx.doi.org/10.1175/2010WAF2222425.1
10. Kouadio, Y.K., Servain, J., Machado, L.T. and Lentini, C.A.D (2012) Heavy Rainfall Episodes in the Eastern Northeast Brazil Linked to Large-Scale Ocean-Atmosphere Conditions in the Tropical Atlantic. Advances in Meteorology, 2012, 1-16.http://dx.doi.org/10.1155/2012/369567
11. Kousky, V.E. (1979) Frontal influence on Northeast Brazil. Monthly Weather Review, 107, 1140-1153. http://dx.doi.org/10.1175/1520-0493(1979)107<1140:FI

ONB>2.0.CO;2

12. Mishra, S.K., Rao, V.B. and Franchito, S.H. (2007) Genesis of the Northeast Brazil Upper-Tropospheric Cyclonic Vortex: A Primitive Equation Barotropic Instability Study. Journal of the Atmospherical Sciences, 64, 1379-1392. http://dx.doi.org/10.1175/JAS3893.1
13. Souza, E.B., Kayano, M.T. and Ambrizzi, T. (2005) Intraseasonal and Submonthly Variability within Autmn Rainy Regime over the Eastern Amazon/Northeast Brazil and Associated Mechanisms. Theoretical and Applied Climatology, 81, 177-191.http://dx.doi.org/10.1007/s00704-004-0081-4 [Paper reference:4]
14. Hastenrath, S. and Heller, L. (1977) Dynamcs of Climatic Hazards in Northeast Brazil. Quarterly Journal of the Royal Meteorological Society, 107, 77-92.http://dx.doi.org/10.1002/qj.49710343505
15. Saravanan, R. and Chang, P. (2000) Interaction between Tropical Atlantic variability and El Ni-o-Southern Oscillation, Journal of Climate, 13, 2177-2194. http://dx.doi.org/10.1175/1520-0442(2000)013<2177:IBTAVA>2.0.CO;2
16. Andreoly, R.V. and Kayano, M.T. (2006) Tropical Pacific and South Atlantic Effects on Rainfall Variability over Northeastern Brazil. International Journal Climatology, 26, 1895-1912. http://dx.doi.org/10.1002/joc.1341
17. Andreoli, R.V. and Kayano, M.T. (2007) A importância relativa do Atlântico Tropical Sul e Pacífico leste na variabi- lidade de precipitação do Nordeste do Brasil. Revista Brasileira de Meteorologia, 22, 63-74. http://dx.doi.org/10.1590/S0102-77862007000100007
18. Kayano, M.T., Oliveira, C.P. and Andreoli, R.V. (2009) Interannual Relations between South American Rainfall and Tropical Sea Surface Temperature Anomalies before and after 1976. International Journal of Climatology, 29, 1439-1448.http://dx.doi.org/10.1002/joc.1824
19. Sterl, A. and Hazeleger, W. (2003) Coupled Variability and Air-Sea Interaction in the South Atlantic Ocean. Climate Dynamics, 21, 559-571.http://dx.doi.org/10.1007/s00382-003-0348-y [Paper reference:3]
20. Oliveira, P.T., Lima, K.C. and Silva, C.M.S. (2013) Synoptic Environment Associated with Heavy Rainfall Events on the Coastland of Northeast Brazil. Advances in Geosciences, 35, 73-78. http://dx.doi.org/10.5194/adgeo-35-73-2013
21. Silva, C.M.S., Silva, A., Oliveira, P. and Lima, K.C. (2014) Dynamical Downscaling of the Precipitation in Northeast Brazil with a Regional Climate Model during Contrasting Years. Atmospheric Science Letters, 15, 50-57. http://dx.doi.org/10.1002/asl2.468
22. Grimm, A.M. and Tedeschi, R.G. (2009) Enso and Extreme Rainfall Events in South America. Journal of Climate, 22, 1589-1609. http://dx.doi.org/10.1175/2008JCLI2429.1
23. Amorim, A.C.B, Chaves, R.R. and Spyrides, M.H.C. (2012) Precipitação no leste do nordeste brasileiro (lneb) e epi- sódios anômalos no oceano atlântico tropical. 17th Congresso Brasileiro De Meteorologia, Gramado, Anais XVII do Congresso

brasileiro de meteorologia, September 2012, Gramado. [Paper reference:3]

24. Huffman, G., Adler, R.F., Arkin, P., Chang, A., Ferraro, R., Gruber, A., et al. (1997) The Global Precipitation Climatology Project (GPCP) Combined Precipitation Dataset. Bulletin of the American Meteorological Society, 78, 5-20. http://dx.doi.org/10.1175/1520-0477(1997)078<0005:TGPCPG>2.0.CO;2

25. Kalnay, E., Kanamitsu, M., Kistler, R., Collins, W., Deaven, D., Gandin, L., et al. (1996) The NCEP/NCAR 40-Year Reanalysis Project. Bulletin of the American Meteorological Society, 77, 437-471. http://dx.doi.org/10.1175/1520-0477(1996)077<0437:TNYRP>2.0.CO;2

26. Dee, D.P., Uppala, S.M., Simmons, A.J., Berrisford, P., Poli, P., Kobayashi, S., et al. (2010) The ERA-Interim Reanalysis: Configuration and Performance of the Data Assimilation System. Quarterly Journal of the Royal Meteorological Society, 137, 553-597.http://dx.doi.org/10.1002/qj.828

27. Smith, T.M. and Reynolds, R.W. (1998) A High-Resolution Global Sea Surface Temperature Climatology for the 1961-90 Base Period. Journal of Climate, 11, 3320-3323. http://dx.doi.org/10.1175/1520-0442(1998)011<3320:AHRGSS>2.0.CO;2

28. Rao, V.B., de Lima, M.C. and Franchito, S.H. (1993) Seasonal and Interannual Variations of Rainfall over Eastern Northeast Brazil. Journal of Climate, 6, 1754-1763.http://dx.doi.org/10.1175/1520-0442(1993)006<1754:SAIVOR>2.0.CO;2

29. Rao, V.B., Chapa, S.R. and Franchito, S.H. (1999) Decadal Variation of Atmosphere-Ocean Interaction in the Tropical Atlantic and Its Relationship to the Northeast-Brazil Rainfall. Journal of the Meteorological Society of Japan, 77, 63- 75.

30. Moura, G.B.A., Aragão, J.O.R., Melo, J.S.P., Silva, A.P.N., Giongo, P.R. and Lacerda, F.F. (2009) Relação entre a pre- cipitação do leste do Nordeste do Brasil e a temperatura dos oceanos. Revista Brasileira de Engenharia Agrícola e Ambiental, 13, 462-469.http://dx.doi.org/10.1590/S1415-43662009000400014

31. Centro de Previsão de Tempo e Estudos Climáticos (CPTEC) (2012) El-Niño e La-Niña.http://enos.cptec.inpe.br/

32. Bezerra, A.C.N. (2006) Aspectos da circulação atmosférica de grande escala sobre o norte e nordeste do Brasil re- lacionados com a temperatura da superfície do mar. Dissertation, Universidade de Federal de Campina Grande, Centro de Tecnologia e Recursos Naturais—Unidade Acadêmica de Ciências Atmosféricas, Campina Grande.

11

CHAPTER

EVALUATION OF SATELLITE AND REANALYSIS SOIL MOISTURE PRODUCTS USING GROUND-BASED MEASUREMENTS

Jian Peng, Jonathan Niesel, Alexander Loew, Shiqiang Zhang and Jie Wang

Keywords: Validation, *in situ* soil moisture, satellite-based soil moisture, reanalysis soil moisture, Southwest China

INTRODUCTION

Soil moisture (SM) plays an important role in the interactions between the atmosphere and the land surface, and has been identified as an essential climate variable by the Global Climate Observing System (GCOS) [1]. Soil moisture connects the energy and water fluxes due to its control on the partitioning of surface net energy into latent and sensible heat fluxes, as well as the partitioning of precipitation into infiltration and runoff [2,3]. Therefore, the knowledge of soil moisture dynamics is important for climate change research, flood and drought monitoring, weather forecasting, as well as water resource management [4,5].

Satellite remote sensing has been recognized as a powerful tool to retrieve soil moisture due to the improvements of sensor technologies and retrieval algorithms [6] using either active [7] or passive microwave [8] remote sensing observations. Until now, several global satellite-based soil moisture products have been released. These products include the Advanced Microwave Scanning Radiometer E for the Earth

Citation: Peng J., Niesel J., Loew A., Zhang S., Wang J., Evaluation of Satellite and Reanalysis Soil Moisture Products Using Ground-Based Measurements, doi:10.3390/rs71115729.

observing system (AMSR-E) [9], the Advanced Scatterometer (ASCAT) [10,11], the Soil Moisture and Ocean Salinity (SMOS) [12], and the European Space Agency's Climate Change Initiative (ESA CCI) soil moisture product [13,14,15]. These satellite-based soil moisture products provide an unprecedented opportunity to investigate the interactions between land and atmosphere and their effects on climate change. However, comprehensive evaluations of these soil moisture products are required before using them. Many evaluation studies have been conducted either based on *in situ* measurements or model simulations. Most of these studies are performed either on a regional scale, for example in Europe [16,17,18,19], Australia [20,21], Africa [22,23], and the United States [24,25], or at the global scale [15,26,27]. However, few studies were conducted in China so far. In order to utilize the satellite-based soil moisture products for practical applications in China, it is necessary to validate them against *in situ* measurements in China.

The objective of this research is to evaluate four satellite-based soil moisture products and one reanalysis product with *in situ* measurements over Yunnan, China (Figure 1). The validation activity is of great importance, because this region is very sensitive and vulnerable to climate change due to its unique geophysical position and climate. Over the last three decades, frequent droughts and floods occurred in this region, with the most extreme drought occurring from winter 2009 to spring 2010 [28,29]. According to the Ministry of Civil Affairs, the drought resulted in a loss of $2.5-billion due to agricultural damage, and about 9.65 million people have had a shortage of drinking water [30]. Therefore, accurate long-term satellite-based soil moisture products are urgently required to assess drought conditions and the links between increasing temperature, soil moisture variations, and vegetation productivity. The following section briefly introduces the study area,*in situ* measurements, and soil moisture products. The evaluation methods are described in Section 3. Results of the evaluation are then presented and discussed in Section 4. Finally, conclusions are summarized in Section 5.

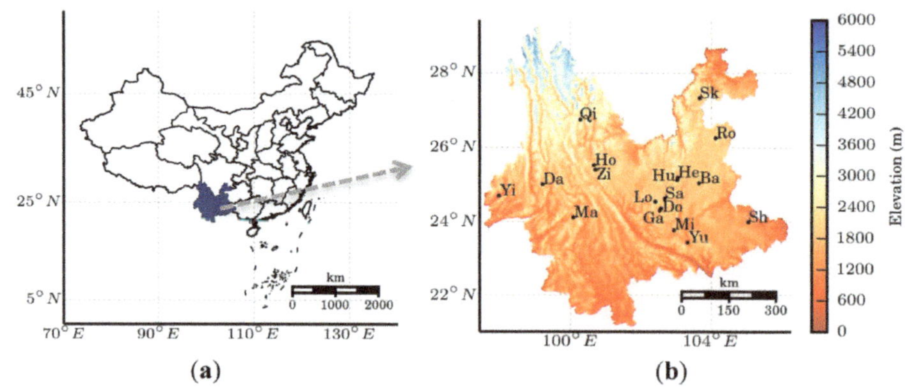

Figure 1. Study area in Southwest China (**a**) and locations of *in situ* soil moisture stations (**b**).

STUDY AREA

The Yunnan province is located in Southwest China (Figure 1), covering an area of roughly 394,000 km^2. Mountainous topography and special karst geomorphology characterize this region, with an elevation gradually decreasing from the northwest to southeast, and an average elevation of 1980 m above sea level [31]. This region has a subtropical climate with dry winters and wet summers. The precipitation is spread unevenly over the year and the region, and is largely influenced by the Tibetan plateau, the tropical Indian monsoon and the eastern Asia monsoon. As a result, about half of the precipitation occurs between June and August. The annual precipitation ranges from 600 mm in valleys to 1700 mm in mountainous regions [32,33]. Due to its special climate and topographical conditions, this region suffers from frequent droughts. The most severe and sustained drought occurred in 2010. It started in the winter of 2009 and became particularly severe in the spring of 2010, leading to significant ecological and economic damages [34]. Numerous streams and small reservoirs dried up, which had a severe impact on the supply of drinking water and the agricultural production [30,35]. Therefore, long-term and spatially distributed soil moisture information is required for this region.

MATERIALS AND METHODS

Satellite-Based and Reanalysis Products

The satellite products used in this study include four soil moisture products as well as one precipitation product. In addition, one soil moisture reanalysis product is also analyzed. In the following subsections, these datasets are briefly introduced.

AMSR-E

The AMSR-E onboard the Aqua satellite is a passive microwave sensor. The instrument operated between May 2002 and October 2011. It provided measurements at six different frequencies. Data is available for descending (1:30 a.m.) and ascending (1:30 p.m.) orbits [36]. Several algorithms have been developed for retrieving soil moisture from the AMSR-E measurements. One of them is the land parameter retrieval model (LPRM) [9]. The corresponding soil moisture product has been developed and released by Vrije Universiteit Amsterdam (VUA) together with NASA. The LPRM product provides global soil moisture on a regular 0.25° global grid derived from C-band and X-band observations. The observations from the X-band are used only for soil moisture estimation when the C-band observations are affected by radio frequency interference (RFI). Based on the observations from these two bands at two overpass times, the LPRM product contains four soil moisture products (C-band and X-band soil moisture in both, in descending and ascending modes). In this study, all these products were compared with *in situ* measurements to investigate their performance using the period from 2008 to 2011. The soil moisture data were masked if the vegetation optical depth (VOD) was higher than 0.8, as suggested by Rebel*et al.* [37] and Lei *et al.* [38].

ASCAT

The ASCAT is a scatterometer onboard the Meteorological Operation-A (MetOp-A) satellite, which was launched in October 2006 and scans the Earth's surface in descending (9:30 a.m.) and ascending (9:30 p.m.) orbits [39]. It operates in the C-band (5.3 GHz) in vertical polarization with a spatial resolution of 25 or 50 km [40]. The soil moisture is derived from the backscatter measurements of ASCAT using a change detection algorithm [11]. In this study, the soil moisture product provided by the European Organization for the Exploitation of Meteorological Satellites (EUMETSAT) with the version of TU-Wien-WARP 5.5 was used. It provides a relative soil moisture saturation degree (using the top 2 cm of the soil column) with values ranging from 0% (dry) to 100% (wet) at spatial resolution of 25 km. To be consistent with *in situ* soil moisture measurements and comparable with other soil moisture products, the ASCAT soil moisture was rescaled into volumetric soil moisture (unit: m^3/m^3) using the soil porosity estimated from Land Data Assimilation System (LDAS) data [41]. The soil moisture products from 2008 to 2012 in both descending and ascending modes were analyzed in this study. The soil moisture data were masked if the processing flags were not equal to 0 (best quality).

ESA CCI

The ESA CCI soil moisture product is a unique 35-year (1978–2013) long satellite-based soil moisture record produced within the framework of the ESA Climate Change Initiative [42]. The CCI soil moisture was generated by merging several passive and active soil moisture products [13,43]. The involved soil moisture products are derived from Scanning Multichannel Microwave Radiometer (SMMR), Special Sensor Microwave/Imager (SSM/I), Tropical Rainfall Measuring Missions (TRMM) Microwave Imager (TMI), ASMR-E, scatterometers (SCAT) and ASCAT. The CCI soil moisture has three products, which differ in the data sources used: passive microwave only, active microwave only, and merged passive-active product [15]. In this study, the merged passive-active product (version 02.1) is analyzed for the period from 2008 to 2012. The dataset has spatial resolution of 0.25° on a daily basis.

SMOS

The SMOS satellite, launched in November 2009, is dedicated to soil moisture monitoring with an L-band (1.4 GHz) passive radiometer. It provides global coverage every three days for both descending (6:00 p.m.) and ascending (6:00 a.m.) orbits with a spatial resolution of about 40 km [44]. The advantage of the L-band is that it has greater sensing depth of up to 5 cm of the soil. The soil moisture is derived using the L-band microwave emission of the biosphere (L-MEB) model [45,46]. In this study, the SMOS level 3 daily soil moisture product was obtained from the Centre Aval de Traitement des Données SMOS (CATDS). As is the case for AMSR-E and ASCAT, both descending and ascending soil moisture products were used for analysis from the years from 2009 to 2012. According to Al-Yaari *et al.* [47], the soil moisture data are masked out if the Data Quality Index (DQX) is larger than 0.06, or the DQX is equal to 0 (filled value), or the RFI is larger than 30%.

ERA-Interim

ERA-Interim is a global atmospheric reanalysis product produced by the ECMWF [48]. It is based on a sequential data assimilation scheme. It covers the period from 1979 until the present. The soil moisture data from ERA-Interim are provided at four depths (0–7, 7–28, 28–100, and 100–289 cm) [48]. In this study, the daily averaged 0.25° upper layer (0–7 cm) of soil moisture products covering the years from 2008 to 2012 were evaluated.

TRMM Precipitation Data

The TRMM precipitation data was used in this study as an alternative to the station precipitation from *in situ*measurements, because it has been extensively validated against ground-based measurements around the world, including the Yunnan province [33]. Details of the TRMM precipitation dataset can be found in Kummerow *et al.* [49] and Iguchi *et al.*[50]. The daily 0.25° TRMM precipitation products (3B42 V7) were used in this study for the same period as indicated in the other dataset (2008–2012).

In Situ Soil Moisture Measurements

There are a total of 18 soil moisture stations used in the current study. The geographical locations and details of these stations are presented in Figure 1 and Table 1. The elevations of these stations range from 847 m to 2293 m, and the land cover types are mainly grassland (4 stations), forest (3 stations), cropland (7 stations), and bare soil (4 stations). The soil moisture contents over these stations were measured at three different depths (10 cm, 20 cm and 40 cm) using frequency domain reflectometry (FDR) sensors. Data was obtained from the Yunnan Hydrology and Water Resources Bureau. The measurements at a 10 cm depth are considered to best represent the soil layer accessible by the satellite observations. We therefore use this layer as a reference in this study.

Table 1. Descriptions of the *in situ* soil moisture stations used in the study

Station Name	Short Name	Land Use	Latitude (°)	Longitude (°)	Elevation (m)
Daciping	Da	Cropland	25.02	99.20	1685
Zijin	Zi	Bare soil	25.42	100.70	1958
Hongqi	Ho	Grassland	25.53	100.67	2044
Yingjiang	Yi	Bare soil	24.70	97.95	847
Yuguopu	Yu	Cropland	23.45	103.32	1283
Mianyangchong	Mi	Forest	23.78	102.93	1433
Sanduo	Sa	Grassland	24.64	102.67	1967
Longwangtang	Lo	Forest	24.55	102.40	2099
Heyou	He	Cropland	25.20	103.03	1968
Huangjiapo	Hu	Grassland	25.13	103.02	2140
Qiaotou	Qi	Cropland	26.77	100.27	2293

Mayidui	Ma	Cropland	24.12	100.07	1344
Bamaochong	Ba	Grassland	25.05	103.63	1857
Rongfeng	Ro	Bare soil	26.26	104.11	1973
Shaba	Sh	Forest	24.00	105.05	1415
Dongfeng	Do	Bare soil	24.37	102.57	1653
Gaocang	Ga	Cropland	24.32	102.52	1650
Sankeshu	Sk	Cropland	27.33	103.67	1909

Evaluation Strategies

In order to evaluate the performances of the satellite-based and reanalysis soil moisture products, they were first compared directly with the *in situ* soil moisture. Established skill scores, like the correlation coefficient (R), the BIAS, the root mean square difference (RMSD), and the unbiased RMSD (ubRMSD) were used to quantify the differences between various soil moisture products and measured soil moisture [51,52]. In addition, the comparisons for soil moisture anomalies were also performed, because the seasonal variations of soil moisture may influence the correlation statistics [53]. The soil moisture anomalies were computed similar to Dorigo *et al.* [15], using the equation below to avoid the seasonal effects on correlations. The anomaly calculation is based on a sliding window of five weeks. The difference from the mean is calculated if there are at least five measurements available during this time window.

$$Anom(t) = SM(t) - \overline{SM(t-17:t+17)} \quad (1)$$

where SM(t) is the soil moisture at day t, the overbar indicates the temporal mean for the five weeks period. Based on the above equation, the soil moisture anomalies of *in situ* soil moisture, satellite-based, and reanalysis products were calculated.

RESULTS AND DISCUSSION

Direct Comparison of Soil Moisture

Pre-Selection of Best Performed AMRS-E, ASCAT, and SMOS

From direct comparisons with *in situ* soil moisture, it is found that the AMSR-E X-band has better performance than the C-band with lower RMSD and ubRMSD values. It implies that the C-band observations might be influenced by the RFI in this region. Regardless of the frequency band, ascending and descending soil moisture products also have different accuracies due to the diurnal variations of the land surface conditions. On one hand, overpass times at night and in the early morning are more suitable for soil moisture retrieval from passive microwave satellites, because the isothermal conditions at night can minimize the errors associated with surface temperature [54]. On the other hand, the soil emission is influenced by the diurnal variation of the water in the vegetation canopy [55]. The attenuation of soil emission caused by water in the vegetation canopy is minimized during the daytime due to loss of water through evapotranspiration [56]. Therefore, both daytime and nighttime retrievals have certain advantages. For AMSR-E, many studies such as Brocca *et al.* [17] have found that ascending soil moisture is more reliable than descending soil moisture. However, in other studies such as Griesfeller *et al.* [57], descending was found to be better than ascending. This demonstrates that the performance of different AMSR-E soil moisture products depends on the surface condition of the study area. In this study, it is found that ascending AMSR-E performs better than descending AMSR-E, which might be due to the heavily vegetated characteristics of the study area. Similar results have been reported by Lei *et al.* [38], who found that the advantage of the nighttime overpass is reduced as the vegetation cover increases and the AMSR-E ascending retrieval is superior over heavily vegetated areas in the United States. Therefore, the descending and ascending soil moisture products should be evaluated before applying them in different regional studies. Compared to AMSR-E, the differences between ascending and descending retrievals are smaller for SMOS. It may be due to the smaller soil moisture temperature differences between ascending/descending overpass times of SMOS than that of AMSR-E. The Faraday rotation effect in the ionosphere is also expected to be minimized at SMOS ascending overpass time [38]. Although SMOS ascending performs slightly better than descending, the sample days of SMOS ascending soil moisture are less than descending soil moisture in the current study. Therefore, the SMOS descending soil moisture is used for the following analyses. Compared to AMSR-E and SMOS, the ASCAT presents quite similar performance for ascending and descending soil moisture. This is partly because the scatterometer-based ASCAT does not require accurate estimates of surface temperature [38]. It is worth noting that the ASCAT descending soil moisture has slightly better correlation values than ascending soil moisture in the current study area. The statistics of the above comparisons are provided in Appendix Figure A1. In the following analyses, only the best performed AMSR-E ascending X band (AMSR-E_A_X), ASCAT descending (ASCAT_D), and SMOS descending (SMOS_D) are considered in comparisons with *in situ* measurements.

Temporal Evolution of Soil Moisture and Precipitation

Figure 2 shows the time series patterns of all the soil moisture products, *in situ* measurements and rainfall for the average values over all stations. All the products except for AMSR-E and SMOS correspond well with rainfall, with the soil moisture increasing during rainfall events and decreasing after rainfall events. Additionally, the temporal dynamics of *in situ* soil moisture are captured well by CCI SM and ERA-Interim. It can also be seen that CCI SM and ERA-Interim, which are in line with *in situ* measurements, capture the spring droughts in 2010. Compared to *in situ* soil moisture, the ASCAT has higher seasonal variability. The poor performance of AMSR-E and SMOS might be due to the effects of RFI over this region, because both AMSR-E and SMOS are reported to correlate well with ground-based measurements over other regions, such as America, West Africa, and Europe [23,25,26].

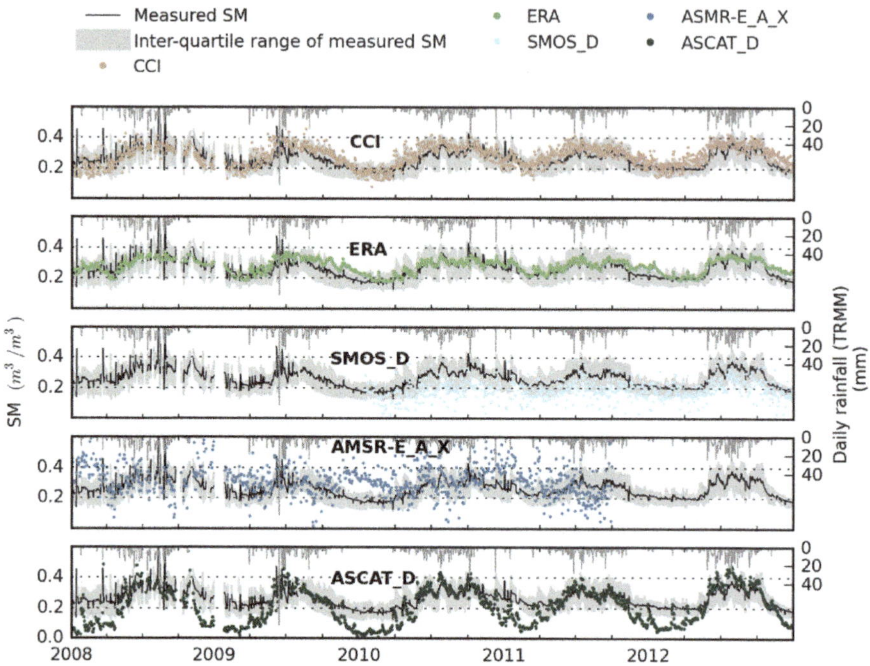

Figure 2. Temporal evolution of the averaged *in situ* measurements, soil moisture products, and precipitation at network scale: *in situ* mean soil moisture (black solid line) and its inter-quartile range (grey shaded areas); European Space Agency's Climate Change Initiative (CCI), Interim Reanalysis (ERA), Soil Moisture and Ocean Salinity (SMOS), Advanced Microwave Scanning Radiometer for the Earth observing system (AMSR-E)_A_X and Advanced Scatterometer (ASCAT)_D (points with different color); and precipitation (top of each panel). The soil moisture has the unit of m^3/m^3(left y-axis), while precipitation has the unit of mm (right y-axis).

Validation of the Soil Moisture against in Situ Measurements

To quantitatively evaluate these soil moisture products, Figure 3 shows the comparison statistics between different soil moisture products and *in situ* measured soil moisture at each station. In general, the ASCAT, CCI SM, and ERA-Interim have similar R values. However, in terms of RMSD and ubRMSD, the best performance was found with the ERA-Interim, followed by CCI SM and ASCAT. It can also be seen that the accuracies of AMSR-E and SMOS are much worse than the expected accuracy of 0.06 $m^3 \cdot m^{-3}$ for AMSR-E and 0.04 $m^3 \cdot m^{-3}$ for SMOS. In theory, the SMOS should provide reliable soil moisture product due to the use of L-band measurements. As stated before, the poor performance found here might be caused by the effects of RFI. Similar results have been reported by Cho *et al.* [58] and Zeng *et al.* [59].

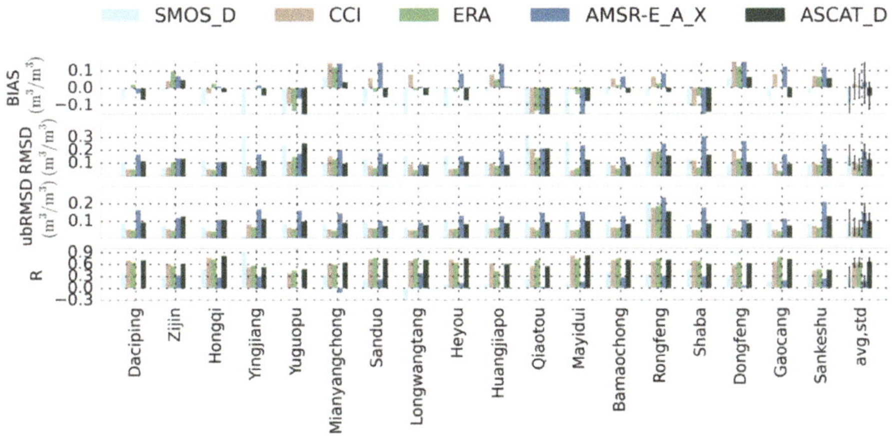

Figure 3. Bar plots of the comparison results between soil moisture products and *in situ* measurements at each station.

In addition to the validation at each individual station, the comparisons between averaged *in situ* soil moisture over all stations and corresponding averaged soil moisture products were also conducted. This comparison method is expected to reduce the uncertainties introduced by scale differences between the *in situ* point and satellite pixel [60]. Table 2 presents the statistical scores of the comparison results at the network scale. Compared to the results at station scale, the comparisons at the network scale generally show the same tendency but better scores. The ERA-Interim outperforms other soil moisture products with lowest RMSD and ubRMSD.

Table 2. Statistics of comparisons between averaged *in situ* soil moisture and soil moisture products at network scale

Products	**R**	**BIAS (m^3/m^3)**	**RMSD (m^3/m^3)**	**ubRMSD (m^3/m^3)**	**Sample Days**
CCI	0.723	0.019	0.050	0.046	1621
ERA	0.782	0.014	0.037	0.035	1692
AMSR-E_A_X	0.086	0.067	0.152	0.136	965
ASCAT_D	0.778	−0.053	0.096	0.080	1686
SMOS_D	0.443	−0.074	0.096	0.062	657

Comparison of Soil Moisture Anomalies

Temporal Evolution of Soil Moisture and Precipitation

The above results give an overview of the evaluation of soil moisture absolute values. To remove the influences of the seasonal cycle in the comparison, the soil moisture anomalies were calculated based on Equation (1). Figure 4 shows the anomaly time series of the soil moisture and rainfall products. The CCI SM, ASCAT, and ERA-

Interim generally represent the variations of *in situ* measurements. As expected, AMSR-E and SMOS have large variations compared to *in situ* soil moisture. Together with the analysis of the absolute soil moisture time series, these results suggest that *in situ* soil moisture temporal variations can be well represented by the satellite-based soil moisture and reanalysis datasets.

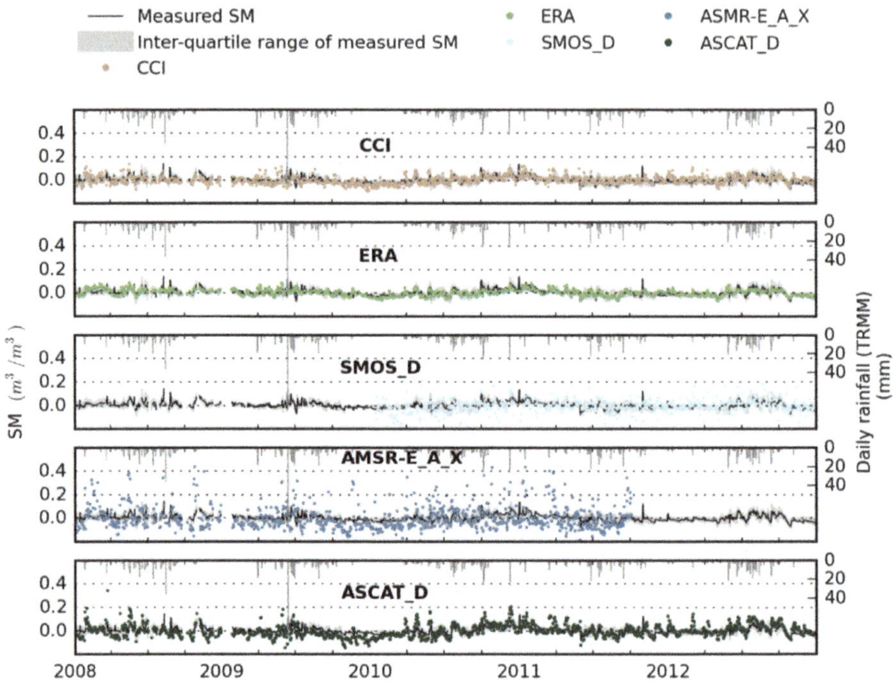

Figure 4. Temporal evolution of the anomalies from averaged *in situ* measurements, soil moisture products, and precipitation at the network scale: *in situ* mean soil moisture (black solid line) and its inter-quartile range (grey shaded areas); CCI, ERA, SMOS, AMSR-E_A_X, and ASCAT_D (points with different colors); and precipitation (top of each panel). The soil moisture has the unit of m^3/m^3 (left y-axis), while precipitation has the unit of mm (right y-axis).

Validation of the Soil Moisture Anomalies

Figure 5 presents the bar plots of the comparison of soil moisture anomalies at each station. Similar to the results from the comparison of absolute soil moisture, the CCI SM, ASCAT, and ERA-Interim have similar performance in terms of R values. As expected, the R values are smaller than that of absolute comparison due to the removal of seasonal cycles. Similar results have been reported by Albergel *et al.* [16] and Brocca *et al.* [17]. However, the RMSD, ubRMSD, and BIAS values of anomaly comparisons are smaller than that of the absolute comparisons. The ERA-Interim has the smallest RMSD, ubRMSD, and BIAS. Similar performance was found with CCI SM, followed by ASCAT. The AMSR-E and SMOS have the worst performance.

The statistic scores of the comparison of soil moisture anomalies at a network scale are summarized in Table 3. Due to the decreasing of uncertainties of scale difference, the scores of network scale are improved compared to that of the station scale. In comparison to the scores of absolute soil moisture (Table 2), the correlations of all the products worsen, while RMSD, ubRMSD, and BISA get smaller. In general, the ERA-Interim has the smallest RMSD and ubRMSD, which indicates that the ERA-Interim agrees best with *in situ* soil moisture. The performance of CCI SM is comparable with the ERA-Interim soil moisture.

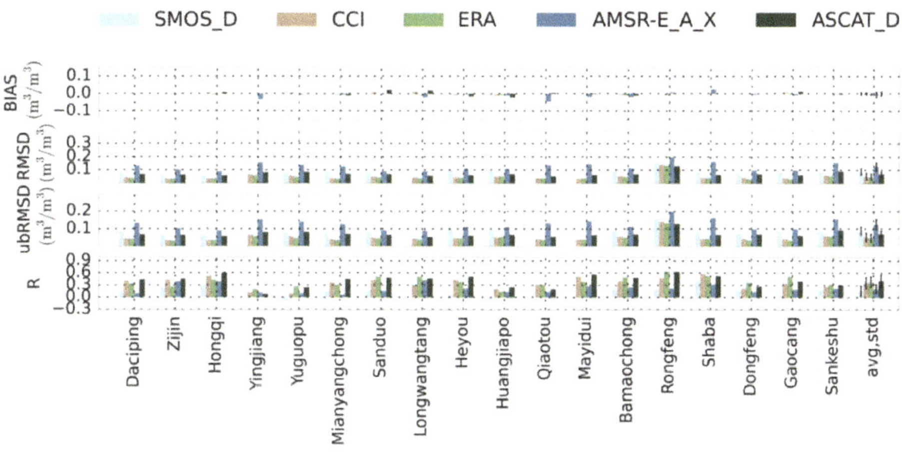

Figure 5. Bar plots of the results from anomalies comparison between soil moisture products and *in situ* soil moisture at each station.

Table 3. Statistics of the anomalies comparison between averaged *in situ* soil moisture and soil moisture products at network scale

Products	R	BIAS (m^3/m^3)	RMSD (m^3/m^3)	ubRMSD (m^3/m^3)	**Sample Days**
CCI	0.480	-0.002	0.030	0.030	1621
ERA	0.634	-0.003	0.023	0.023	1692
AMSR-E_A_X	0.222	-0.007	0.101	0.101	962
ASCAT_D	0.564	-0.004	0.044	0.044	1686
SMOS_D	0.303	-0.004	0.054	0.054	656

Seasonal and Land Cover Analysis

Furthermore, the seasonal analysis was conducted to identify any seasonal variations of the dataset accuracies. The study period is separated into four seasons September/October/November (autumn), December/January/February (winter), March/April/May (spring), and June/July/August (summer) respectively. It can be seen from Figure 6 that all products have small variations over different seasons in terms of R, RMSD, and ubRMSD values. This indicates that the accuracy of these soil moisture products is stable in different seasons. In addition, the influence of land use on the statistical scores was also examined. The stations are divided into four land use groups: bare soil (4), cropland (7), forest (3), and grassland (4). The results were shown in Figure 7. In general, different land use categories have similar performance in terms of statistical scores for all the products except SMOS. SMOS has very low R value over cropland. These results suggest that the performance of soil moisture products is generally insensitive to land use.

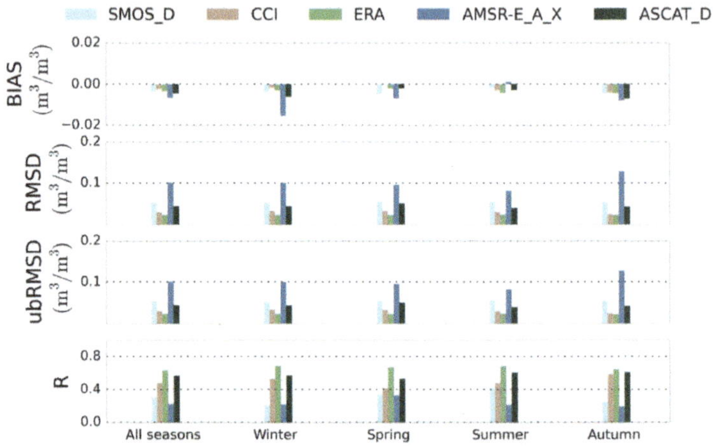

Figure 6. Bar plots of the comparison results between soil moisture products and *in situ* measurements over different seasons.

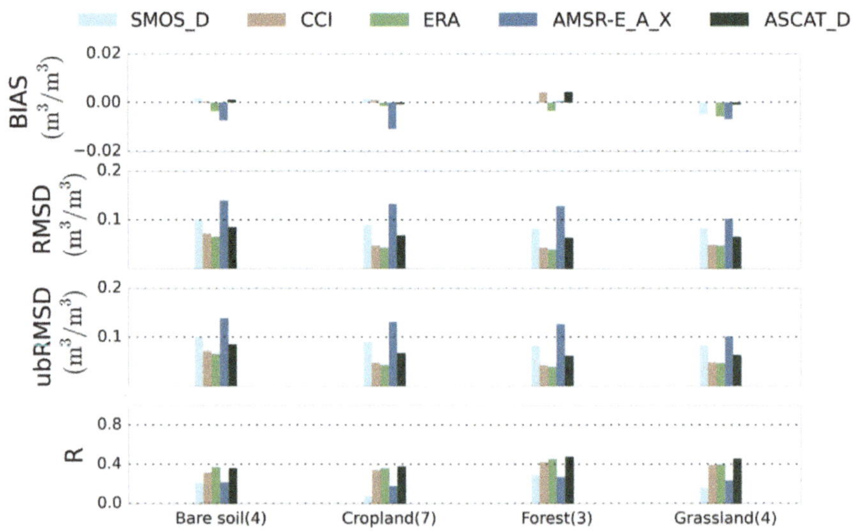

Figure 7. Bar plots of the comparison results between soil moisture products and *in situ* measurements over different land cover types.

Discussion

This study examined the ability of several soil moisture products to represent the surface soil moisture conditions over Yunnan, China. The results highlight the performance of ERA-Interim and CCI SM. From the time series analysis of both absolute values and anomalies, it is found that ERA-Interim and CCI SM can capture the variations of *in situ* soil moisture and respond well to the rainfall events. From quantitative comparison with *in situ* measurements, the ERA-Interim has the lowest RMSD and ubRMSD. The CCI SM has comparable performance to ERA-Interim. The accuracy levels are also similar to that reported by validation studies over other regions [59,61,62]. Additionally, the seasons and land use have little effect on the performance of these products. This indicates that both ERA-Interim and CCI SM have the potential to benefit the local hydrological applications such as drought monitoring, and water planning and management.

The unfavorable results of AMSR-E and SMOS are very likely due to the affects of RFI. The RFI disturbs the natural microwave emission observed by SMOS, leading to the unreliable estimation of soil moisture. The impacts of RFI on SMOS and AMSR-E have been found in Western Europe, Northeast Asia, and Middle East [58,63]. The present study further indicates the strong impacts of RFI on SMOS and AMSR-E in Southwest China.

Although most of the soil moisture can represent the surface soil moisture, there are still several issues that need to be addressed in future studies: (I) the scale mismatch between satellite pixel and *in situ* point; (II) the mismatch between the satellite penetration depth and *in situ* measurement depth; (III) the inaccuracies of the input data (such as land surface temperature and soil texture) for the soil moisture retrieval; and (IV) the improvement of the retrieval algorithms.

In addition to the improvement of soil moisture retrieval accuracy, there is also a need to improve the spatial resolution of the soil moisture product. The current global soil moisture products normally have spatial resolutions of tens of kilometers. However, many regional hydrological applications require a spatial resolution of 1–10 km [64,65]. The optical/thermal infrared (TIR) sensors can provide complementary information of soil moisture at higher spatial resolutions. Therefore, the synergistic use of microwave and optical/thermal data to estimate soil moisture at high spatial resolution should be another direction in future study.

CONCLUSIONS

Validation of satellite-based soil moisture products is essential at both the regional and global scale. The present study evaluated four satellite-based and one reanalysis soil moisture product over an intensive soil moisture network in Southwest China. The results show that these products can capture well the seasonal variations of *in situ* soil moisture. Larger deviations were observed in particular for AMSR-E and SMOS products. The SMOS underestimates soil moisture, while AMRS-E overestimates soil moisture. The biases of AMSR-E and SMOS are very likely due to the strong radio frequency interference (RFI) that affects the measurements in this region. Quantitative validations show that ERA-Interim (R = 0.782, ubRMSD = 0.035 m^3/m^3) and CCI SM (R = 0.723, ubRMSD = 0.046 m^3/m^3) outperform the other products. In addition, these soil moisture products were found to provide similar performance across the year and for different landcover types. No systematic differences in soil moisture accuracy could be identified throughout the different seasons. Due to their long-term coverage (1978–2013 for CCI SM, 1979—present for ERA-Interim), the CCI SM and ERA-Interim dataset have the potential to provide valuable information for hydrological applications and water resource management applications in the future.

ACKNOWLEDGMENTS

The authors would like to acknowledge Yunnan Hydrology and Water Resources Bureau for providing the *in situ* soil measurements. The authors also would like to thank the Integrated Climate Data Centre (ICDC), Centre Aval de Traitement des Données SMOS (CATDS), the European Space Agency (ESA), and the European Centre for Medium-Range Weather Forecasts (ECMWF) for providing the satellite-based and reanalysis soil moisture products. Additionally, Gitta Lasslop was acknowledged for reviewing the first version of the manuscript. Authors are also grateful to the three anonymous reviewers for their constructive comments and suggestions that helped improving the manuscript. This research was supported by the Cluster of Excellence CliSAP (EXC177), University of Hamburg, funded through the German Science Foundation (DFG), and the ESA CCI program (CMUG), and the project of Yunnan Social Development (2012CA021).

AUTHOR CONTRIBUTIONS

Jian Peng designed the research, analyzed the results, and wrote the manuscript. Jonathan Niesel performed the research. Alexander Loew designed and supervised the research. Shiqiang Zhang and Jie Wang processed the *in situ* measurements.

APPENDIX

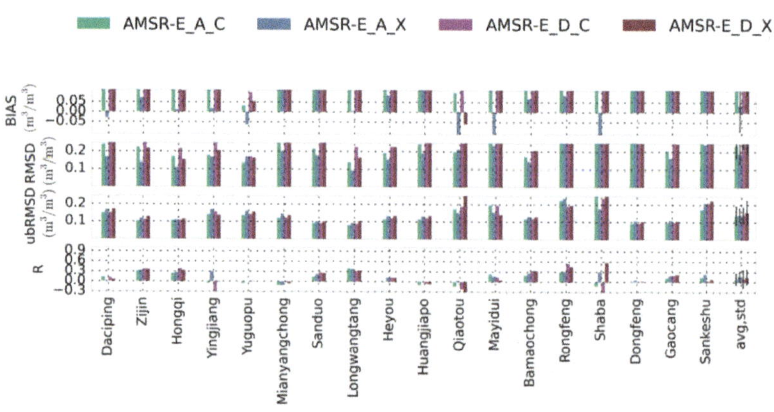

(a)

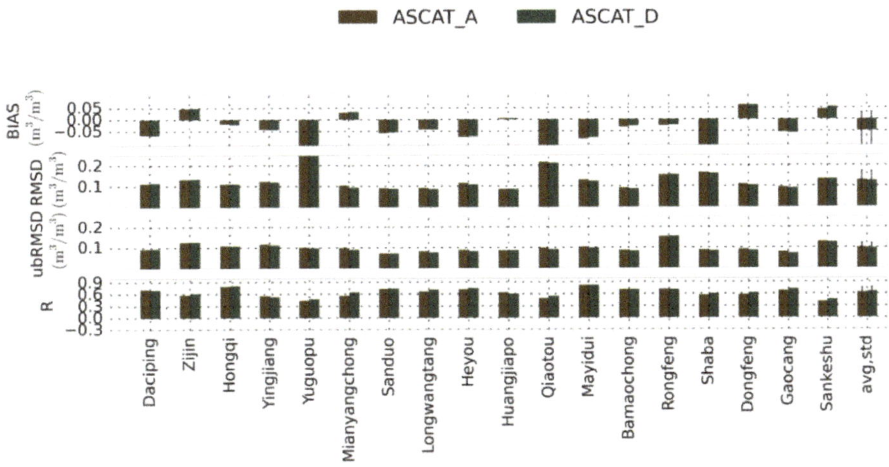

(b)

(c)

Figure A1. Bar plots of the comparison results between *in situ* measurements and soil moisture products: (**a**) AMSR-E; (**b**) ASCAT; (**c**) SMOS.

REFERENCES

1. GCOS. Available online: https://www.wmo.int/pages/prog/gcos/Publications/gcos-154.pdf (accessed on 23 November).
2. Porporato, A.; Daly, E.; Rodriguez-Iturbe, I. Soil water balance and ecosystem response to climate change. *Am. Naturalist* **2004**, *164*, 625–632. [PubMed]
3. Vereecken, H.; Huisman, J.; Pachepsky, Y.; Montzka, C.; van der Kruk, J.; Bogena,

H.; Weihermüller, L.; Herbst, M.; Martinez, G.; Vanderborght, J. On the spatio-temporal dynamics of soil moisture at the field scale. *J. Hydrol.* **2014**, *516*, 76–96.

4. Seneviratne, S.I.; Corti, T.; Davin, E.L.; Hirschi, M.; Jaeger, E.B.; Lehner, I.; Orlowsky, B.; Teuling, A.J. Investigating soil moisture–climate interactions in a changing climate: A review. *Earth-Sci. Rev.* **2010**, *99*, 125–161.
5. Loew, A.; Holmes, T.; de Jeu, R. The european heat wave 2003: Early indicators from multisensoral microwave remote sensing? *J. Geophys. Res.: Atmos. (1984–2012)* **2009**, *114*.
6. Wagner, W.; Naeimi, V.; Scipal, K.; de Jeu, R.; Martínez-Fernández, J. Soil moisture from operational meteorological satellites. *Hydrogeol. J.* **2007**, *15*, 121–131.
7. Loew, A.; Ludwig, R.; Mauser, W. Derivation of surface soil moisture from envisat asar wide swath and image mode data in agricultural areas. *IEEE Trans. Geosci. Remote Sens.* **2006**, *44*, 889–899.
8. Njoku, E.G.; Entekhabi, D. Passive microwave remote sensing of soil moisture. *J. Hydrol.* **1996**, *184*, 101–129.
9. Owe, M.; de Jeu, R.; Holmes, T. Multisensor historical climatology of satellite-derived global land surface moisture. *J. Geophys. Res. Earth Surf.* **2008**, *113*, F01002.
10. Naeimi, V.; Scipal, K.; Bartalis, Z.; Hasenauer, S.; Wagner, W. An improved soil moisture retrieval algorithm for ERS and METOP scatterometer observations. *IEEE Trans. Geosci. Remote Sens.* **2009**, *47*, 1999–2013.
11. Wagner, W.; Lemoine, G.; Rott, H. A method for estimating soil moisture from ERS scatterometer and soil data. *Remote Sens. Environ.* **1999**, *70*, 191–207.
12. Kerr, Y.H.; Waldteufel, P.; Wigneron, J.P.; Martinuzzi, J.; Font, J.; Berger, M. Soil moisture retrieval from space: The soil moisture and ocean salinity (SMOS) mission. *IEEE Trans. Geosci. Remote Sens.* **2001**, *39*, 1729–1735.
13. Liu, Y.Y.; Parinussa, R.M.; Dorigo, W.A.; de Jeu, R.A.M.; Wagner, W.; van Dijk, A.I.J.M.; McCabe, M.F.; Evans, J.P. Developing an improved soil moisture dataset by blending passive and active microwave satellite-based retrievals. *Hydrol. Earth Syst.Sci.* **2011**, *15*, 425–436.
14. Wagner, W.; Dorigo, W.; de Jeu, R.; Fernandez, D.; Benveniste, J.; Haas, E.; Ertl, M. Fusion of active and passive microwave observations to create an essential climate variable data record on soil moisture. *ISPRS Ann. Photogramm. Remote Sens. Spat. Inf. Sci.* **2012**, *I-7*, 315–321.
15. Dorigo, W.A.; Gruber, A.; de Jeu, R.A.M.; Wagner, W.; Stacke, T.; Loew, A.; Albergel, C.; Brocca, L.; Chung, D.; Parinussa, R.M.; *et al.* Evaluation of the ESA CCI soil moisture product using ground-based observations. *Remote Sens. Environ.* **2015**, *162*, 380–395.
16. Albergel, C.; Rüdiger, C.; Carrer, D.; Calvet, J.-C.; Fritz, N.; Naeimi, V.; Bartalis, Z.; Hasenauer, S. An evaluation of ascat surface soil moisture products with *in situ* observations in southwestern france. *Hydrol. Earth Syst. Sci.* **2009**, *13*, 115–

124.

17. Brocca, L.; Hasenauer, S.; Lacava, T.; Melone, F.; Moramarco, T.; Wagner, W.; Dorigo, W.; Matgen, P.; Martínez-Fernández, J.; Llorens, P.; *et al.* Soil moisture estimation through ASCAT and AMSR-E sensors: An intercomparison and validation study across europe. *Remote Sens. Environ.* **2011**, *115*, 3390–3408.
18. Sánchez, N.; Martínez-Fernández, J.; Scaini, A.; Pérez-Gutiérrez, C. Validation of the SMOS l2 soil moisture data in the remedhus network (Spain). *IEEE Trans. Geosci. Remote Sens.* **2012**, *50*, 1602–1611.
19. Fascetti, F.; Pierdicca, N.; Pulvirenti, L.; Crapolicchio, R.; Muñoz-Sabater, J. A comparison of ASCAT and SMOS soil moisture retrievals over europe and northern africa from 2010 to 2013. *Int. J. Appl. Earth Obs. Geoinf.* **2015**. in press.
20. Draper, C.S.; Walker, J.P.; Steinle, P.J.; de Jeu, R.A.; Holmes, T.R. An evaluation of AMSR–E derived soil moisture over australia. *Remote Sens. Environ.* **2009**, *113*, 703–710.
21. Su, C.-H.; Ryu, D.; Young, R.I.; Western, A.W.; Wagner, W. Inter-comparison of microwave satellite soil moisture retrievals over the murrumbidgee basin, southeast Australia. *Remote Sens. Environ.* **2013**, *134*, 1–11.
22. Gruhier, C.; Rosnay, P.D.; Hasenauer, S.; Holmes, T.; Jeu, R.D.; Kerr, Y.; Mougin, E.; Njoku, E.; Timouk, F.; Wagner, W. Soil moisture active and passive microwave products: Intercomparison and evaluation over a Sahelian site. *Hydrol. Earth Syst. Sci.* **2010**, *14*, 141–156.
23. Louvet, S.; Pellarin, T.; al Bitar, A.; Cappelaere, B.; Galle, S.; Grippa, M.; Gruhier, C.; Kerr, Y.; Lebel, T.; Mialon, A.; *et al.* SMOS soil moisture product evaluation over West-Africa from local to regional scale. *Remote Sens. Environ.* **2015**,*156*, 383–394.
24. Al Bitar, A.; Leroux, D.; Kerr, Y.H.; Merlin, O.; Richaume, P.; Sahoo, A.; Wood, E.F. Evaluation of SMOS soil moisture products over continental US using the SCAN/SNOTEL network. *IEEE Trans. Geosci. Remote Sens.* **2012**, *50*, 1572–1586.
25. Collow, T.W.; Robock, A.; Basara, J.B.; Illston, B.G. Evaluation of SMOS retrievals of soil moisture over the central United States with currently available *in situ* observations. *J. Geophys. Res. Atmos.* **2012**, *117*.
26. Al-Yaari, A.; Wigneron, J.P.; Ducharne, A.; Kerr, Y.; de Rosnay, P.; de Jeu, R.; Govind, A.; Al Bitar, A.; Albergel, C.; Muñoz-Sabater, J.; *et al.* Global-scale evaluation of two satellite-based passive microwave soil moisture datasets (SMOS and AMSR-E) with respect to land data assimilation system estimates. *Remote Sens. Environ.* **2014**, *149*, 181–195.
27. Albergel, C.; Dorigo, W.; Reichle, R.H.; Balsamo, G.; de Rosnay, P.; Muñoz-Sabater, J.; Isaksen, L.; de Jeu, R.; Wagner, W. Skill and global trend analysis of soil moisture from reanalyses and microwave remote sensing. *J. Hydrometeorol.***2013**, *14*, 1259–1277.
28. Long, D.; Shen, Y.; Sun, A.; Hong, Y.; Longuevergne, L.; Yang, Y.; Li, B.; Chen,

L. Drought and flood monitoring for a large karst plateau in southwest China using extended grace data. *Remote Sens. Environ.* **2014**, *155*, 145–160.

29. Xu, K.; Yang, D.; Xu, X.; Lei, H. Copula based drought frequency analysis considering the spatio-temporal variability in southwest China. *J. Hydrol.* **2015**, *527*, 630–640.
30. Qiu, J. China drought highlights future climate threats. *Nature* **2010**, *465*, 142–143. [PubMed]
31. Fan, Z.-X.; Bräuning, A.; Thomas, A.; Li, J.-B.; Cao, K.-F. Spatial and temporal temperature trends on the Yunnan plateau (southwest China) during 1961–2004. *Int. J. Climatol.* **2011**, *31*, 2078–2090.
32. Li, Y.-G.; He, D.; Hu, J.-M.; Cao, J. Variability of extreme precipitation over Yunnan province, China 1960–2012. *Int. J. Climatol.* **2015**, *35*, 245–258.
33. Abbas, S.; Nichol, J.; Qamer, F.; Xu, J. Characterization of drought development through remote sensing: A case study in central Yunnan, china. *Remote Sens.* **2014**, *6*, 4998–5018.
34. Yang, J.; Gong, D.; Wang, W.; Hu, M.; Mao, R. Extreme drought event of 2009/2010 over southwestern China. *Meteorol. Atmos. Phys.* **2012**, *115*, 173–184.
35. Zhang, L.; Xiao, J.; Li, J.; Wang, K.; Lei, L.; Guo, H. The 2010 spring drought reduced primary productivity in southwestern China. *Environ. Res. Lett.* **2012**, *7*, 045706.
36. Njoku, E.G.; Jackson, T.J.; Lakshmi, V.; Chan, T.K.; Nghiem, S.V. Soil moisture retrieval from AMSR-E. *IEEE Trans. Geosci. Remote Sens.* **2003**, *41*, 215–229.
37. Rebel, K.T.; de Jeu, R.A.M.; Ciais, P.; Viovy, N. A global analysis of soil moisture derived from satellite observations and a land surface model. *Hydrol. Earth Syst. Sci.* **2012**, *16*, 833–847.
38. Lei, F.; Crow, W.; Shen, H.; Parinussa, R.; Holmes, T. The impact of local acquisition time on the accuracy of microwave surface soil moisture retrievals over the contiguous United States. *Remote Sens.* **2015**, *7*, 13448–13465.
39. Figa-Saldaña, J.; Wilson, J.J.; Attema, E.; Gelsthorpe, R.; Drinkwater, M.; Stoffelen, A. The advanced scatterometer (ASCAT) on the meteorological operational (METOP) platform: A follow on for European wind scatterometers. *Can. J. Remote Sens.* **2002**, *28*, 404–412.
40. Wagner, W.; Bloeschl, G.; Pampaloni, P.; Calvet, J.-C.; Bizzarri, B.; Wigneron, J.-P.; Kerr, Y. Operational readiness of microwave remote sensing of soil moisture for hydrologic applications. *Nord. Hydrol.* **2007**, *38*, 1–20.
41. Reynolds, C.A.; Jackson, T.J.; Rawls, W.J. Estimating soil water-holding capacities by linking the food and agriculture organization soil map of the world with global pedon databases and continuous pedotransfer functions. *Water Resour. Res.* **2000**, *36*, 3653–3662.
42. Hollmann, R.; Merchant, C.J.; Saunders, R.; Downy, C.; Buchwitz, M.; Cazenave, A.; Chuvieco, E.; Defourny, P.; de Leeuw, G.; Forsberg, R.; *et al.* The ESA climate change initiative: Satellite data records for essential climate variables.*Bull. Am.*

Meteorol. Soc. **2013**, *94*, 1541–1552.

43. Liu, Y.Y.; Dorigo, W.A.; Parinussa, R.M.; de Jeu, R.A.M.; Wagner, W.; McCabe, M.F.; Evans, J.P.; van Dijk, A.I.J.M. Trend-preserving blending of passive and active microwave soil moisture retrievals. *Remote Sens. Environ.* **2012**, *123*, 280–297.
44. Kerr, Y.H.; Waldteufel, P.; Wigneron, J.-P.; Delwart, S.; Cabot, F.O.; Boutin, J.; Escorihuela, M.-J.; Font, J.; Reul, N.; Gruhier, C. The SMOS mission: New tool for monitoring key elements ofthe global water cycle. *IEEE Proc.* **2010**, *98*, 666–687.
45. Wigneron, J.-P.; Kerr, Y.; Waldteufel, P.; Saleh, K.; Escorihuela, M.-J.; Richaume, P.; Ferrazzoli, P.; de Rosnay, P.; Gurney, R.; Calvet, J.-C. L-band microwave emission of the biosphere (L-MEB) model: Description and calibration against experimental data sets over crop fields. *Remote Sens. Environ.* **2007**, *107*, 639–655.
46. Kerr, Y.H.; Waldteufel, P.; Richaume, P.; Wigneron, J.P.; Ferrazzoli, P.; Mahmoodi, A.; Al Bitar, A.; Cabot, F.; Gruhier, C.; Juglea, S.E. The SMOS soil moisture retrieval algorithm. *IEEE Trans. Geosci. Remote Sens.* **2012**, *50*, 1384–1403.
47. Al-Yaari, A.; Wigneron, J.P.; Ducharne, A.; Kerr, Y.H.; Wagner, W.; de Lannoy, G.; Reichle, R.; Al Bitar, A.; Dorigo, W.; Richaume, P.; *et al.* Global-scale comparison of passive (SMOS) and active (ASCAT) satellite based microwave soil moisture retrievals with soil moisture simulations (Merra-Land). *Remote Sens. Environ.* **2014**, *152*, 614–626.
48. Dee, D.P.; Uppala, S.M.; Simmons, A.J.; Berrisford, P.; Poli, P.; Kobayashi, S.; Andrae, U.; Balmaseda, M.A.; Balsamo, G.; Bauer, P.; *et al.* The Era-interim reanalysis: Configuration and performance of the data assimilation system. *Q. J. R. Meteorol. Soc.* **2011**, *137*, 553–597.
49. Kummerow, C.; Simpson, J.; Thiele, O.; Barnes, W.; Chang, A.T.C.; Stocker, E.; Adler, R.F.; Hou, A.; Kakar, R.; Wentz, F.; *et al.* The status of the tropical rainfall measuring mission (TRMM) after two years in orbit. *J. Appl. Meteorol.* **2000**,*39*, 1965–1982.
50. Iguchi, T.; Kozu, T.; Meneghini, R.; Awaka, J.; Okamoto, K.I. Rain-profiling algorithm for the TRMM precipitation radar. *J. Appl. Meteorol.* **2000**, *39*, 2038–2052.
51. Willmott, C.J. Some comments on the evaluation of model performance. *Bull. Am. Meteorol. Soc.* **1982**, *63*, 1309–1313.
52. Entekhabi, D.; Reichle, R.H.; Koster, R.D.; Crow, W.T. Performance metrics for soil moisture retrievals and application requirements. *J. Hydrometeorol.* **2010**, *11*, 832–840.
53. Scipal, K.; Drusch, M.; Wagner, W. Assimilation of a ERS scatterometer derived soil moisture index in the ECMWF numerical weather prediction system. *Adv. Water Resour.* **2008**, *31*, 1101–1112.
54. Jackson, T.J.; Cosh, M.H.; Bindlish, R.; Starks, P.J.; Bosch, D.D.; Seyfried, M.;

Goodrich, D.C.; Moran, M.S.; Jinyang, D. Validation of advanced microwave scanning radiometer soil moisture products. *IEEE Trans. Geosci. Remote Sens.* **2010**,*48*, 4256–4272.

55. Saleh, K.; Wigneron, J.-P.; de Rosnay, P.; Calvet, J.-C.; Escorihuela, M.J.; Kerr, Y.; Waldteufel, P. Impact of rain interception by vegetation and mulch on the L-band emission of natural grass. *Remote Sens. Environ.* **2006**, *101*, 127–139.
56. Rowlandson, T.L.; Hornbuckle, B.K.; Bramer, L.M.; Patton, J.C.; Logsdon, S.D. Comparisons of evening and morning SMOS passes over the Midwest United States. *IEEE Trans. Geosci. Remote Sens.* **2012**, *50*, 1544–1555.
57. Griesfeller, A.; Lahoz, W.A.; Jeu, R.A.M.D.; Dorigo, W.; Haugen, L.E.; Svendby, T.M.; Wagner, W. Evaluation of satellite soil moisture products over Norway using ground-based observations. *Int. J. Appl. Earth Obs. Geoinf.* **2015**. in press.
58. Cho, E.; Choi, M.; Wagner, W. An assessment of remotely sensed surface and root zone soil moisture through active and passive sensors in northeast Asia. *Remote Sens. Environ.* **2015**, *160*, 166–179.
59. Zeng, J.; Li, Z.; Chen, Q.; Bi, H.; Qiu, J.; Zou, P. Evaluation of remotely sensed and reanalysis soil moisture products over the Tibetan plateau using *in situ* observations. *Remote Sens. Environ.* **2015**, *163*, 91–110.
60. Su, Z.; Rosnay, P.; Wen, J.; Wang, L.; Zeng, Y. Evaluation of ecmwf's soil moisture analyses using observations on the Tibetan plateau. *J. Geophys. Res. Atmos.* **2013**, *118*, 5304–5318.
61. Pratola, C.; Barrett, B.; Gruber, A.; Kiely, G.; Dwyer, E. Evaluation of a global soil moisture product from finer spatial resolution sar data and ground measurements at Irish sites. *Remote Sens.* **2014**, *6*, 8190–8219.
62. Peng, J.; Niesel, J.; Loew, A. Evaluation of soil moisture downscaling using a simple thermal based proxy—The remedhus network (Spain) example. *Hydrol. Earth Syst. Sci. Discuss.* **2015**, *12*, 8505–8551.
63. Oliva, R.; Daganzo, E.; Kerr, Y.H.; Mecklenburg, S.; Nieto, S.; Richaume, P.; Gruhier, C. SMOS radio frequency interference scenario: Status and actions taken to improve the RFI environment in the 1400–1427-MHZ passive band.*IEEE Trans. Geosci. Remote Sens.* **2012**, *50*, 1427–1439.
64. Piles, M.; Camps, A.; Vall-llossera, M.; Corbella, I.; Panciera, R.; Rudiger, C.; Kerr, Y.H.; Walker, J. Downscaling SMOS-derived soil moisture using modis visible/infrared data. *IEEE Trans. Geosci. Remote Sens.* **2011**, *49*, 3156–3166.
65. Peng, J.; Loew, A.; Zhang, S.; Wang, J.; Niesel, J. Spatial downscaling of satellite soil moisture data using a vegetation temperature condition index. *IEEE Trans. Geosci. Remote Sens.* **2016**, *54*, 558–566.

12
CHAPTER

ESTIMATING TOTAL DISCHARGE IN THE YANGTZE RIVER BASIN USING SATELLITE-BASED OBSERVATIONS

Vagner G. Ferreira, Zheng Gong, Xiufeng He, Yonglei Zhang and Samuel A. Andam-Akorful

Keywords: Water balance, evaporation, hydrogeodesy, climate changes, GRACE, TRMM, GLDAS

INTRODUCTION

The ability of Gravity Recovery and Climate Experiment (GRACE) to detect continental water-storage variations has been proven during recent years. The terrestrial water-storage is virtually a measure of total water content in surface stores, soil layers, ice (including snow), groundwater reservoirs and biomass (which is negligible in most cases) [1]. GRACE has been used for estimating regional water-storage variations in a number of locations, for instance: the Amazon River basin [2–4]; the Ganges River basin [5]; the Congo River basin [6]; the Orinoco River basin [7]; and the Yangtze River (Chang Jiang) basin [8]. Other important studies from GRACE related to the monitoring of water-storage are: groundwater withdrawal in India [9]; contributions of glaciers and ice caps to sea-level rise [10,11]; monitoring the mass balance of Antarctica [12] and Greenland [13–15]; and Alaskan permafrost groundwater storage changes [16]. In this study, we focus on the Yangtze River basin, which is an important region in China in terms of culture, society and economy, and it plays an important role in the ecological environmental conservation of China [17].

Citation: Ferreira, V.G.; Gong, Z.; He, X.; Zhang, Y.; AndamAkorful, S.A. Estimating Total Discharge in the Yangtze River Basin Using Satellite-Based Observations. Remote Sens. 2013, 5, 3415-3430. doi:10.3390/rs5073415.

Hu et al. [8], for example, compared the seasonal water-storage over the Yangtze basin derived from GRACE data and two hydrologic models: the Climate Prediction Center (CPC) and the Global Land Data Assimilation (GLDAS). Their results showed good agreement (7 mm of equivalent water height) in terms of the differences in annual amplitudes between GRACE and the model predictions using fifteen months of GRACE spherical harmonic solutions, from April 2002 to December 2003. Furthermore, Wang et al. [18] investigated the ability of GRACE to monitor the water systems area (a set of five sub-basins) of the Three Gorges reservoir by comparing the inversion results from GRACE with the results of the CPC model. They found that the root mean square error (RMSE) is 21 mm for the total water-storage changes. Zhao et al. [19] used the first release of the Delft Institute of Earth Observation and Space Systems (DEOS) Mass Transport (DMT-1) model, based on GRACE data, to analyze water-storage changes in the Yangtze basin. Their results showed that the water-storage of the Yangtze basin has a large and statistically significant increase, 7 ± 1.6 mm/yr, over the period of February 2003 to May 2008. Huang et al. [20] considered the soil moisture and snow water equivalent as an estimation of water-storage and they found that the Yangtze basin is drying up. The results were based on data obtained from Interim Reanalysis Data (ERA-Interim) and Noah model from GLDAS for the period between 1979 and 2010.

In [21], Huang et al. examined the changes in the water-storage of the Yangtze basin over a period of approximately seven years using monthly gravity fields of GRACE and water level measurements. Because of the limited resolution of the GRACE satellite data, they concluded that no changes could be detected in the water-storage capacity, owing to the water impoundment of the Three Gorges Dam (TGD). However, Wang et al. [22] applied a novel approach for isolating the signal from mass changes for the water impoundment of the TGD. The TGD's contribution from water-storage changes was isolated by using the WaterGAP Global Hydrology Model (WGHM) and the residual (GRACE-WGHM) was compared to in situ measurements of volume changes. Wang et al. [22] concluded that GRACE can detect the mass shift and retrieve the amplitudes of large surface water-storage changes in a concentrated area that is smaller than GRACE's spatial resolution.

All of the aforementioned studies outline the potential of GRACE for investigating water-storage and its changes within the Yangtze basin. However, there have been efforts to incorporate other data sets in estimating total basin discharge [23] at a river basin scale, for example, atmospheric models that predict precipitation minus evaporation (we use evaporation to describe all processes of vaporization). Syed et al. [24] have used satellite measurements of variations in continental water-storage from the GRACE mission to present first estimates of monthly freshwater discharge from the entire Pan-Arctic for the period 2003–2005. The methodology published in [23,24] has been used by Syed et al. [25] to estimate the monthly freshwater discharge from continents, drainage regions, and global land for the period of 2003–2005. In [26], Seo et al. used a novel approach, which avoids influences from uncertainties in the estimation of atmospheric moisture flux, in order to evaluate the global fresh water discharge by solving the water balance equation over the oceans.

To the best of our knowledge, there has been only one application, carried out by Syed et al. [25], that estimates the total freshwater discharge from Yangtze basin for the three-year period from 2003 to 2005. Their results show a correlation coefficient of 0.92 between the annual cycles of the observed streamflow and estimated discharges for the Yangtze basin. Syed et al. [23] pointed out that the limitation in the use of the terrestrial water budget equation is the high uncertainties of evaporation. However, this has never been tested before. Hence, the aims of the present study are to estimate the total discharge of the Yangtze basin by using the terrestrial water budget equation and to assess if the water-storage (as derived from GRACE) can be considered negligible in this estimation. To achieve these goals, we applied precipitation data from: the Tropical Rainfall Measuring Mission (TRMM); data on water-storage changes derived from the latest Level-02 (Release-05) GRACE data from three processing centers (i.e., Center for Space Research—CSR, Jet Propulsion Laboratory—JPL, and GeoForschungsZentrum—GFZ); and evaporation predictions from four versions of GLDAS version 1 (CLM, Mosaic, Noah and VIC) [27]. The results are compared with a time series of in situ streamflow data.

METHODS AND DATA SOURCES

Datasets

Study Area and In Situ Discharge of Yangtze River

The Yangtze River basin lies within the subtropical zone in China [20], see Figure 1a. The Yangtze River originates in the highlands of the east Qinghai-Tibet Plateau. It owes its streamflow to the glaciers of the Dangla Mountain Range. Besides the glaciers from plateau, the Yangtze River receives discharge from numerous tributaries and lakes, Lake Poyang in particular, till it finally reaches the East China Sea at Shanghai. In the present study, daily streamflow observations for the Yangtze basin were obtained from the Datong gauging station from the Yangtze River Estuary Survey Bureau of Hydrology and Water Resource, Ministry of Water Resources. Datong is located near the Yangtze Estuary and measures the contribution from an upstream area of approximately 1.7×10^6 km^2 (Figure 1b). Owing to tidal effects, it is not possible to measure streamflow from a station at the mouth of the Yangtze River (Datong is the tidal limit of the estuary). In order to convert the observed streamflow (m^3/s) at the Datong station into daily net surface runoff rate (mm/day) per unit area over Yangtze basin, the drainage area size is required ($\sim 1.7 \times 10^6$ km^2). Thus, monthly surface runoff rate (R in mm/month) for the basin was computed as the sum of the daily surface runoff rate.

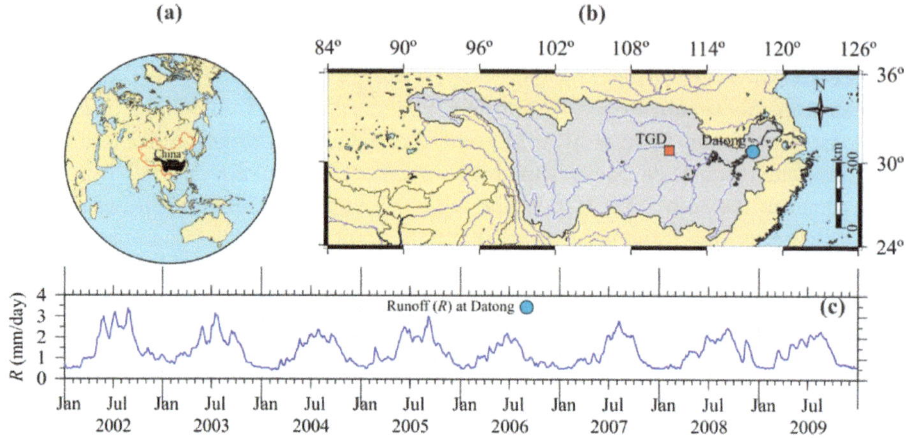

Figure 1. (**a**) Location of the study area in China; (**b**) Yangtze River basin (shadowed portion with an area of approximately 1.8×10^6 km^2), Three Gorges Dam (TGD) and the Datong Hydrologic Station. The graphical scale is related to the parallel 30°; (**c**) Daily net surface runoff (R) of Datong Hydrologic station.

Precipitation Data

In this study, we used the global monthly accumulated rain grids supplied by the TRMM, as Level-3 V7 products, more specifically the TRMM 3B43. TRMM is a joint satellite mission of Goddard Space Flight Centre (GSFC), from the National Aeronautics and Space Administration (NASA), and the Japan Aerospace Exploration Agency (JAXA) [28]. The TRMM Multi-satellite Precipitation Analysis (TMPA) was designed to combine all available precipitation datasets from different satellite sensors and monthly surface rain gauge data to provide an estimate of precipitation at spatial resolution of 0.25°(arc-degree) [28]. Since the end of 1997, the TRMM has provided monthly rainfall rates for tropical and subtropical regions. Owing to the availability of the discharge and GRACE products at the time of the study, the time period used is limited from January 2003 to December 2009 (total 84 months).

Hydrological Models

GRACE-derived values of water-storage anomalies are of potential importance as stand-alone quantities or when combined with other data types, for example, land surface models that offer detailed estimates of distributed hydrological fluxes and storages. The Global Land Data Assimilation System (GLDAS) is generating a series of land surface state (e.g., soil moisture and surface temperature) and flux (e.g., evaporation and sensible heat flux) products simulated by four land surface models (CLM, Mosaic, Noah and VIC) [27,29–31]. We used four versions (CLM, Mosaic Noah, and VIC) of the GLDAS Version 1 (GLDAS-1) 1.0° resolution. This was necessary because of the lack of a spatial evaporation data set necessary to estimate the total discharge by using the terrestrial water budget.

GRACE Level 2 Products

The GRACE observations are processed at: the Center for Space Research (CSR), University of Texas; the Jet Propulsion Laboratory (JPL); the GeoForschungsZentrum (GFZ); the Centre National d'Études Spatiales (CNES); the Delft Institute of Earth Observation and Space Systems (DEOS), Delft University of Technology; and at a few other institutions. The final results, known as Level 2 (L2) products, are usually monthly geopotential solutions expressed in terms of spherical harmonic coefficients (truncated at certain degree and order d/o), which are widely used to study mass changes in the Earth system. Furthermore, each center follows different data processing methodologies, which might cause some differences in the solutions [32]. Thus, the new Release-05 (RL05) L2 products from the three GRACE project processing centers (CSR, JPL, GFZ) were used for this study. Whiles the JPL and CSR still recommend that users replace C_{20} estimates from Satellite Laser Ranging (SLR), the GFZ recommends that users maintain the RL05 estimates of C_{20}. As in Chen et al. [11], GRACE-estimatedC_{20} coefficients for the JPL and CSR were replaced by the values derived from SLR [33]. The monthly degree 1 coefficients for all solutions were used from Swenson et al. [34]. The data for this study include 84 CSR-RL05 (d/o 60), 84 JPL-RL05 (d/o 90, some are of d/o 60) and 84 GFZ-RL5 (d/o 90) GRACE monthly solutions, covering the period from January 2003 to January 2010. The missing period is June 2003 for the three processing centers. The RL05 products for 2002 are not available yet (6 June 2013).

Methodology

Computation of Water-Storage Variations from GRACE

The sets of coefficients from JPL and GFZ were truncated at d/o 60 to ensure compatibility with the ones from CSR. This limit fixes the spatial resolution ρ at approximately 334 km at the equator ($\rho \approx \pi nR\rho \approx \pi nR$ with R = 6,371 km is the radius of the Earth and n is the degree). Each monthly Stokes's coefficients data set was reduced by the individual long-term mean (the difference between coefficients of a month t and the mean gravity field obtained as the time average of the available coefficients). Thus, only the time-variable component of the change in surface mass can be recovered (for details see [35]). The GRACE-based land water solution, computed with the methodology described in Wahr et al. [35], provides the water-storage anomaly values δS (i.e., deviations from a reference value), usually expressed in terms of equivalent water height (or equivalent water thickness).

GRACE gravity fields at high order coefficients exhibit a high level of noise which is known as "stripes" in the spatial domain [36,37]. Therefore, in order to obtain coherent results, it is necessary to remove these stripes in post-processing by reducing correlated errors with minimal impact on the real signal. These correlations can be reduced, using either an empirical method based on a polynomial fit [38], or an a priori synthetic model of the observation geometry [39]. In this study, the polynomial fit scheme filter suggested by Chen et al. [40] was applied to the residual Stokes coefficients. For residual Stokes' coefficients with orders 6 and above, a least squares fit of degree 4 polynomial was removed from the even and odd degree coefficient pairs [41]. In Chen et al. [41], they call this procedure the de-correlation filter; hereafter, abbreviated as P4M6. After P4M6 filtering, a 300-km Gaussian low-pass filter was applied to further suppress the remaining short-wavelength errors [38].

A regional average of the water-storage anomalies was computed by defining a mask with a perimeter shown by the solid red line in Figure 2 of [42]. Following Klees et al. [43], the spatial smoothing reduces the noise and also introduces a bias in the estimated monthly mean water-storage. In order to reduce this, we applied a scale factor, k, of 1.0442 required to restore the amplitudes for the Yangtze basin. We used the methodology from Landerer and Swenson (cf. Section 4.1 of [44]) to estimate the scale factor k. The water-storage estimates from Noah driven GLDAS were first converted to spherical harmonic coefficients and truncated up to d/o 60. These spherical harmonic coefficients were then used to estimate the regional water-storage by using the mask in Figure 2 and the two step filtering scheme (i.e., P4M6 and Gaussian 300-km). The scale factor was then obtained through a least squares minimization as [44]:

$$v = \sum_{i=1}^{84} (\delta S_i^u - k \delta S_i^f)^2 \qquad (1)$$

where δS^u and δS^f is the unfiltered (true) and filtered, respectively, water-storage time series from January 2003 to December 2009 (84 months). It is important to note that the scale factor does not match the GRACE-derived water-storage to those of GLDAS; it only gives the relative signal attenuation and restores the signal to its "original" form. Thus, when working with other gridded datasets, one only needs to scale the GRACE signals with the gain factor for consistent comparisons [44].

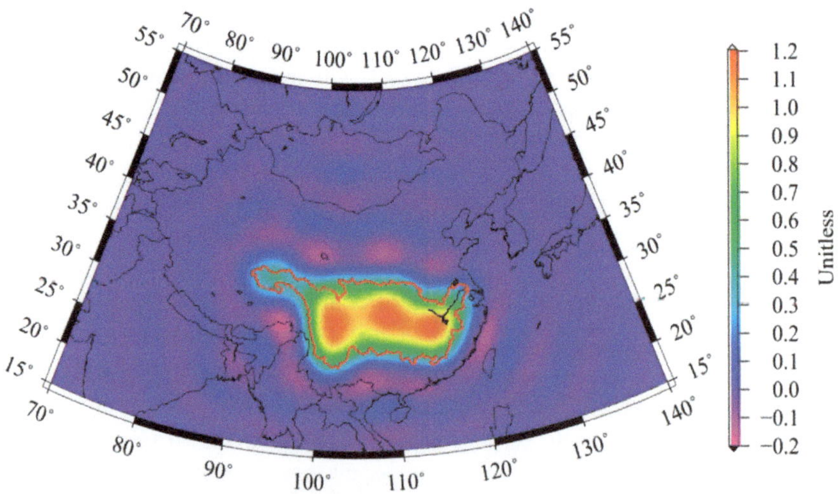

Figure 2. The average function (mask) adopted for the Yangtze basin analysis. The solid red line represents the perimeter of the Yangtze basin.

Computation of Total Discharge using Terrestrial Water Budget

Water budget methods are based on the principle of conservation of mass, applied to some part of the hydrologic cycle. Figure 3 shows a simplified diagram to represent the water cycle where only the land surface is considered. Over a land surface of area Ω (cf. Figure 3), the mean evaporation rate, E, can be expressed in terms of the water balance equation as follows [45]:

$$E = P + [(Q_{ri} + Q_{gi}) - (Q_{ro} + Q_{go})] - \frac{dS}{dt} \qquad (2)$$

where P is the areal mean rate of precipitation; Q_{ri} is the total surface inflow, Q_{ro} is the total surface outflow, Q_{gi} is the total groundwater inflow, and Q_{go} is the total groundwater outflow rates, all per unit area; and S is the water volume stored per unit area.

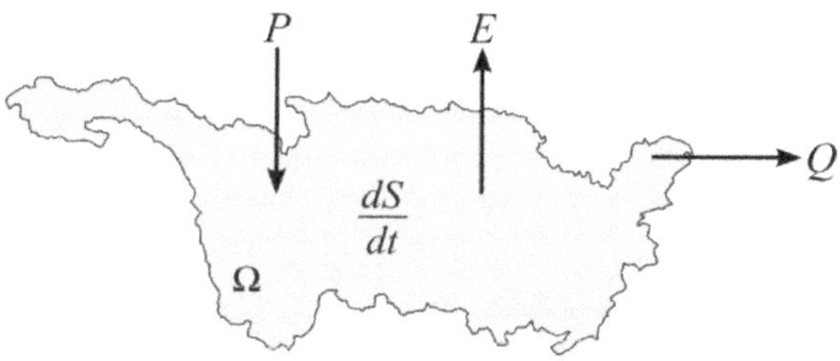

Figure 3. A simplified scheme for the relation of quantities in the land water budget.

If the area Ω s a natural river basin, bounded by natural divides, the outflow terms (Q_{ro} and Q_{go}) are generally larger than inflow terms (Q_{ri} and Q_{gi}). Thus, $Q = (Q_{ro} + Q_{go}) - (Q_{ri} + Q_{gi})$ represents the total basin discharge. Generally, these hydrological variables are expressed in terms of water mass (mm of equivalent water height) or pressure (kg/m^2) per day.

Equation (2) can be solved directly for Q as:

$$Q=P-E-\Delta S \qquad (3)$$

where the variables in Equation (3) are the monthly values of total discharge, precipitation, evaporation, and

$$\Delta S=S(t_2)-S(t_1) \qquad (4)$$

is the water-storage variation between times t_1 and t_2 (in which the subscripts 1 and 2 refer to the beginning and the end of the month). Note that here, ΔS in Equation (3) is the derivative called "water-storage changes". With regard to the total discharge estimated using the GRACE-derived water-storage changes over Yangtze basin, it is important to mention that this value includes the total surface inflows and outflows, total groundwater inflows and outflows, as well as tidal inflows and outflows [23].

Given that the difference between S and δS is a constant value (mean of the study period), the following equation can be derived from numerical differentiation using the center difference (two-sided difference):

$$\Delta S_i \approx \frac{1}{2}(\delta S_{i+1} - \delta S_{i-1}) \tag{5}$$

where ΔS_i is the approximation of the water mass variations of month i necessary in Equation (3) for estimates of the total basin discharge Q. As mentioned in Section 2.1.4, the RL05 of L2 products (Stokes's coefficients) are missing for 2002. To calculate the water-storage change (ΔS) for January 2003, the water-storage anomaly (δS) of December 2002 is necessary. Thus, the water-storage changes were calculated during the period February 2003 to December 2009. This time span was adopted because the two records (GRACE and observed streamflow) overlap for the period January 2003 to December 2009. Monthly water-storage anomalies for the missing period of June 2003 were interpolated based on values corresponding both to the previous and following months [36].

Furthermore, as reported by Syed et al. [23], the difficulty in the direct application of Equation (3) is the high uncertainty of E (evaporation). To estimate evaporation over the Yangtze basin we used the four versions of GLDAS (Section 2.1.3). As in Xue et al. [46], we assessed the relative quality of the GLDAS-estimated evaporation by using a proxy of E from water balance method. Several authors (e.g., [36,47,48]) have used evaporation predicted by the GLDAS to validate results derived from a combined approach of using GRACE and other data sets. In Rodell et al. [47], an RMSE of 0.83 mm/day (~24.9 mm/month) was found over the Mississippi River basin; in Ramillien et al. [36], 0.53 mm/day (~15.9 mm/month) was determined over the Yangtze basin; and Cesanelli and Guarracino [48] found 0.83 mm/day (~24.9 mm/month) over the Salado basin in Argentina. Typically, they reach values of approximately 28.8 mm/month, which seems to be the current level of accuracy of GRACE solutions in terms of the water-storage changes for large river basins. For comparison with the GRACE-derived water-storage changes, TRMM precipitation data and the four versions (CLM, Mosaic, Noah and VIC) of GLDAS-estimated evaporation were used. In this regard, we did not apply a filter scheme as for GRACE because the water-storage anomalies were rescaled (cf. sub-Section 2.2.1).

RESULTS AND DISCUSSION

Overall, the results presented in Figure 4a show that the GRACE-derived water-storage changes by using the RL05 of L2 products (i.e., spherical harmonic coefficients) from the three different processing centers (i.e., CSR, GFZ and JPL) are in good agreement. Even without replacing the GFZ-estimated C_{20} by those derived from SLR (cf. [11]). We performed a cross-correlation between the three time series of water-storage changes. We found a correlation coefficient of 0.96 between CSR and GFZ, 0.98 between CSR and JPL, and 0.95 between GFZ and JPL, which are significant at the 95% confidence level. All GRACE solutions show comparable root mean square (RMS) signals between 22.32 mm (CSR), 23.04 mm (GFZ) and 21.57 mm (JPL). The four evaporation products (Figure 4b) show similar seasonal behavior over the Yangtze basin over the period of study (January 2003 to December 2009). However, a relative comparison in terms of the amplitude between CLM-estimated evaporation and the other three models (Mosaic,

Noah and VIC) shows that CLM has the lowest value. We decided to check this by using the proxy of evaporation from water balance method considering E ≈ P – R, i.e., ΔS = 0, whereR is the observed streamflow. We found that the Mosaic, VIC and Noah estimated evaporation have the worst results in terms of bias (22.54, 16.81 and 12.90 mm/month, respectively) and RMSE (33.85, 29.37 and 26.29 mm/month, respectively). Thus, it seems that the CLM-estimated evaporation has the best performance for the basin with a bias of 0.55 mm/month and an RMSE of 20.89 mm/month.

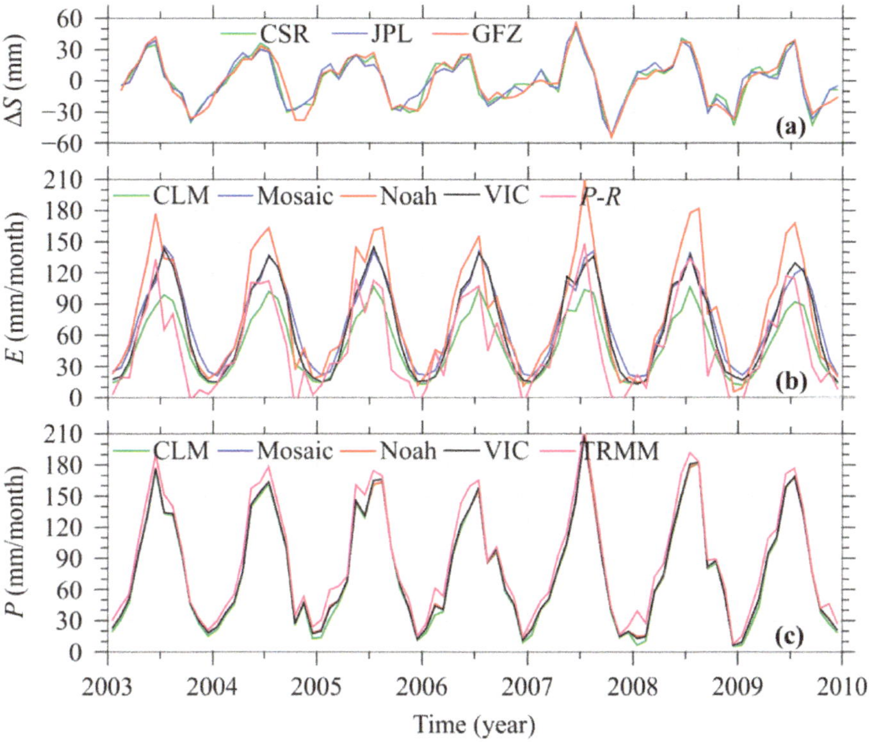

Figure 4. (**a**) Monthly Gravity Recovery and Climate Experiment (GRACE)-derived water-storage changes from the three different processing centers (Center for Space Research (CSR), Jet Propulsion Laboratory (JPL) and GeoForschungsZentrum (GFZ)); (**b**) Monthly evaporations from four versions of Global Land Data Assimilation System (GLDAS) (CLM, Mosaic, Noah and VIC) and those estimated by P – R; (**c**) Tropical Rainfall Measurement Mission (TRMM) precipitation and those estimated by GLDAS (CLM, Mosaic, Noah and VIC).

In [46], Xue et al. pointed out that the uncertainties in the evaporation products come from various sources such as meteorological and surface cover data as well as the algorithm used. Details of these forcing data are provided in [27]. Here, we investigated the time series of the forcing data in terms of precipitation over Yangtze basin. The Figure 4c shows that the CLM, Mosaic, Noah and VIC estimated precipitation all have similar fluctuations among them, but slight differences to those of TRMM. There are negative biases for all four versions of GLDAS precipitation products, for CLM it is −11.64 mm/

month, for both Mosaic and Noah it is approximately −9.32 mm/month, and for VIC it is −9.36 mm/month. It seems that the GLDAS precipitation has a dry bias over Yangtze basin; this will need to be investigated as well as the systematic error from other input variables in GLDAS (e.g., downward shortwave radiation forcing).

The Taylor diagram [49] (Figure 5) presents the results of statistical comparisons between the observed streamflow (Ref.) and: GRACE-estimated discharge by using different process centers (CSR, GFZ and JPL) combined with evaporation predicted from four versions of GLDAS (CLM, Mosaic, Noah and VIC); by considering the water-storage changes equal a zero ($\Delta S = 0$, i.e., $Q \approx P - E$). In general, the RMSE between GRACE-based discharge and observed streamflow are less than 25 mm/month (~0.83 mm/day). The lowest RMSEs are found with JPL, CSR and GFZ by combing evaporation predicted from CLM which are approximately 14.30, 14.62 and 16.02 mm/month with correlation coefficients of 0.74, 0.71 and 0.62, respectively. Additionally, the residuals between observed streamflow and derived discharge by considering water-storage changes equal a zero provide RMSEs of 20.98, 33.96, 26.36, and 29.51 for CLM, Mosaic, Noah and VIC, respectively. For some reasons, not explored in this study, CLM-estimated evaporation delivers the best results over Yangtze basin. The overall agreement between the observed streamflow and estimated discharge is better than that of P − E. An improvement of approximately 32% between the best estimation by using GRACE (i.e., JPL/CLM) and the best estimation by using P − E is noted in terms of RMSE. This means that GRACE-derived water-storage changes seem to play an important role in the water balance at a seasonal time scales. The annual accumulated ΔS between 2004 and 2009 reveals that the largest value was 27.98 mm in 2008 and the smallest was 3.86 mm in 2009 which is equivalent to approximately 5% and 1% of the annual accumulated P − E, respectively.

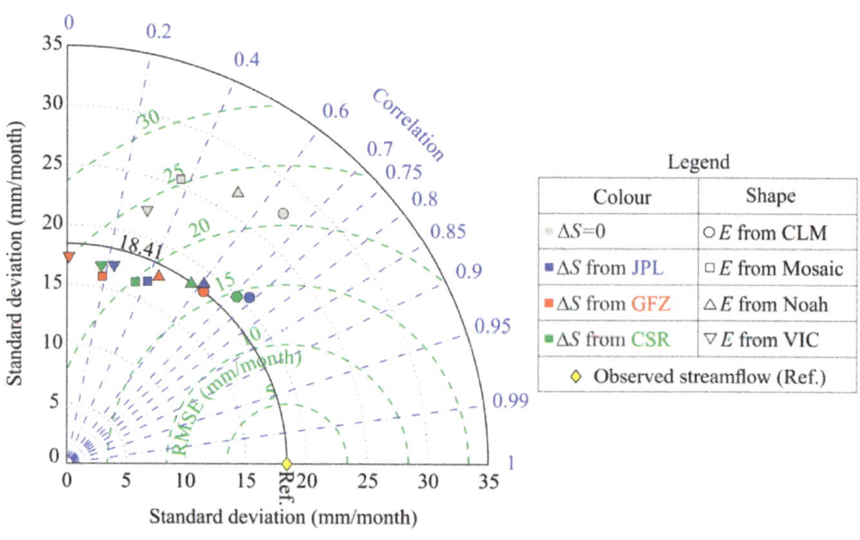

Figure 5. Taylor's diagram of statistical comparison between the time series of observed streamflow (reference) at Datong hydrological station and estimated discharge rates.

Because the results using coefficients from CSR, GFZ, and JPL are statistically identical over Yangtze basin, for the remainder of this study we utilized only the total discharge estimated by JPL/CLM. The same holds for CLM-estimated evaporation due to its good performance in terms of the bias and RMSE. Overall, the results presented in Figure 6a show that the estimated total basin discharge (black line) and observed streamflow (blue line) are in consensus for the Yangtze basin. These results for the Yangtze basin are comparable to that of Syed et al. in [25], who applied a different methodology and data sets (atmospheric moisture storage and divergence from two available global reanalysis products (National Centers for Environmental Protection and National Center for Atmospheric Research NCEP–NCAR). Their results for the total discharge derived from GRACE–NCEP–NCAR for the Yangtze basin (see Figure 2 of Syed et al. [25]) indicate seasonal cycles with magnitudes of approximately 119 km^3 (~66 mm considering the Yangtze basin's area equal 1.8×10^6 km^2) peak-to-peak with a correlation coefficient of 0.92. As such, our results show that this range is not consistent with the total variation in the estimated total basin discharge time series (~47 mm) in Figure 6b and similarly, we found a correlation coefficient of 0.93. The difference between both results (~19 mm) could be associated with a difference in the methods of GRACE data processing (e.g., RL03 vs. RL05), coupled land-atmosphere water mass balance (here we used only land), the duration of the study periods, and the size of the study area. Additionally, Syed et al. [24] mentioned that the error in atmospheric moisture flux from reanalysis affects the accuracy of the estimated total discharge. For comparison, Figure 6balso shows that the variability of the estimated discharges using P − E (red line) and the observed streamflow are in weak agreement in terms of the amplitude and phase within the Yangtze basin.

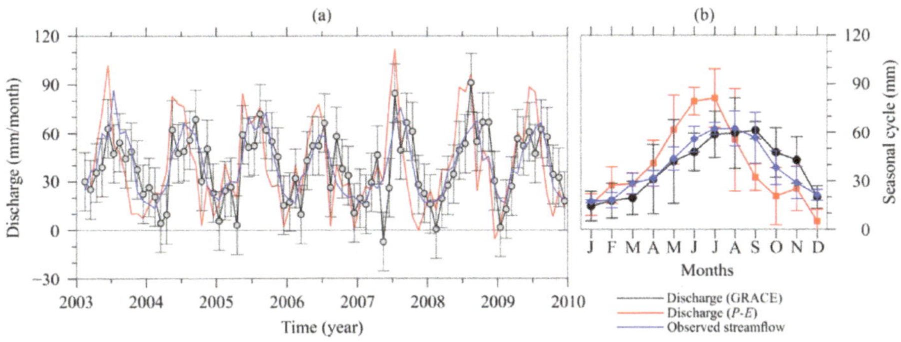

Figure 6. (**a**) Monthly estimated Yangtze total basin discharge (black line) and observed streamflow (blue line) as well as estimated discharge with those of P − E (red line). The error bars in GRACE-derived discharge were calculated by using the Equation (7) of [36] and Equation (28) of [42] with a 95% confidence; (**b**) Comparison of the seasonal cycles of estimated discharge with those of P − E and observed streamflow. The error bars represent the standard deviation for the monthly mean values.

The variability of the estimated discharge by using P − E (red line in Figure 6a) often exceeds that shown by the observed streamflow. In contrast, the annual cycles of the GRACE-derived discharge can in part explain the majority of the discrepancies in terms of the amplitude and phase. For the Yangtze basin, the inclusion of GRACE-estimated water-storage changes leads to a better representation of the discharge. The magnitude of the annual amplitude for P − E discharge is 32.0 mm/month, whereas is it only 23.4 mm/month for the observed streamflow and 22.0 mm/month for the GRACE-derived results (Table 1). Additionally, estimated low flows are similar for GRACE-based discharge and those from P − E, however, lower than observed streamflow (Figure 6a). The phase shift between the time series (Figure 6) can be verified in Table 1 at both annual and semi-annual time scales. For the case of GRACE-estimated discharge and observed streamflow this phase difference is approximately 0.51 ± 0.22 months (generally, observed streamflow precedes the GRACE-estimated discharge). We observed a time lag between precipitation and observed streamflow of approximately 0.8 ± 0.12 months. We also performed a comparison with the P − E estimated discharges where a phase lag between the observed streamflow and predicted discharge is approximately −1.11 ±0.18 months (generally, the observed streamflow lags that predicted by P − E).

Table 1. Coefficients for least squares best fit over the time window of February 2002 to December 2009 with confidence level of 95%

Variables	Annual		Semi-Annual	
	Amplitude (mm/month)	**Phase (month)**	**Amplitude (mm/month)**	**Phase (month)**
Obs. streamflow	23.4 ±1.3	6.8 ±0.1	2.7 ±1.3	2.8 ±0.9
Discharge (GRACE)	22.0 ±2.2	7.3 ±0.2	2.5 ±2.2	6.6 ±1.7
Discharge (P – E)	32.0 ±2.4	5.6 ±0.1	8.4 ±2.4	11.8 ±0.5

The results from this study are comparable with previous GRACE-based study on discharge [23,25], the noted inconsistencies are perhaps explained by the fundamental differences between the two methodologies and data sets. The Yangtze basin was chosen as the study region owing to the availability of the observed streamflow data. However, it is worth mentioning that the discharge of a river (streamflow) is a hydrological flux that represents the spatially integrated upstream hydrological process [50]. As such, the evaluation of the GRACE-estimated discharge would be more realistic if the estimated discharge is properly routed through the river basin as, e.g., in Zaitchik et al. [50]. This is one limitation of our study; we thus plan to repeat the validation in future work by using, e.g., the source-to-sink routing scheme [50]. Furthermore, the runoff originating downstream of the Datong gauging station is unmeasured, but as the comparisons were made in mm/month, the error which we can made due to the different river basins is reduced. Overall, the results are encouraging and show that the method could be applied to any large drainage basins.

CONCLUSIONS

We assessed the total discharge of the Yangtze basin on a monthly time scale from

February 2003 to December 2009 by using the terrestrial water budget approach. This method uses water-storage changes derived from the Gravity Recovery and Climate Experiment (GRACE) measurements; precipitation measured by, e.g., the Tropical Rainfall Measurement Mission (TRMM); and evaporation, e.g., simulated by the Global Land Data Assimilation System (GLDAS). We found that the estimated discharge of the Yangtze basin was correlated well with the observed streamflow (0.74) and showed an RMSE of 14.30 mm/month (~0.48 mm/day) (by using the GRACE-JPL derived water-storage changes and GLDAS-CLM simulated evaporation). The GRACE-derived results showed that the inclusion of water-storage changes improved the results against precipitation minus evaporation ($Q \approx P - E$) in approximately 32% in terms of the RMSE. This means that the water-storage changes, as derived from GRACE's measurements, play a crucial role in estimate discharge. In light of these findings, we believe that our results will help Hydrologists and Oceanographers improve their understanding of the energy cycle between the Yangtze River and the East China Sea. Future work will entail refining our results by exploiting accurate evaporation data and the evaluation of the estimated discharge by properly routed through the Yangtze River basin.

ACKNOWLEDGMENTS

Vagner G. Ferreira acknowledges the support of grant 2012B01314 from the Hohai University. Zheng Gong acknowledges the State Key Development Program of Basic Research of China (2010CB429001), the Program for New Century Excellent Talents in University (Grant No. NCET-12-0841), and the National Natural Science Foundation of China (Grant No. 51009062 and Grant No. 51179067). We also thank the satellite teams from the TRMM mission, the science teams from the GRACE processing centers (CSR, JPL, GFZ) for providing the monthly gravity fields. The GLDAS data used in this study were acquired as part of the mission of NASA's Earth Science Division and were archived and distributed by the Goddard Earth Sciences (GES) Data and Information Services Center (DISC). The work of these engineers and scientists is the basis for our investigations. The spherical harmonic expansions of the GLDAS and the mask function in Figure 2 were performed by using the FORTRAN code dragon. We would also like to thank the six anonymous reviewers and the associated editor for their helpful comments in enhancing the first submission of this manuscript.

REFERENCES

1. Cazenave, A.; Chen, J. Time-variable gravity from space and present-day mass redistribution in the Earth system.Earth Planet. Sci. Lett 2010, 298, 263–274.
2. Chen, J.L.; Wilson, C.R.; Tapley, B.D.; Yang, Z.L.; Niu, G.Y. 2005 drought event in the Amazon River basin as measured by GRACE and estimated by climate models. J. Geophys. Res 2009, 114, B05404.
3. Frappart, F.; Ramillien, G.; Ronchail, J. Changes in terrestrial water storage versus rainfall and discharges in the Amazon basin. Int. J. Climatol. 2013.
4. Pokhrel, Y.N.; Fan, Y.; Miguez-Macho, G.; Yeh, P.J.-F.; Han, S.-C. The role of groundwater in the Amazon water cycle: 3. Influence on terrestrial water storage

computations and comparison with GRACE. J. Geophys. Res.-Atmos 2013, 118, 3233–3244.

5. Schmidt, R.; Schwintzer, P.; Flechtner, F.; Reigber, C.; Güntner, A.; Döll, P.; Ramillien, G.; Cazenave, A.; Petrovic, S.; Jochmann, H.; et al. GRACE observations of changes in continental water storage. Global Planet. Chang 2006, 50, 112–126.
6. Crowley, J.W.; Mitrovica, J.X.; Bailey, R.C.; Tamisiea, M.E.; Davis, J.L. Land water storage within the Congo Basin inferred from GRACE satellite gravity data. Geophys. Res. Lett 2006, 33, L19402.
7. Syed, T.H.; Famiglietti, J.S.; Rodell, M.; Chen, J.; Wilson, C.R. Analysis of terrestrial water storage changes from GRACE and GLDAS. Water Resour. Res 2008, 44, W02433.
8. Hu, X.; Chen, J.; Zhou, Y.; Huang, C.; Liao, X. Seasonal water storage change of the Yangtze River basin detected by GRACE. Sci. China Ser. D 2006, 49, 483–491.
9. Rodell, M.; Velicogna, I.; Famiglietti, J.S. Satellite-based estimates of groundwater depletion in India. Nature 2009, 460, 999–1002.
10. Jacob, T.; Wahr, J.; Pfeffer, W.T.; Swenson, S. Recent contributions of glaciers and ice caps to sea level rise. Nature 2012,482, 514–518.
11. Chen, J.L.; Wilson, C.R.; Tapley, B.D. Contribution of ice sheet and mountain glacier melt to recent sea level rise. Nat. Geosci 2013, 6, 549–552.
12. Velicogna, I. Increasing rates of ice mass loss from the Greenland and Antarctic ice sheets revealed by GRACE.Geophys. Res. Lett 2009, 36, L19503.
13. Luthcke, S.B.; Zwally, H.J.; Abdalati, W.; Rowlands, D.D.; Ray, R.D.; Nerem, R.S.; Lemoine, F.G.; McCarthy, J.J.; Chinn, D.S. Recent Greenland ice mass loss by drainage system from satellite gravity observations. Science 2006, 314, 1286–1289.
14. Chen, J.L.; Wilson, C.R.; Tapley, B.D. Interannual variability of Greenland ice losses from satellite gravimetry. J. Geophysi. Res 2011, 116, B07406.
15. Bergmann, I.; Ramillien, G.; Frappart, F. Climate-driven interannual ice mass evolution in Greenland. Global Planet. Chang. 2012, 82–83, 1–11.
16. Muskett, R.R.; Romanovsky, V.E. Alaskan permafrost groundwater storage changes derived from GRACE and ground measurements. Remote Sens 2011, 3, 378–397.
17. Wang, W.; Xing, W.; Yang, T.; Shao, Q.; Peng, S.; Yu, Z.; Yong, B. Characterizing the changing behaviours of precipitation concentration in the Yangtze River Basin, China. Hydrol. Process. 2012.
18. Wang, H.-S.; Wang, Z.-Y.; Yuan, X.-D.; Wu, P.; Rangelova, E. Water storage changes in Three Gorges water systems area inferred from GRACE time-variable gravity data. Chin. J. Geophys 2007, 50, 650–657.
19. Zhao, Q.-L.; Liu, X.-L.; Ditmar, P.; Siemes, C.; Revtova, E.; Hashemi-Farahani, H.; Klees, R. Water storage variations of the Yangtze, Yellow, and Zhujiang river

basins derived from the DEOS Mass Transport (DMT-1) model. Sci. China Earth Sci 2010, 54, 667–677.

20. Huang, Y.; Salama, M.S.; Krol, M.S.; van der Velde, R.; Hoekstra, A.Y.; Zhou, Y.; Su, Z. Analysis of long-term terrestrial water storage variations in Yangtze River basin. Hydrol. Earth Syst. Sci 2012, 9, 11487–11520.
21. Huang, H.; Song, D.S.; Yun, H.S.; Lee, D.H.; Cho, J.M. Water level change caused from three gorges dam construction in Yangtze River Basin. Proceedings of the 11th International Coastal Symposium, Szczecin, Poland, 9–14 May 2011; pp. 1672–1675.
22. Wang, X.; de Linage, C.; Famiglietti, J.; Zender, C.S. Gravity Recovery and Climate Experiment (GRACE) detection of water storage changes in the Three Gorges Reservoir of China and comparison with in situ measurements. Water Resour. Res 2011, 47, W12502.
23. Syed, T.H.; Famiglietti, J.S.; Chen, J.; Rodell, M.; Seneviratne, S.I.; Viterbo, P.; Wilson, C.R. Total basin discharge for the Amazon and Mississippi River basins from GRACE and a land-atmosphere water balance. Geophys. Res. Lett 2005,32, L24404.
24. Syed, T.H.; Famiglietti, J.S.; Zlotnicki, V.; Rodell, M. Contemporary estimates of Pan-Arctic freshwater discharge from GRACE and reanalysis. Geophys. Res. Lett 2007, 34, L19404.
25. Syed, T.H.; Famiglietti, J.S.; Chambers, D.P. GRACE-based estimates of terrestrial freshwater discharge from basin to continental scales. J. Hydromet 2009, 10, 22–40.
26. Seo, K.-W.; Waliser, D.E.; Tian, B.; Famiglietti, J.S.; Syed, T.H. Evaluation of global land-to-ocean fresh water discharge and evapotranspiration using space-based observations. J. Hydrol 2009, 373, 508–515.
27. Rodell, M.; Houser, P.R.; Jambor, U.; Gottschalck, J.; Mitchell, K.; Meng, C.-J.; Arsenault, K.; Cosgrove, B.; Radakovich, J.; Bosilovich, M.; et al. The global land data assimilation system. Bull. Am. Meteorol. Soc 2004, 85, 381–394.
28. Huffman, G.J.; Adler, R.F.; Bolvin, D.T.; Gu, G.; Nelkin, E.J.; Bowman, K.P.; Hong, Y.; Stocker, E.F.; Wolff, D.B. The TRMM Multisatellite Precipitation Analysis (TMPA): Quasi-global, multiyear, combined-sensor precipitation estimates at fine scales. J. Hydrometeorol 2007, 8, 38–55.
29. Koster, R.D.; Suarez, M.J. Modeling the land surface boundary in climate models as a composite of independent vegetation stands. J. Geophys. Res 1992, 97, 2697.
30. Liang, X.; Lettenmaier, D.P.; Wood, E.F.; Burges, S.J. A simple hydrologically based model of land surface water and energy fluxes for general circulation models. J. Geophys. Res 1994, 99, 14415.
31. Ek, M.B. Implementation of Noah land surface model advances in the national centers for environmental prediction operational mesoscale Eta model. J. Geophys. Res 2003, 108, 8851.
32. Klees, R.; Liu, X.; Wittwer, T.; Gunter, B.C.; Revtova, E.A.; Tenzer, R.; Ditmar,

P.; Winsemius, H.C.; Savenije, H.H.G. A comparison of global and regional GRACE models for land hydrology. Surv. Geophys 2008, 29, 335–359.

33. Cheng, M.; Tapley, B.D. Variations in the Earth's oblateness during the past 28 years. J. Geophys. Res 2004, 109, B09402.

34. Swenson, S.; Chambers, D.; Wahr, J. Estimating geocenter variations from a combination of GRACE and ocean model output. J. Geophys. Res 2008, 113, B08410.

35. Wahr, J.; Molenaar, M.; Bryan, F. Time variability of the Earth's gravity field: Hydrological and oceanic effects and their possible detection using GRACE. J. Geophys. Res 1998, 103, 30205–30229.

36. Ramillien, G.; Frappart, F.; Güntner, A.; Ngo-Duc, T.; Cazenave, A.; Laval, K. Time-variations of the regional evapotranspiration rate from GRACE satellite gravimetry. Water Resour. Res 2006, 42, W10403.

37. Chen, J.L.; Wilson, C.R.; Famiglietti, J.S.; Rodell, M. Spatial sensitivity of the Gravity Recovery and Climate Experiment (GRACE) time-variable gravity observations. J. Geophys. Res 2005, 110, B08408.

38. Swenson, S.; Wahr, J. Post-processing removal of correlated errors in GRACE data. Geophys. Res. Lett 2006, 33, L08402.

39. Kusche, J. Approximate decorrelation and non-isotropic smoothing of time-variable GRACE-type gravity field models. J. Geodesy 2007, 81, 733–749.

40. Chen, J.L.; Wilson, C.R.; Tapley, B.D.; Grand, S. GRACE detects coseismic and postseismic deformation from the Sumatra-Andaman earthquake. Geophys. Res. Lett 2007, 34, L13302.

41. Chen, J.L.; Wilson, C.R.; Tapley, B.D.; Longuevergne, L.; Yang, Z.L.; Scanlon, B.R. Recent La Plata basin drought conditions observed by satellite gravimetry. J. Geophy. Res 2010, 115, D22108.

42. Swenson, S.; Wahr, J. Methods for inferring regional surface-mass anomalies from Gravity Recovery and Climate Experiment (GRACE) measurements of time-variable gravity. J. Geophys. Res. 2002.

43. Klees, R.; Zapreeva, E.A.; Winsemius, H.C.; Savenije, H.H.G. The bias in GRACE estimates of continental water storage variations. Hydrol. Earth Syst. Sci 2007, 11, 1227–1241.

44. Landerer, F.W.; Swenson, S.C. Accuracy of scaled GRACE terrestrial water storage estimates. Water Resour. Res 2012,48, W04531.

45. Brutsaert, W. Hydrology: An Introduction; Cambridge University Press: New York, NY, USA, 2005; p. 142.

46. Xue, B.-L.; Wang, L.; Li, X.; Yang, K.; Chen, D.; Sun, L. Evaluation of evapotranspiration estimates for two river basins on the Tibetan Plateau by a water balance method. J. Hydrol 2013, 492, 290–297.

47. Rodell, M.; Famiglietti, J.S.; Chen, J.; Seneviratne, S.I.; Viterbo, P.; Holl, S.; Wilson, C.R. Basin scale estimates of evapotranspiration using GRACE and other observations. Geophys. Res. Lett 2004, 31, L20504.

48. Cesanelli, A.; Guarracino, L. Estimation of regional evapotranspiration in the extended Salado Basin (Argentina) from satellite gravity measurements. Hydrogeol. J 2011, 19, 629–639.

49. Taylor, K.E. Summarizing multiple aspects of model performance in a single diagram. J. Geophys. Res 2001, 106, 7183–7192.

50. Zaitchik, B.F.; Rodell, M.; Olivera, F. Evaluation of the global land data assimilation system using global river discharge data and a source-to-sink routing scheme. Water Resour. Res 2010, 46, W06507.

13
CHAPTER

DIATOM PHENOLOGY IN THE SOUTHERN OCEAN: MEAN PATTERNS, TRENDS AND THE ROLE OF CLIMATE OSCILLATIONS

Mariana A. Soppa, Christoph Völker and Astrid Bracher

Keywords: Diatom, phenology, phytoplankton, SAM, ENSO, ocean colour, Southern Ocean

INTRODUCTION

Phytoplankton in many parts of the world ocean follows a distinct seasonal pattern, and the timing, or phenology, of events in that seasonal cycle, such as the spring bloom, strongly depends on physical influences, as the depth of the surface mixed layer, ice cover or temperature [1,2,3,4]. Changes in the phytoplankton phenology are therefore sensitive indicators of environmental change [5]. Furthermore, they may have a large effect on the functioning of marine ecosystems [6] by affecting the temporal coincidence between parts of the ecosystem that influence each other, such as in predator-prey interactions [7,8].

Numerous studies have investigated the mean phytoplankton bloom phenology patterns in the Southern Ocean and the conditions that drive these blooms at large (e.g., entire Southern Ocean [3,4,9,10], global [5]) and over local scales (e.g., South Georgia, [11], marginal ice zone, [12]). The biomass, timing, magnitude and duration of blooms vary from one year to the next and studies have suggested that part of this variability is

Citation: Soppa, M.A., Völker, C.; Bracher, A. Diatom Phenology in the Southern Ocean: Mean Patterns, Trends and the Role of Climate Oscillations. Remote Sens. 2016, 8, 420. doi:10.3390/rs8050420.

linked to the climate oscillations El Niño-Southern Oscillation (ENSO) and Southern Annular Mode (SAM) [5,13,14,15,16].

ENSO [17] and SAM [18] dominate the climate variability on interannual timescales in the tropics and in the Southern Ocean, respectively. ENSO is associated with anomalous high (low) sea surface temperature (SST) in the east Pacific during a positive (negative) phase, called El Niño (La Niña) [19]. SAM, also known as the Antarctic Oscillation, consists of anomalous low (high) pressure at high latitudes and high (low) pressure at mid latitudes during a positive (negative) phase. A positive SAM phase is characterized by a strengthening and shift of the westerly winds towards the pole. Studies have also shown that in the Southern Ocean El Niño results in anomaly patterns in ocean and atmosphere that are similar to a negative phase of SAM, and *vice versa* [15,20,21]. As a result, ENSO and SAM effects on bloom phenology may be amplified when El Niño (La Niña) and negative (positive) SAM events coincide. The influence of these climate oscillations on the phytoplankton bloom phenology has been observed for example, in the Antarctic Peninsula [14,15] and the Ross Sea [13]. Moreover, the significant correlation between chlorophyll-a (Chla) and the Southern Annular Mode (SAM) is a strong evidence of a relationship between phytoplankton phenology and these natural climate oscillations [22].

Diatoms are the most diverse [23,24] and abundant eukaryotic phytoplanktonic group in the global oceans with an extremely important role as major marine primary producers. They occur in a wide range of environments due to several abilities. Diatoms can under nutrient or light stress: migrate in the water column by controlling their buoyancy, store nutrients in the central vacuoles for later use, reduce iron requirements and maintain symbiosis with nitrogen fixing cyanobacteria [23,24]. In addition, diatoms produce thick cell walls, spines and toxins to avoid grazers [24,25] and the rapid mass sinking events are considered a seeding strategy to overcome periods adverse to growth conditions [26,27]. They shape the biogeochemistry of the oceans by being responsible for much of the vertical flux of carbon out of the surface layer [28,29]; specially in the Southern Ocean they dominate production almost everywhere [30]. Variations in the timing, duration and biomass of diatom blooms in the Southern Ocean could therefore lead to important consequences for marine biogeochemical cycles.

In the Southern Ocean, the biogeochemical role of diatoms differs between iron-limited and iron replete regions. In iron-replete regions such as the Patagonian shelf, weakly silicified diatoms prevail. These diatoms have high growth rates and form high biomass blooms that drive the carbon pump (carbon sinkers) [28]. In iron-limited regions, diatoms are not less important. Here, the community of diatoms is dominated by species with thick silica shells for grazer protection [29]. When these diatoms are grazed, the organic biomass is recycled near the surface, but the sinking out of the thick shells sequesters silicon, resulting in loss of silicon (silica sinkers) but retention of nitrogen (N) and phosphorous (P) at surface. Part of the frustules dissolves during sinking and increase silicic acid in the Circumpolar Deep Water. Part is buried into the sediments forming the opal belt, the major global biogenic silica accumulation [25,28,31,32].

Despite the importance of diatoms in the Southern Ocean, a study on their phenology using remotely sensed data has not been published so far. This study discusses the mean patterns, trends and role of climate oscillations in the diatom bloom phenology regimes over 1997–2012 (covering from September to April). We extend previous studies on phytoplankton bloom phenology in the Southern Ocean by: (i) looking specifically at the concentration of Chla in diatoms; (ii) examining the different characteristics of the phenology (iii) using a new merged satellite Chla product with improved spatial and temporal coverage than the data sets based only on one sensor; (iv) investigating trends and (v) investigating if the interannual variability of the diatom phenology could be modulated by the large scale climate oscillations ENSO and SAM.

DATA AND METHODS

Satellite Data

We used 15 years (September 1997 to April 2012) of the level 3 Chla data (ESACCI-OC-L3S product, 4 km, version 1.0) from the Ocean Colour Climate Change Initiative (OC-CCI) [33]. The OC-CCI project is an European effort to produce high quality ocean colour products by combining data from the MERIS, MODIS-Aqua and SeaWiFS sensors. The data processing improves limitations of ocean colour remote sensing in polar regions due to low solar elevation and frequent cloud cover. Radiometric contamination by sun glint, thin clouds or heavy aerosol plumes are removed from the MERIS with the POLYMER algorithm [34], while the SeaWiFS and MODIS data are processed for atmospheric correction with the algorithm of [35]. The global validation of the Chla product with *in situ* High-Performance Liquid Chromatography Chla data presented relative errors lower than 30% for most of the Chla range, except for concentrations lower than 0.1 mg·m−3 [36]. More details on the project and processing steps can be found in http://www.esa-oceancolour-cci.org/, where the Chla data are also available. In our study, we calculated weekly averages of Chla (a week is defined as 8 days averages) from daily data onto a 15 min spatial grid for the area south of 50∘ S. To avoid coastal waters, we removed the three grid cells closest to the coastline.

Diatom abundance was derived by applying the regional abundance based model developed by [37] to the weekly Chla data, hereinafter referred to as Diatom-Chlorophyll-a (Dia-Chla). The model of [37] was developed based on the work of [38] which uses satellite-derived Chla together with empirical relationships between Chla and fraction of Phytoplankton Functional Types (PFTs, e.g., diatoms, dinoflagellates), tuned using *in situ* phytoplankton pigment measurements, to derive the PFTs. [37] revised the parametrizations of [38] for diatoms using a large global *in situ* dataset of phytoplankton pigments, particularly with more samples collected in the Southern Ocean, and to take account of the information on the penetration depth. The authors observed that the relationship between Chla and the fraction of diatom in the Southern Ocean is different from the global one. Therefore, a regional model was developed for the Southern Ocean to retrieve the concentration of diatoms from Chla (log10Diatom = 1.1559log10Chla − 0.2901). This regional model improves the retrieval of diatoms in the Southern Ocean by 40% compared to the original global model of [38], but more validation should be done on the model and satellite Dia-Chla product. The advantage of using ocean colour data to study phytoplankton phenology is the high temporal, compared to *in situ* measurements, and spatial resolution, compared to model outputs, that allows to investigate the full development of the bloom and in different regions simultaneously in time. However, we are aware that still knowledge gaps exist that might affect the phenology studies and where further investigations are needed. One limitation is that some regions of the Southern Ocean present a deep Chla maximum (~ 60–90 m) which is not seen by the sensor (e.g., southern Indian and Pacific sectors of the Southern Ocean [39]). This implies that subsurface blooms deeper than the penetration depth are not accounted for in the satellite data. Nevertheless, the deep Chla maximum may also result from photoacclimation to reduced light levels and nutrient availability and may not represent an increase in the organic carbon content [40,41]. A second limitation is that due to gaps in satellite data the right timing of the bloom may be missed [1,42]. Racault *et al.* [42] have shown that if 40% of data are missing in an annual time-series, the RMSE and bias in estimates of timing of peak are ~30 and 10 days respectively. Data gaps due to clouds, high solar zenith angle and sea ice for example also reduce the length of the time series and the significance of the statistical tests. We have attempted to minimize this error by filling the gaps by linear temporal interpolation and through the use of the use of OC-CCI merged satellite Chla product. For example, in January 2003 the number of observations per pixel including all three sensors (SeaWiFS, MODIS and MERIS) is 89 observations (averaged over Southern Ocean). Only MERIS data processed with the POLYMER algorithm represents more than half of these observations, 56% of this total. Even using the OC-CCI Chla product, 64% of the time series were interpolated in 2003–2004 for instance and on average 34% of the values in the time series were interpolated.

Fronts Position

We used the weekly position of the Polar Front (PF) available at http://ctoh.legos.obs-mip.fr/applications/mesoscale/southern-ocean-fronts. This product is based on sea level anomalies observed in altimetry data and climatological mean sea level from historical data and ARGO profiles [43]. The mean position of the PF was calculated from 1997 to 2012, for the months of September to April, the same period that was used to describe the diatom bloom phenology (see Section 2.5). In addition, we included the mean position of the Southern Antarctic Circumpolar Current Front (SACCF) in our analysis (no temporal resolved product available). The SACCF position is derived from historical hydrographic data of the Southern Ocean until 1990 [44].

Maximum Sea Ice Extent

To delineate the seasonal ice zone, we used the September sea ice extent data of [45] for the Southern Ocean, made available by the National Snow and Ice Data Center (NSIDC) atftp://sidads.colorado.edu/DATASETS/NOAA/G02135/shapefiles/. The maximum sea ice extent for each year was binned into longitude bins of 1 degree. Coordinates were automatically extracted from the sea ice extent data and mean position was calculated for each longitude bin in the entire period (1997 to 2012).

Climate Indices

To investigate if the Dia-Chla phenology in the Southern Ocean is influenced by ENSO and SAM climate oscillations, we used two indices: the Multivariate El Niño Southern Oscillation index (MEI) and the Antarctic Oscillation (AAO) index. The MEI, available at http://www.esrl.noaa.gov/ psd/enso/mei/#loadings, is based on six variables (cloudiness, sea surface temperature, sea-level pressure, surface air temperature and the zonal and meridional components of the surface wind) over the tropical Pacific from 30° N to 30° S [46]. Positive MEI values can characterize El Niño events while negative values indicate La Niña events. The Antarctic Oscillation (AAO) index, available athttp://www.cpc.ncep.noaa.gov/products/precip/CWlink/daily_ao index/ aao/monthly. aao.index.b79.current.ascii.table, is based on the first principal component of monthly mean pressure anomalies at 700 mb for the region south of 30° S [47]. Positive/negative phases of the Southern Annular Mode (SAM) are associated with positive/negative values of AAO, respectively.

Annual ENSO and SAM indices were calculated by averaging their respective indices from September of the previous year to April of the following year, the same period used to estimate the phenological indices. During the 1997–2012 period there were six El Niño years (1997/1998, 2002/2003, 2003/2004, 2004/2005, 2006/2007, 2009/2010), eight La Niña years (1998/1999, 1999/2000, 2000/2001, 2005/2006, 2007/2008, 2008/2009, 2010/2011, 2011/2012), seven years of a positive phase of SAM (1998/1999, 1999/2000, 2001/2002, 2007/2008, 2008/2009, 2010/2011, 2011/2012) and four of a negative phase (2000/2001, 2002/2003, 2003/2004, 2009/2010) (Figure 1 and Table 1).

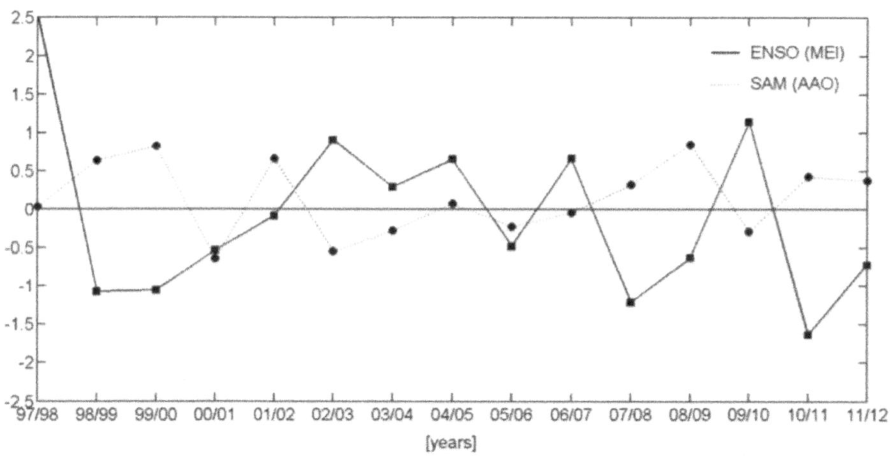

Figure 1. Time series (dimensionless) of the annual Multivariate ENSO Index (representing the ENSO, solid line) and the Antarctic Oscillation index (representing the SAM, dashed line).

Table 1. List of ENSO and SAM events. El Niño and positive SAM events are represented by + and La Niña and negative SAM by –

Events	ENSO	SAM
1997/1998	+	
1998/1999	–	+
1999/2000	–	+
2000/2001	–	–
2001/2002		+
2002/2003	+	
2003/2004	+	–
2004/2005	+	
2005/2006	–	
2006/2007	+	
2007/2008	–	+
2008/2009	–	+
2009/2010	+	–
2010/2011	–	+
2011/2012	–	+

Phenological Indices

We assessed the diatom phenology using a threshold method initially developed based on the total chlorophyll-a concentration [48]. We note that there are different methods to estimate the bloom phenology. Here we use a robust and widely applied method to investigate phytoplankton phenology from ocean color data [5,9,48,49,50,51].

Phytoplankton blooms start (defining the bloom start date - BSD) when the Chla value exceeds a value of 5% above the median [48] and remains above this threshold for at least two consecutive weeks [9] (Table 2 and Figure 2). To isolate primary blooms from secondary blooms, we first identified the maximum Dia-Chla of the time series and then looked backwards in time to find the bloom start date [51]. The bloom end date (BED) was determined as the first week when Dia-Chla level fell below the threshold. The period between bloom start date and end date defines the total bloom duration (BD). Within this period the Dia-Chla reaches a maximum (CM) at the date of Dia-Chla maximum (CMD). The sub-periods before and after the maximum determine the bloom growth duration (BGD) and bloom decline duration (BDD), respectively. During the growth duration, the average (CAV) and integrated Dia-Chla values (CI) are calculated. In addition, the amplitude of the bloom (CA) is determined as the difference between maximum and threshold Dia-Chla value.

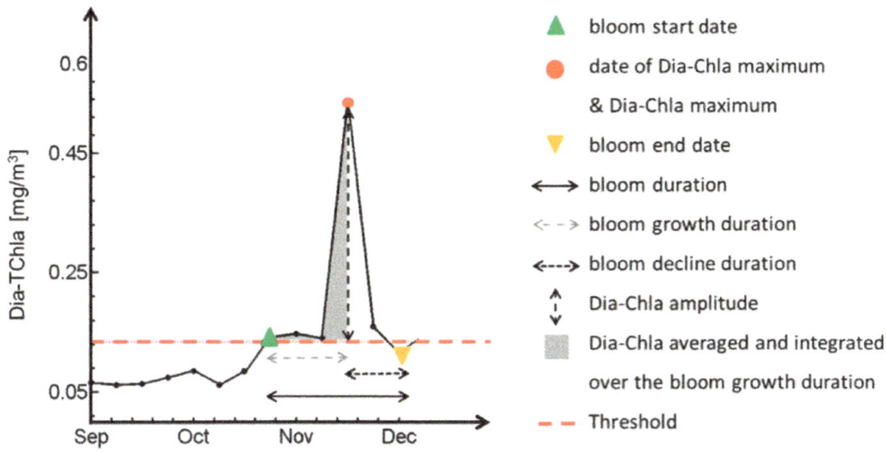

Figure 2. Schematic of the indices used to describe the diatom phenology.

Table 2. Phenological indices as used in this study

Index	**Abbreviation**	**Unit**
Bloom Start Date	BSD	Week
Date of Dia-Chla Maximum	CMD	Week
Bloom End Date	BED	Week
Bloom Duration	BD	Week
Bloom Growth Duration	BGD	Week
Bloom Decline Duration	BDD	Week
Dia-Chla Amplitude	CA	mg·m−3
Dia-Chla Maximum	CM	mg·m−3
Dia-Chla averaged over BGD	CAV	mg·m−3
Dia-Chla integrated over BGD	CI	mg·m−3

Using these indices, we analyzed the phenology of the entire time series (1997 to 2012), from September to April of the following year (e.g., September 2002–April 2003). Before computing the phenological indices, the time series were linearly interpolated in time to fill gaps less than 3 weeks in length [49]. After the temporal interpolation, if there were remaining gaps of more than two weeks between the date of Dia-Chla maximum and the estimated bloom start or bloom end date, these phenological indices were not calculated to avoid erroneous detection of the bloom timing. This led to slightly different data coverage of the phenological indices. The best data coverage is achieved for date of Dia-Chla maximum, Dia-Chla maximum and amplitude of the bloom.

Statistical Analysis

The mean spatial patterns were obtained by averaging the 15 years of phenological indices. To examine the interannual variability of the diatom phenology, we estimated the trends, correlations with ENSO and SAM indices and composite maps of the anomaly of the phenological indices. The analyses were performed using the standardized anomaly data. Standardized anomalies (dimensionless and hereafter termed as anomalies) were produced by subtracting the average (15-yr) from the annual phenology data (e.g., 2002–2003) and dividing by the standard deviation (15-yr), pixel by pixel. Trends were investigated with non-parametric Kendall's tau test with Sen's method at the 95% confidence level for each grid cell (only when 100% of data were available). End-point bias was not accounted for. The correlation between the climate indices and anomalies of the phenological indices was determined using Spearman correlation. Partial correlations were used to study the influence of both oscillations separately, for example, by considering the relationship between SAM and Dia-Chla maximum after removal of the effect of ENSO [21], since the correlation between the annual ENSO and SAM indices is −0.58 (p-value = 0.03). Composite maps of the anomalies of the phenological indices were computed by averaging the anomalies from the different phases (positive/negative) of ENSO and SAM. Using composite maps we investigated the dominant patterns of the anomalies associated with the different phases and oscillations [52]. Unfortunately, because of the short length of our time series it was not possible to distinguish between amplified (e.g., El Niño coincided with negative phase of SAM) and non-amplified years (Table 1).

RESULTS AND DISCUSSION

Mean Patterns

The spatial patterns of the diatom phenological indices averaged over the 15-year period are presented in Figure 3 to Figure 5, together with the corresponding latitudinal variation in Figure 6. Overall, the spatial patterns are associated to the location of the contour of the SACCF and of the maximum sea ice extent. This association is particularly clear for the bloom start date, maximum date, growth duration and total duration of the diatom blooms. The diatom blooms start and reach their maximum earlier north of the SACCF–outside the seasonal ice zone (Figure 3). The spatial pattern of the bloom start date is consistent with [9] and mainly follows the increase in light availability [5,9]. On the other hand, the end of the bloom is more likely related to the exhaustion of nutrients [25,26,28]. In the South Georgia region the exhaustion of silicate is thought to be a limiting factor for the end of the spring diatom bloom [11]. Grazing pressure is thought to control the diatom species composition and biomass, rather than the end of the diatom bloom [28]. In the seasonal ice zone, the start of the bloom is driven by light as well as water column stability; as the sea ice retreats, the melting of ice increases the stratification of the water column which favors to maintain the phytoplankton in the euphotic zone [12]. The end of the bloom occurs when the mixed layer deepens due to wind forcing, which dilutes the phytoplankton in the water column [12] and can bring them to lower light levels.

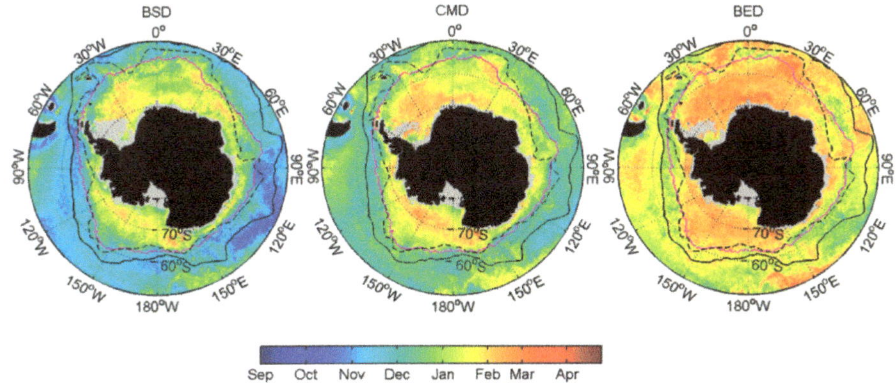

Figure 3. Spatial distribution of the mean diatom phenology in 1997–2012: (**left**) bloom start date–BSD, (**center**) date of Dia-Chla maximum–CMD, (**right**) bloom end date–BED. Grey areas represent missing data. Black solid lines show the mean position of the Polar Front [43] over 1997–2012. Dashed lines show the Southern Antarctic Circumpolar Front [44]. Purple line displays the median position of the maximum sea ice extent over 1997–2012 [45].

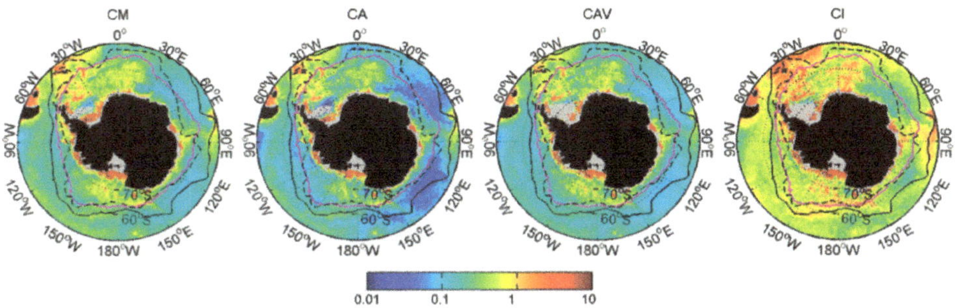

Figure 5. Same as Figure 3, but for Dia-Chla maximum (CM), Dia-Chla amplitude (CA), Dia-Chla average (CAV) and Dia-Chla integrated over the growth duration (CI). Units are in mg·m−3.

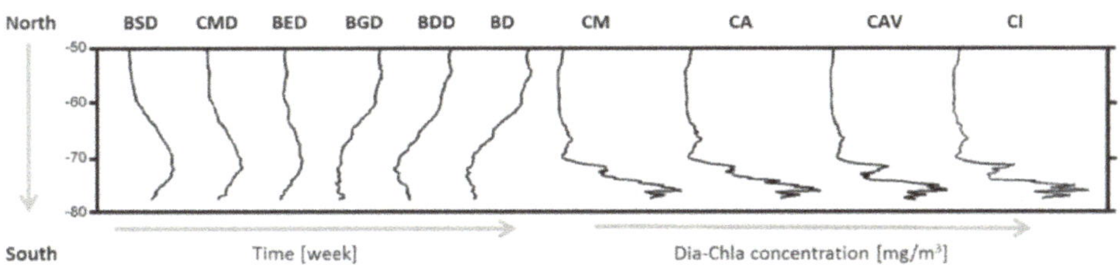

Figure 6. Schematic representation of the latitudinal variability (longitudinal average) of the phenological indices: bloom start date (BSD), date of Dia-Chla maximum (CMD), bloom end date (BED), bloom growth duration (BGD), bloom decline duration (BDD), bloom duration (BD), Dia-Chla maximum (CM), Dia-Chla amplitude (CA), Dia-Chla averaged BGD (CAV), Dia-Chla integrated over BGD (CI).

Particularly notable is the early start of the diatom blooms in the waters surrounding Antarctica in December (light green), caused by the opening of areas free of ice around the continent. Arrigo *et al.* [53] showed that in the Amundsen polynya small areas free of ice occur throughout the year and that their size increases with three factors: advection of sea ice offshore, increase in temperature and melting of ice. These factors, combined with an increase in solar radiation and water column stability, as shown by [12], are linked to the earlier bloom start date in these waters surrounding Antarctica as compared to other regions of the seasonal ice zone.

The duration of the diatom blooms is shorter south of the SACCF, in the seasonal ice zone, and *vice versa* (Figure 4). Outside this region it forms a belt of higher values (longer duration) around the Polar Front (PF), particularly between 30∘W and 120∘ E. Previous phenology studies (e.g., [3,4,5,9]) based on total Chla data, which includes all PFTs, reported durations of phytoplankton blooms, which were longer than the durations of the diatom blooms observed in the present study. Specifically, the average duration of the blooms for the regions 50∘ S–60∘ S and 60∘ S–70∘ S was 8.3 and 6.5 weeks for the diatom blooms, while total chla blooms were shown to range between 14 and 11 weeks [5], respectively.

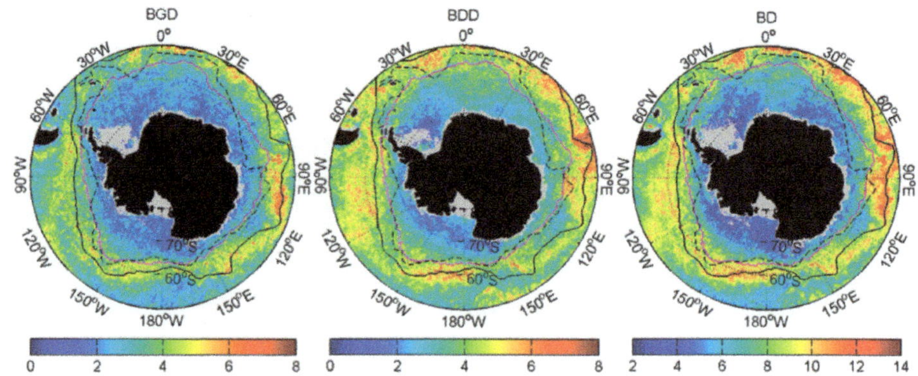

Figure 4. Same as Figure 3, but for bloom growth duration (BGD), bloom decline duration (BDD) and total duration (BD) of the diatom blooms. Units are in weeks.

The duration of the bloom in the seasonal ice zone results from a combination of factors influencing the growth and decline phases of the bloom, mainly light and stability of the water column, while nutrients are less important [12]. The belt of 'longer lasting' blooms outside the seasonal ice zone is likely linked to a complex inter-play of different forcings: longer light periods and deeper mixed layers [54] that enhance the supply of nutrients at surface as well as reduce the grazing pressure by zooplankton [55]. The mixed layer depth is deeper in the vicinity the fronts; around 100 m in the summer and up to 400 m in the winter [54].

The deepening of the mixed layer in the winter together with diapycnal diffusion replenishes the surface with nutrients from subsurface waters, including iron [56]. It is known that iron is a limiting nutrient in the surface waters of the Southern Ocean controlling phytoplankton growth, particularly in the open ocean. This micronutrient is rapidly depleted by spring blooms. In late spring and summer, phytoplankton relies on the pelagic recycling until the following deepening of the mixed layer in autumn [56]. Open ocean diatoms have the ability to reduce their requirement of iron [23] which can help to sustain their blooms for longer periods. Other important factors controlling the duration of the bloom are the increasing of grazing pressure and algal viruses [28,55].

The relationship of the biomass indices with the fronts is not as evident as for the other indices (Figure 5). The spatial distribution shows that more intense diatom blooms (higher biomass) occur in coastal regions, in the seasonal ice zone and in the Atlantic sector of the Southern Ocean. Diatom blooms around Antarctica can be considered as more efficient blooms, with short duration and high biomass. Sokolov and Rintoul [57] have shown that at a broader scale the distribution of Chla is mainly controlled by the upwelling of nutrients via Ekman transport while the upwelling associated with bathymetric features is responsible for the magnitude and duration of the blooms. In regions where the Antarctic Circumpolar Front (ACC) interacts with the topography, the nutrient supply is enhanced [57] leading to higher Chla and consequently, higher amplitude of the blooms. This occurs for example in the Pacific Antarctic Ridge (see Figure 7 in [57]). The enrichment from the coastal and shelf sediments close to islands (e.g., Kerguelen, Crozet and South Georgia Islands) are also important sources of nutrients, especially iron [58,59,60].

The latitudinal variability displays, from north to south, a progressive delay in the start, maximum and end date of the diatom blooms until about 73∘ S (Figure 6). The opposite is observed for the duration; there is a decrease in the growth, decline and total duration of the diatom blooms from north to south. South of 73∘ S the trend is reversed except by the growth duration which holds at about the same duration. The biomass indices present similar latitudinal variations, but are rather small until the first peak at ~67∘ S, followed by two steep peaks at ~72∘ S and at ~76∘ S, and then decreasing towards the south.

The computation of the phenological indices reported here may not only be affected by the data drawbacks mentioned in the methods (missing the deep Chla maximum, gaps), but also satellite pixels contaminated by white caps, sea ice or sun glint can potentially affect the results. Ocean colour data in polar oceans are known to have issues related to sea ice, low sun elevation, clouds and polar aerosols [61]. Adjacency effect and sea ice contaminated pixels can lead to overestimation of the satellite Chla and standard SeaWiFS and MODIS flags may not remove all impacted pixels [62,63]. The extent of these issues in the timing of the diatom blooms has not been yet quantified and needs more attention. By using the OC-CCI Chla product in this study, where POLYMER is used for the MERIS atmospheric correction, we believe to work with the best long term dataset available for the Southern Ocean.

Interannual Variability

Trends

Because among the phenological indices only the Dia-Chla maximum, Dia-Chla maximum and amplitude indices are gap free, trends could only be determined for those indices. Trends in the Dia-Chla amplitude are very similar to Dia-Chla maximum and not shown.

Coherent patches of significant positive and negative trends were detected for the date of Dia-Chla maximum and Dia-Chla maximum (Figure 7). For example, in the region between the Malvinas and South Georgia Islands (Figure 7, green star) there is a trend towards an earlier maximum of the bloom leading to higher biomass (Dia-Chla maximum). However, the opposite relationship where a later start of the bloom leads to an increase in biomass can also be observed (e.g., in the region south of 60∘ S and between 120∘ E and 150∘ E) (Figure 7, black star). Although we could not estimate trends in the bloom start date and bloom end date, we can expect a similar pattern to the ones detected for date of Dia-Chla maximum since these indices are highly correlated.

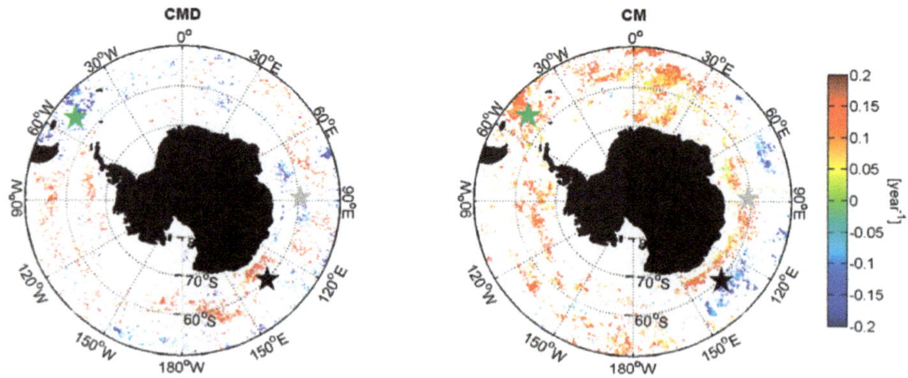

Figure 7. Trends of the annual standardized anomalies of date of Dia-Chla maximum (CMD) and Dia-Chla maximum (CM). Reddish colour indicates a positive trend and bluish indicates a negative trend. Only statistically significant trends ($p < 0.05$) are shown. The stars highlight the regions between Malvinas and South Georgia Islands (green) and south of 60∘ S between 120∘ E to 150∘ E (black) and 60∘ E to 120∘ E (grey).

These observations combined with recent studies on the trends in sea surface temperature [64] and sea ice cover [65] over the last three decades, suggest a link between these two variables and the diatom phenology. For example, in the region south of 60∘ S and from 60∘ E to 120∘ E (Figure 7, grey star) the earlier date of Dia-Chla maximum and the increased Dia-Chla maximum coincide with the observed increase trend in SST and decrease in sea ice cover (earlier sea ice melt).

Compared to literature, the spatial distribution of trends in Dia-Chla maximum are similar to trends in total Chla from SeaWiFS reported by [66] and [67] for the 1997–2007 and 1997–2010 periods, respectively. Recent decadal trends (1998–2012) in diatom concentration have been investigated by [68] based on model results, where large areas in the Southern Ocean with positive trends have been observed as in this study (e.g., off the Patagonian shelf), but also positive trends between the 60∘ E and 150∘ E north of 60∘ S whereas we observed negative trends. Furthermore, the general increase in Dia-Chla maximum observed here coincide with regions where [68] observed a shallowing of the MLD and an increase in silicate, iron and nitrate.

Relationships with ENSO and SAM

To further explore the interannual variability of the diatom phenology, we examined the relationship of the annual phenological indices with ENSO and SAM. The correlation maps are presented in Figure 8 for date of Dia-Chla maximum and Dia-Chla maximum as representative of the date indices and biomass indices, respectively. Significant positive (negative) correlations indicate that the anomalies are in (out of) phase with ENSO and SAM. Consistent areas in the duration indices are less evident for the duration indices.

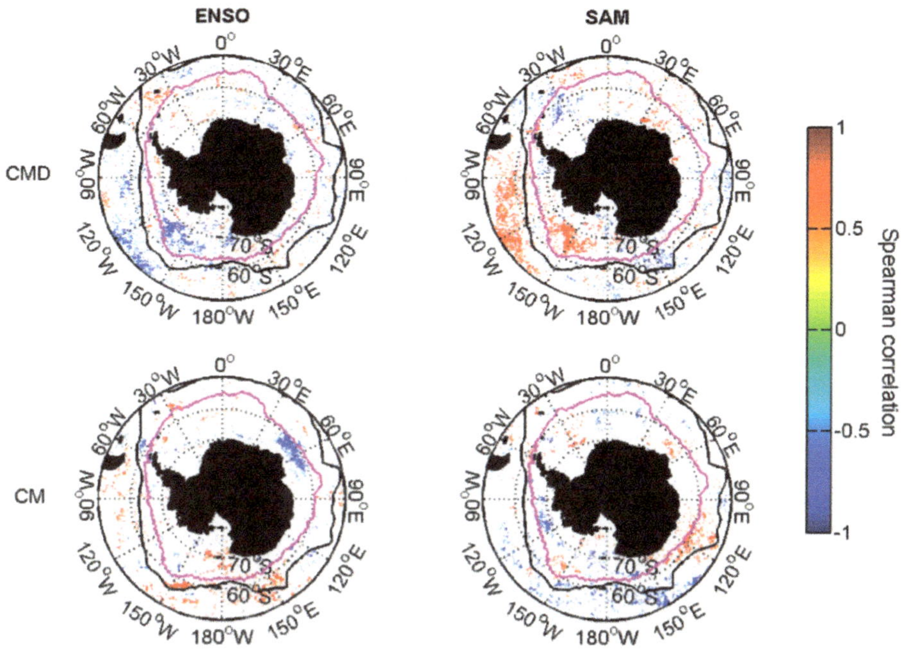

Figure 8. Correlation coefficients of the standardized anomalies of date of Dia-Chla maximum (CMD) and Dia-Chla maximum (CM) *vs.* ENSO (MEI) and SAM (AAO) indices. Only statistically significant trends ($p < 0.05$) are shown. Black lines show the mean position of the Polar Front [43] over 1997–2012. Purple line displays the median position of the maximum sea ice extent [45] over 1997–2012.

Several areas show significant correlation between ENSO and SAM and the diatom phenology. The correlation coefficients for ENSO are opposite to that of SAM. For example, the date of Dia-Chla maximum in the sector of the seasonal ice zone between 120∘ W and 150∘ W is negatively correlated with ENSO and positively correlated with SAM. Moreover, the patterns in El Niño (La Niña) years and negative (positive) SAM are similar. These results are in line with observations of the sea ice concentration, SST, Chla and wind speed and direction in the Southern Ocean [20,21]. Smith *et al.* [15] also observed that high Chla biomass offshore the Western Antarctic Peninsula region was associated La Niña and/or positive SAM events. Hence, we can expect the spatial patterns of the anomalies of phenological indices during El Niño (La Niña) years and negative (positive) phase of SAM to resemble each other.

The most remarkable feature in the correlation maps of the date of Dia-Chla maximum can be seen in the Pacific Sector (90○ W to 150○ W), north of the PF and south of the maximum sea ice extent. This is consistent with patterns observed by [52,69] for earlier periods, 1982–1998 and 1980–1999 and using satellite and model data, respectively. For the same region, Kwok and Comiso [52] observed an increase in SST and a decline in sea ice concentration associated with El Niño. Lefebvre*et al.* [69] showed that the winter sea ice concentration decreases in negative SAM events. As a result, an earlier start, maximum and end of the bloom can be expected in El Niño or negative SAM events.

The Dia-Chla maximum displays less significant correlations, but the general pattern of Dia-Chla maximum is consistent with the correlations of satellite Chla and SAM presented by [22] for an earlier period (1997–2004). The observed lower Dia-Chla maximum at 60○ E during El Niño (negative correlation) can be linked to lower SST in El Niño years, as shown by [52] (see Figure 6 in [52]). In contrast, the general increase in diatom concentration between 50○ S and 70○ S during positive SAM event showed by [70] was not observed in our results. The authors used a coupled ecosystem-general circulation model and lagged correlations (4 months) to investigate the relationship and these might be the cause for disagreement between the results, as well as the different periods analyzed in their study (1948–2010) and in the present study (1997–2012).

Because SAM and ENSO are not linearly independent at interannual time scales during the austral summer season [21,71], we expect that some of the variability we observed related to SAM may be influenced by ENSO, or *vice versa*. This was in part confirmed by the partial correlations (Figure S2 in the supplement), but the differences between the correlations and partial correlations are in general small. Higher differences were observed between the date of Dia-Chla maximum and MEI. The correlations between the date of Dia-Chla maximum and SAM, and Dia-Chla maximum and MEI or SAM did not change. One possible reason for not observing differences is that the short time series used here might not allow to distinguish the influence of the respective oscillations, reinforcing the need for continuous and longer ocean colour records.

The composite maps of the anomalies of bloom start date, Dia-Chla maximum and bloom duration are shown in Figure 9and Figure 10 and Figure S3 in the supplement, respectively, and provide insight into the magnitude of the anomalies during the ENSO and SAM events. In the seasonal ice zone there are two regions with inverse patterns and high anomalies of bloom start date: the Weddell Sea region (white dashed box) and the sector between 120○ W and 180○ W (white box), north of 70○ S. In the Weddell Sea, El Niño/negative SAM years are characterized by later start, shorter duration and slightly higher diatom biomass, which are likely a response of more extensive ice cover in these years [52,69]. In the sector between 120○ W and 180○ W the pattern presents opposite sign.

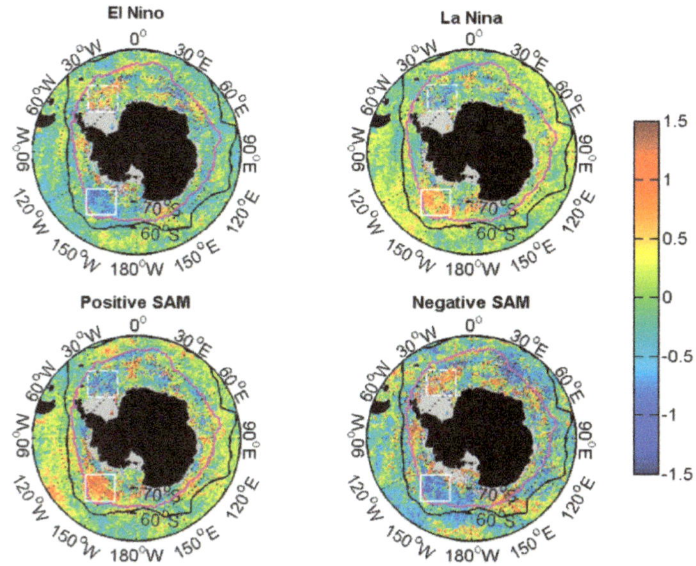

Figure 9. Composites of bloom start date (BSD) standardized anomalies during El Niño ($N = 6$), La Niña ($N = 8$), positive SAM ($N = 7$) and negative SAM ($N = 4$) years. Grey areas represent missing data. Black lines show the mean position of the Polar Front [43] over 1997–2012. Purple line displays the median position of the maximum sea ice extent [45] over 1997–2012. The white boxes depict the Weddell Sea region (dashed) and the sector between 120◦ W and 180◦ W.

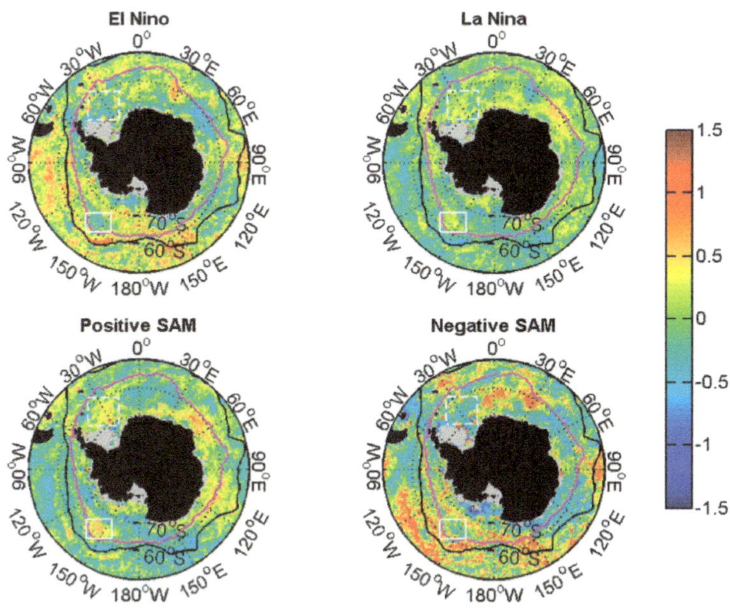

Figure 10. Same as Figure 9 but for Dia-Chla maximum.

CONCLUSIONS

Over the last decade, the phenology of phytoplankton blooms in the Southern Ocean has been examined using satellite-derived estimates of Chla [3,9,10]. However, by looking at the total Chla provided by satellite no information about the phytoplankton community composition (and changes in) is provided [72]. In this study we were able to look specifically at the diatom biomass by using a satellite-derived diatom concentration. We investigated the mean spatial and temporal patterns of diatom phenology and their interannual variability. We find a clear correspondence between ENSO and SAM and the phenology of diatoms, as revealed by the correlation and the anomaly composite maps. The influence of the climate oscillations varies depending on the region. It is also evident that ENSO and SAM have opposite effects in the diatom phenology. These results emphasize the influence of climate oscillations on the diatom phenology in the Southern Ocean.

A next step would be to investigate in more detail the relationship between climate oscillations, environmental variables and diatom phenology. Such investigation is not straight forward and requires a comprehensive dataset of, at least, weekly temporal resolved data including not only information on SST and PAR, as usual used in phytoplankton phenology studies since these variables are freely available from remote sensing, but also information on e.g., water column mixing, sea ice concentration, dissolved iron and silicate and grazing pressure. Combining remote sensing and model data can help to explain the missing link between climate oscillations, environmental anomalies and diatom bloom phenology.

Last, the knowledge of other phytoplankton types forming blooms in the Southern Ocean, mainly haptophytes, is essential to understand phytoplankton community shift and the factors controlling it. *E. huxleyi* are known to occur along the 'Great Calcite Belt' and dense blooms are often observed at the shelf break and off the Patagonian shelf after the spring bloom of diatoms. Moreover, while it is generally accepted that diatoms dominate the spring bloom in the Southern Ocean [26] this might not be true everywhere as the spring bloom of *P. antarctica* in the Ross Sea [73].

ACKNOWLEDGMENTS

We would like to acknowledge the Ocean Colour CCI project for the production and distribution of the chlorophyll data. This work was partially supported by the Alfred-Wegener-Institute, the Total Foundation Project "Phytoscope" and the ESA-SEOM SY-4SCI SYNERGY project "SynSenPFT".

AUTHOR CONTRIBUTIONS

Mariana Soppa led the study, performed data analysis and wrote the manuscript. Christoph Völker and Astrid Bracher were involved in discussions and assisted in writing the manuscript.

REFERENCES

1. Kahru, M.; Brotas, V.; Manzano-sarabia, M.; Mitchell, B. Are phytoplankton blooms occurring earlier in the Arctic?*Glob. Change Biol.* **2011**, *17*, 1733–1739.
2. Ardyna, M.; Babin, M.; Gosselin, M.; Devred, E.; Rainville, L.; Tremblay, J.É. Recent Arctic Ocean sea ice loss triggers novel fall phytoplankton blooms. *Geophys. Res. Lett.* **2014**, *41*, 6207–6212.
3. Carranza, M.M.; Gille, S.T. Southern Ocean wind-driven entrainment enhances satellite chlorophyll-a through the summer. *J. Geophys. Res.: Oceans* **2015**, *120*, 304–323.
4. Cole, H.S.; Henson, S.; Martin, A.P.; Yool, A. Basin-wide mechanisms for spring bloom initiation: How typical is the North Atlantic? *ICES J. Mar. Sci.: J. Conseil* **2015**, *72*, 2029–2040.
5. Racault, M.F.; Le Quéré, C.; Buitenhuis, E.; Sathyendranath, S.; Platt, T. Phytoplankton phenology in the global ocean.*Ecol. Indic.* **2012**, *14*, 152–163.
6. Edwards, M.; Richardson, A.J. Impact of climate change on marine pelagic phenology and trophic mismatch. *Nature***2004**, *430*, 881–884.
7. Platt, T.; Fuentes-Yaco, C.; Frank, K.T. Marine ecology: Spring algal bloom and larval fish survival. *Nature* **2003**, *423*, 398–399.
8. Koeller, P.; Fuentes-Yaco, C.; Platt, T.; Sathyendranath, S.; Richards, A.; Ouellet, P.; Orr, D.; Skúladóttir, U.; Wieland, K.; Savard, L.; *et al.* Basin-scale coherence in phenology of shrimps and phytoplankton in the North Atlantic Ocean. *Science* **2009**, *324*, 791–793.
9. Thomalla, S.; Fauchereau, N.; Swart, S.; Monteiro, P. Regional scale characteristics of the seasonal cycle of chlorophyll in the Southern Ocean. *Biogeosciences* **2011**, *8*, 2849–2866.
10. Sallée, J.B.; Llort, J.; Tagliabue, A.; Lévy, M. Characterization of distinct bloom phenology regimes in the Southern Ocean. *ICES J. Mar. Sci.: J. Conseil* **2015**, *72*, 1985–1998.
11. Borrione, I.; Schlitzer, R. Distribution and recurrence of phytoplankton blooms around South Georgia, Southern Ocean. *Biogeosciences* **2013**, *10*, 217–231.
12. Taylor, M.H.; Losch, M.; Bracher, A. On the drivers of phytoplankton blooms in the Antarctic marginal ice zone: A modeling approach. *J. Geophys. Res.: Oceans* **2013**, *118*, 63–75.
13. Arrigo, K.R.; van Dijken, G.L. Annual cycles of sea ice and phytoplankton in Cape Bathurst polynya, southeastern Beaufort Sea, Canadian Arctic. *Geophys. Res. Lett.* **2004**.
14. Montes-Hugo, M.; Vernet, M.; Martinson, D.; Smith, R.; Iannuzzi, R. Variability on phytoplankton size structure in the western Antarctic Peninsula (1997–2006). *Deep Sea Res. Part II: Top. Stud. Oceanogr.* **2008**, *55*, 2106–2117.
15. Smith, R.C.; Martinson, D.G.; Stammerjohn, S.E.; Iannuzzi, R.A.; Ireson, K. Bellingshausen and western Antarctic Peninsula region: Pigment biomass and

sea-ice spatial/temporal distributions and interannual variabilty. *Deep Sea Res. Part II: Top. Stud. Oceanogr.* **2008**, *55*, 1949–1963.

16. Alvain, S.; Le Quéré, C.; Bopp, L.; Racault, M.F.; Beaugrand, G.; Dessailly, D.; Buitenhuis, E.T. Rapid climatic driven shifts of diatoms at high latitudes. *Remote Sens. Environ.* **2013**, *132*, 195–201.
17. McPhaden, M.J.; Zebiak, S.E.; Glantz, M.H. ENSO as an integrating concept in earth science. *Science* **2006**, *314*, 1740–1745.
18. Thompson, D.W.; Wallace, J.M. Annular modes in the extratropical circulation. Part I: Month-to-month variability*. *J. Clim.* **2000**, *13*, 1000–1016.
19. Penland, C.; Sun, D.Z.; Capotondi, A.; Vimont, D.J. A brief introduction to El Nino and La Nina. In *Climate Dynamics: Why Does Climate Vary?* Elsevier: Amsterdam, The Netherlands, 2010; pp. 53–64.
20. Lovenduski, N.S. *Impact of the Southern Annular Mode on Southern Ocean Circulation and Biogeochemistry*; University of California: Los Angeles, CA, USA, 2007.
21. Pohl, B.; Fauchereau, N.; Reason, C.; Rouault, M. Relationships between the Antarctic Oscillation, the Madden-Julian Oscillation, and ENSO, and consequences for rainfall analysis. *J. Clim.* **2010**, *23*, 238–254.
22. Lovenduski, N.S.; Gruber, N. Impact of the Southern Annular Mode on Southern Ocean circulation and biology.*Geophys. Res. Lett.* **2005**.
23. Armbrust, E.V. The life of diatoms in the world's oceans. *Nature* **2009**, *459*, 185–192.
24. Kooistra, W.; Gersonde, R.; Medlin, L.K.; Mann, D.G. The origin and evolution of the diatoms: Their adaptation to a planktonic existence. In *Evolution of Primary Producers in the Sea*; Elsevier: Amsterdam, The Netherlands, 2007; pp. 207–249.
25. Smetacek, V. Diatoms and the ocean carbon cycle. *Protist* **1999**, *150*, 25–32.
26. Smetacek, V. Role of sinking in diatom life-history cycles: Ecological, evolutionary and geological significance. *Mar. Biol.* **1985**, *84*, 239–251.
27. Raven, J.; Waite, A. The evolution of silicification in diatoms: Inescapable sinking and sinking as escape? *New Phytol.***2004**, *162*, 45–61.
28. Smetacek, V.; Assmy, P.; Henjes, J. The role of grazing in structuring Southern Ocean pelagic ecosystems and biogeochemical cycles. *Antarct. Sci.* **2004**, *16*, 541–558.
29. Assmy, P.; Smetacek, V.; Montresor, M.; Klaas, C.; Henjes, J.; Strass, V.H.; Arrieta, J.M.; Bathmann, U.; Berg, G.M.; Breitbarth, E.; *et al.* Thick-shelled, grazer-protected diatoms decouple ocean carbon and silicon cycles in the iron-limited Antarctic Circumpolar Current. *Proc. Natl. Acad. Sci. USA* **2013**, *110*, 20633–20638.
30. Rousseaux, C.S.; Gregg, W.W. Interannual variation in phytoplankton primary production at a global scale. *Remote Sens.* **2013**, *6*, 1–19.
31. Tréguer, P.J.; De La Rocha, C.L. The world ocean silica cycle. *Annu. Rev. Mar.*

Sci. **2013**, *5*, 477–501.

32. Tréguer, P.J. The southern ocean silica cycle. *Comptes Rendus Geosci.* **2014**, *346*, 279–286.

33. Sathyendranath, S.; Krasemann, H. Climate Assessment Report: Ocean Colour Climate Change Initiative (OC-CCI) – Phase One. Technical Report, ESA OC-CCI. 2014. Available online: http://www.esa-oceancolour-cci.org/?q=documents(accessed on 9 May 2016).

34. Steinmetz, F.; Deschamps, P.Y.; Ramon, D. Atmospheric correction in presence of sun glint: Application to MERIS.*Opt. Express* **2011**, *19*, 9783–9800.

35. Wang, M.; Gordon, H.R. A simple, moderately accurate, atmospheric correction algorithm for SeaWiFS. *Remote Sens. Environ.* **1994**, *50*, 231–239.

36. Krasemann, H.; Belo-Couto, A.; Brando, V.; Brewin, R.J.; Brockmann, C.; Brotas, V.; Doerffer, R.; Feng, H.; Frouin, R.; Gould, R.; *et al. Product Validation and Inter–Comparison Report*; Technical Report 1, Ocean Colour Climate Change Initiative; Helmholtz-Zentrum Geesthacht: Geesthacht, Germany, 2014.

37. Soppa, M.A.; Hirata, T.; Silva, B.; Dinter, T.; Peeken, I.; Wiegmann, S.; Bracher, A. Global Retrieval of Diatom Abundance Based on Phytoplankton Pigments and Satellite Data. *Remote Sens.* **2014**, *6*, 10089–10106.

38. Hirata, T.; Hardman-Mountford, N.; Brewin, R.; Aiken, J.; Barlow, R.; Suzuki, K.; Isada, T.; Howell, E.; Hashioka, T.; Noguchi-Aita, M.; *et al.* Synoptic relationships between surface Chlorophyll-a and diagnostic pigments specific to phytoplankton functional types. *Biogeosciences* **2011**, *8*, 311–327.

39. Holm-Hansen, O.; Kahru, M.; Hewes, C.D. Deep chlorophyll a maxima (DCMs) in pelagic Antarctic waters. II. Relation to bathymetric features and dissolved iron concentrations. *Mar. Ecol. Progr. Ser.* **2005**, *297*, 71–81.

40. Fennel, K.; Boss, E. Subsurface maxima of phytoplankton and chlorophyll: Steady-state solutions from a simple model. *Limnol. Oceanogr.* **2003**, *48*, 1521–1534.

41. Parslow, J.S.; Boyd, P.W.; Rintoul, S.R.; Griffiths, F.B. A persistent subsurface chlorophyll maximum in the Interpolar Frontal Zone south of Australia: Seasonal progression and implications for phytoplankton-light-nutrient interactions.*J. Geophys. Res.: Oceans* **2001**, *106*, 31543–31557.

42. Racault, M.F.; Sathyendranath, S.; Platt, T. Impact of missing data on the estimation of ecological indicators from satellite ocean-colour time-series. *Remote Sens. Environ.* **2014**, *152*, 15–28.

43. Sallée, J.; Speer, K.; Morrow, R. Southern Ocean fronts and their variability to climate modes. *J. Clim.* **2008**, *21*, 3020–3039.

44. Orsi, A.H.; Whitworth, T.; Nowlin, W.D. On the meridional extent and fronts of the Antarctic Circumpolar Current.*Deep Sea Res. Part I: Oceanogr. Res. Pap.* **1995**, *42*, 641–673.

45. Fetterer, F.; Knowles, K.; Meier, W.; Savoie, M. Sea Ice Index. Boulder, CO: National Snow and Ice Data Center. *Digit. Media* **2002**, *6*. Available online: ftp://

sidads.colorado.edu/DATASETS/NOAA/G02135/shapefiles/ (accessed on 13 May 2016).

46. Wolter, K.; Timlin, M.S. Monitoring ENSO in COADS with a seasonally adjusted principal component index. In Proceedings of the 17th Climate Diagnostics Workshop, Norman, OK, USA, 18–23 October1993; pp. 52–57.
47. Mo, K.C. Relationships between low-frequency variability in the Southern Hemisphere and sea surface temperature anomalies. *J. Clim.* **2000**, *13*, 3599–3610.
48. Siegel, D.; Doney, S.; Yoder, J. The North Atlantic spring phytoplankton bloom and Sverdrup's critical depth hypothesis. *Science* **2002**, *296*, 730–733.
49. Henson, S.A.; Thomas, A.C. Interannual variability in timing of bloom initiation in the California Current System. *J. Geophys. Res.: Oceans* **2007**.
50. Henson, S.A.; Raitsos, D.; Dunne, J.P.; McQuatters-Gollop, A. Decadal variability in biogeochemical models: Comparison with a 50-year ocean colour dataset. *Geophys. Res. Lett.* **2009**.
51. Brody, S.R.; Lozier, M.S.; Dunne, J.P. A comparison of methods to determine phytoplankton bloom initiation. *J. Geophys. Res.: Oceans* **2013**, *118*, 2345–2357.
52. Kwok, R.; Comiso, J. Southern Ocean climate and sea ice anomalies associated with the Southern Oscillation. *J. Clim.***2002**, *15*, 487–501.
53. Arrigo, K.R.; Lowry, K.E.; van Dijken, G.L. Annual changes in sea ice and phytoplankton in polynyas of the Amundsen Sea, Antarctica. *Deep Sea Res. Part II: Top. Stud. Oceanogr.* **2012**, *71*, 5–15.
54. Sallée, J.; Speer, K.; Rintoul, S. Zonally asymmetric response of the Southern Ocean mixed-layer depth to the Southern Annular Mode. *Nat. Geosci.* **2010**, *3*, 273–279.
55. Behrenfeld, M.J.; Doney, S.C.; Lima, I.; Boss, E.S.; Siegel, D.A. Annual cycles of ecological disturbance and recovery underlying the subarctic Atlantic spring plankton bloom. *Glob. Biogeochem. Cycles* **2013**, *27*, 526–540.
56. Tagliabue, A.; Sallée, J.B.; Bowie, A.R.; Lévy, M.; Swart, S.; Boyd, P.W. Surface-water iron supplies in the Southern Ocean sustained by deep winter mixing. *Nat. Geosci.* **2014**, *7*, 314–320.
57. Sokolov, S.; Rintoul, S.R. On the relationship between fronts of the Antarctic Circumpolar Current and surface chlorophyll concentrations in the Southern Ocean. *J. Geophys. Res.: Oceans* **2007**.
58. Blain, S.; Quéguiner, B.; Armand, L.; Belviso, S.; Bombled, B.; Bopp, L.; Bowie, A.; Brunet, C.; Brussaard, C.; Carlotti, F.; *et al.* Effect of natural iron fertilization on carbon sequestration in the Southern Ocean. *Nature* **2007**, *446*, 1070–1074.
59. Planquette, H.; Statham, P.J.; Fones, G.R.; Charette, M.A.; Moore, C.M.; Salter, I.; Nedelec, F.H.; Taylor, S.L.; French, M.; Baker, A.R.; *et al.* Dissolved iron in the vicinity of the Crozet Islands, Southern Ocean. *Deep Sea Res. Part II: Top. Stud. Oceanogr.* **2007**, *54*, 1999–2019.
60. Borrione, I.; Aumont, O.; Nielsdóttir, M.; Schlitzer, R. Sedimentary and

atmospheric sources of iron around South Georgia, Southern Ocean: A modelling perspective. *Biogeosciences* **2014**, *11*, 1981–2001.

61. IOCCG. *Ocean Colour Remote Sensing in Polar Seas*; IOCCG Report Series, No. 16; International Ocean-Colour Coordinating Group: Dartmouth, NS, Canada, 2015.
62. Bélanger, S.; Ehn, J.K.; Babin, M. Impact of sea ice on the retrieval of water-leaving reflectance, chlorophyll a concentration and inherent optical properties from satellite ocean color data. *Remote Sens. Environ.* **2007**, *111*, 51–68.
63. Wang, M.; Son, S.; Shi, W. Evaluation of MODIS SWIR and NIR-SWIR atmospheric correction algorithms using SeaBASS data. *Remote Sens. Environ.* **2009**, *113*, 635–644.
64. Maheshwari, M.; Singh, R.K.; Oza, S.R.; Kumar, R. An investigation of the southern ocean surface temperature variability using long-term optimum interpolation SST data. *ISRN Oceanogr.* **2013**, *2013*, 392632.
65. Maksym, T.; Stammerjohn, S.E.; Ackley, S.; Massom, R. Antarctic sea ice—A polar opposite? *Oceanography* **2012**, *25*, 140–151.
66. Henson, S.A.; Sarmiento, J.L.; Dunne, J.P.; Bopp, L.; Lima, I.D.; Doney, S.C.; John, J.; Beaulieu, C. Detection of anthropogenic climate change in satellite records of ocean chlorophyll and productivity. *Biogeosciences* **2010**, *7*, 621–640.
67. Siegel, D.A.; Buesseler, K.O.; Doney, S.C.; Sailley, S.F.; Behrenfeld, M.J.; Boyd, P.W. Global assessment of ocean carbon export by combining satellite observations and food-web models. *Glob. Biogeochem. Cycles* **2014**, *28*, 181–196.
68. Rousseaux, C.S.; Gregg, W.W. Recent decadal trends in global phytoplankton composition. *Glob. Biogeochem. Cycles***2015**, *29*, 1674–1688.
69. Lefebvre, W.; Goosse, H.; Timmermann, R.; Fichefet, T. Influence of the Southern Annular Mode on the sea ice–ocean system. *J. Geophys. Res.: Oceans* **2004**.
70. Hauck, J.; Völker, C.; Wang, T.; Hoppema, M.; Losch, M.; Wolf-Gladrow, D.A. Seasonally different carbon flux changes in the Southern Ocean in response to the southern annular mode. *Glob. Biogeochem. Cycles* **2013**, *27*, 1236–1245.
71. L'Heureux, M.L.; Thompson, D.W. Observed relationships between the El Niño-Southern Oscillation and the extratropical zonal-mean circulation. *J. Clim.* **2006**, *19*, 276–287.
72. Hopkins, J.; Henson, S.A.; Painter, S.C.; Tyrrell, T.; Poulton, A.J. Phenological characteristics of global coccolithophore blooms. *Glob. Biogeochem. Cycles* **2015**, *29*, 239–253.
73. Smith, W.O., Jr.; Sedwick, P.N.; Arrigo, K.R.; Ainley, D.G.; Orsi, A.H. The Ross Sea in a sea of change. *Oceanography***2012**, *25*, 90–103.

14

CHAPTER

ANALYSIS OF STORM STRUCTURE USING THE TRMM PRECIPITATION RADAR DATA

Ayodeji Richard Balogun, Zechariah Debo Adeyewa

Keywords: Storm Height, Storm Structure, Storm Counts, Histogram

INTRODUCTION

It's documented [1] that a major modeling-based effort is taken to evaluate a sensitivity of their rainfall, and heating results (among other aspects) in partitioning methods that classify precipitation echoes into various types (i.e. stratiform and convective). The convective-stratiform distinction is important because of differences in Z-R (The relationships between reflectivity and rainfall rate are referred to as Z-R Relations) relationships, e.g. [2], which affect not only the radar-based rainfall estimation, but also the latent heat release profile, and hence the energy balances of the tropical atmosphere [3]. Differences in precipitation types may lead to distinctly different vertical profiles of latent heat released to the atmosphere [4- 7], which has implications for the large-scale circulation and climate [8].

There appear to be two primary modes of how the atmosphere is affected by moist processes: Young vigorous convective cells induce an atmospheric response that is two-layered, with air converging into the active convection at low levels and diverging aloft, and the other, weak intermediary and stratiform precipitation areas induce a three-

Citation: A. Balogun and Z. Adeyewa, Analysis of Storm Structure using the Trmm Precipitation Radar Data, Vol. 3 No. 4, 2013, pp. 538-551. doi: 10.4236/acs.2013.34057.

layered response, in which environmental air converges into the weak precipitation area at mid levels and diverges out from it at lower and upper levels [9]. The convective mode thus displays a heating release throughout the depth of the troposphere (associated with precipitation particles growing in the updraft) whereas the stratiform mode exhibits warming (from growth supported by rising air motion) in the upper half of the troposphere and a cooling (caused by melting and evaporation of particles in subsiding air) at lower levels [4- 6].

Storms are classified based on the precipitation types; thus, we have convective storms and stratiform storms. Yet either type of storm structures, which are ambiguous in nature, may neither be classified as convective or stratiform existance [10]. These types are generally termed "unconditioned or unclassified". When air is sufficiently moist and unstable, convective clouds can grow to great heights which develop vigorous updrafts and produce rain, lighting, hail and sometimes flooding. R. A. Houze [11] has noted the importance of high vertical velocities in the convective clouds, which could cause serious flooding, wind and hail damage.

Spaceborne rainfall estimation, which is usually guided by raingauge data and also NWP (Numerical Weather Prediction) output, has evolved dramatically over the last two decades [12]. This is particularly important in Africa where large parts have a poor raingauge record. Spaceborne techniques have been developed based on IR brightness temperatures (e.g. [13]), and 14 GHz radar reflectivities (e.g. [14]). Often a combination of these is used to benefit from specific strength, e.g. IR-based techniques using geostationary satellite data can be used continuously, day and night. Obviously IR-based techniques are much inferior to radar-based techniques, in principle at least, because the anvil of large convective systems is much larger than the radar echoes underneath, and its topography is much more uniform. But large differences exist even between passive microwave and radar-based rainfall estimates, both on instantaneous and cumulative bases [15].

Radar has been used as an essential tool for studies of convective storms for over a generation. Pictures of individual vertical slices through convective storm have been a classic method of gaining insight into storm structure. An individual vertical slice, however, does not necessarily show that the maximum reflectivity at each altitude unless storm is aligned perfectly along the plane of the vertical scan. We define this desired Vertical Profile of Radar Reflectivity (VPRR) as the maximum reflectivity of a cell and a function of height. Because individual cells tend to tilt, the VPRR requires some attention to derive quantitatively in details [16,17]. Such a study is performed, and a classic paper is derived from the VPRR for convective cells in New England. Comparing the VPRR for thunderstorms with rain only, or with rain and hail, or with tornadoes, the storms with hail and tornadoes had greater radar reflectivity at all altitudes, but had the greatest differences in the mid troposphere. One of the principal sensors onboard the Tropical Rainfall Measuring Mission (TRMM) satellite is the Precipitation Radar (PR). Being the first space-borne radar designed to capture more comprehensive structure of rainfall than any space-borne sensor before it, the PR has produced three-dimensional rainfall data from space unprecedented by any previous scientific spacecraft [18,19]. An example of the superior observation capability of PR is its ability (with suitable methodology) to discriminate between stratiform and convective precipitation. Such details are not easily detectable by passive microwave radiometer and visible infrared sensors [20].

The purpose of this work is to analyze storm heights statistics from the TRMM PR, focusing on storms over Africa and the Atlantic Ocean. Histograms of storm heights are used to map regions where storms are dominant and to determine the modal height of the distribution.

DATA AND METHODS

The TRMM Science Data Information System (TSDIS) derived the data utilized in this study from TRMM-PR observation. The PR algorithm version 5, provided by the TRMM Science Team, was used by TSDIS to derive storm height and rainfall characteristics for 5 years. The level 3 data, 3A25, which is the monthly and seasonal observations of the PR, have been used for this study. The 3A25 (monthly product) has two resolutions: a low horizontal resolution (5° × 5° latitude/longitude from 40°N to 40°S) and high horizontal resolution (0.5° × 0.5° latitude/longitude from 37°N to 37°S) [18]. Level 3 products also include histogram data, composite within the 5° × 5° cells for all storms and for convective and stratiform storms separately. Counts of storm heights are recorded in 30 intervals having vertical resolutions of 0.5 km from the surface to 13 km, increasing to 1 km from 13 km to 16 km, ending with a final 4 km bin covering the interval from 16 to 20 km. All storm height values represent height of storm top above the standard geoide an imaginary surface of the earth that coincides with mean sea level over oceans and extend through the continent. The ten regions used in this study, which covers the entire Africa and some part of the Atlantic Ocean are summarized in **Table 1**.

For the sake of brevity, the analyses were conducted on quarterly basis: December-January-February (DJF), Marc-April-May (MAM), June-July-August (JJA) and September-October-November (SON). The region 40°N to 40°S and 20°E to 60°W was structured into a 16 × 16 grid points and the latitude and longitude axes respectively [18,19].

Most TRMM data are written in Hierarchical data format (HDF), containing array of data called scientific data sets. Codes were written to extract and analyze the data set over the region used for this study. The software packages used include Noesys for extraction of data set, transform and excel for plotting. Noesys was used for the computation of seasonal averages of the storm height mean and histograms.

RESULTS AND DISCUSSIONS

Seasonal Distribution of Mean Storm Height

Distribution of Mean Storm Height over Africa

The averages of the total (stratiform, convective and unconditioned storm types) are shown in **Figure 1**, for each of the seasons. It shows the true features of all the storms combined. The effects or presence of other types of storms, apart from the stratiform and convective, is visualized here. In **Figure 1**(a) (DJF), stratiform storm tops are observed in East of West Africa, that is, the Northcentral region in Africa while convective storms are far in the South, the unconditioned is in the same locations as the stratiform, indicating dominance of the stratiform and unclassified types. Because both the stratiform and convective have dominance in West Africa, thus, the average storm tops are concentrated around West Africa, in **Figure 1**(b) (MAM). In **Figure 1**(c) (JJA), convective storm tops are observed in the West through central Africa to the East whereas the stratiform is overshadowed already and the unconditioned is both in the South and North. This indicates the spread of the storms across Africa-South, West and East. The pattern observed in Figure 1(a) (DJF) above also applies to **Figure 1**(d) (SON), because the strong convective activities are dying down giving way for the stratiform, thus, making stratiform and unclassified storm tops to dominate. Slight influences of convective storm tops are observed in the savanna south region.

Table 1. Grid points and locations of selected climatic regions in Africa

REGIONS	Geographical Co-ordinates	Latitudinal Reference Grids	Longitudinal Reference Grids
Desert (North)	20°N - 30°N 10°W - 30°E	3.4	3.10
Semi Desert (North)	15°N - 20°N 10°W - 35°E	5.5	3.10
Brush Grass Savanna/Steppe (north)	10°N - 15°N 10°W - 35°E	6.6	3.11
Deciduous Forest (North)	5°N - 10°N 10°W - 30°E	7.7	3.10
Tropical Rainforest	5°N - 5°S 10°E - 30°E	8.9	7.10
Deciduous Forest (South)	10°S - 15°S 25°E - 35°E	11.11	10.11
Brush Grass Savanna/ Steppe (south)	20°S - 25°S 20°E - 30°E	13.13	9.10
Temperate Grass Land/ Montane Forest	25°S - 30°S 25°E - 30°E	14.14	10.10
Steppe (East)	0° - 10°N 40°E - 45°E	7.8	13.13
Atlantic Ocean	0° - 5°S 10°W - 15°E	9.9	5.6

Regional Distribution of the Mean Storm Heights

Figures 2 compares the stratiform, convective and unclassified storm types on regional basis. In the desert (Sahara), semi-desert (Sahel) and brush grass savanna (North), in **Figure 2**(a) (DJF), unclassified storms have tops greatest while the stratiform is greater than the convective storm tops. The unclassified decreases slightly in deciduous forest (North) and tropical rainforest, giving way for the convective which overshadowed both of them, having tops close to 8 km.

The unclassified gained vertical momentum again from deciduous forest south through brush grass savanna (South), temperate grassland/Montane forest to steppe East with mean heights of storm tops around 7.3 km. The stratiform also follow suite as in the first three regions in the north above. In the Atlantic, Convective storms have greater height of storm tops followed by the unclassified and stratiform. A very similar pattern is observed in **Figure 2**(c) (JJA), as in DJF, except that in the desert, semidesert and brush grass savanna (North), the convective is second to the unclassified unlike in DJF where stratiform is second to unclassified, also in deciduous forest (South) and brush grass savanna (South), the convective storm top heights are greater but are decreasing unlike in DJF where the unclassified is Greater.

Seasonal Distribution of Storm Height Histograms

Seasonal Distribution of Storm Height Histograms over Africa

Here, emphasis will be laid only on histograms at 250 mb (TEJ level) and 700 mb (AEJ level) levels because of the importance of the Tropical Easterly Jet and Africa Easterly Jet in the continent. Simple storm counts in **Figure 3**(a) shows that there are high concentrations of stratiform storms extending from central Africa down to the East of Southern Africa, including Madagascar. Although, the AEJ, which has it southernmost limit at about 3°N, is not observed, the TEJ has its southernmost limit at about 8°S in February. The influence of the TEJ at this latitude could be the cause of the very high concentration of stratiform storms in the Congo (DRC). Another possible cause of the high storms especially in the Madagascar could be the winter Monsoon in the Indian Ocean. Although, the cool, dry air from the land in Asia is poor in moisture content, but on reaching the Madagascar region, it would have been moistened because of the large ocean mass and distance covered before reaching there, thus inducing storms and hence precipitation in the region. In spring (MAM), as shown in **Figure 3**(b), storm activities have already shifted to W/Africa with a strong core around 3°N, this corresponds to the southernmost position of the AEJ. The ITD is already ahead and the Southeasterly have started gaining strength over the landmasses. During the summer, as shown in **Figure 3**(c), the storms at this level have already migrated up to the savanna and Sahel region having the TEJ core at around 12°N, which correspond to the core of the storm. By autumn, in Figure 3(d), storms concentrations are in the central Africa region.

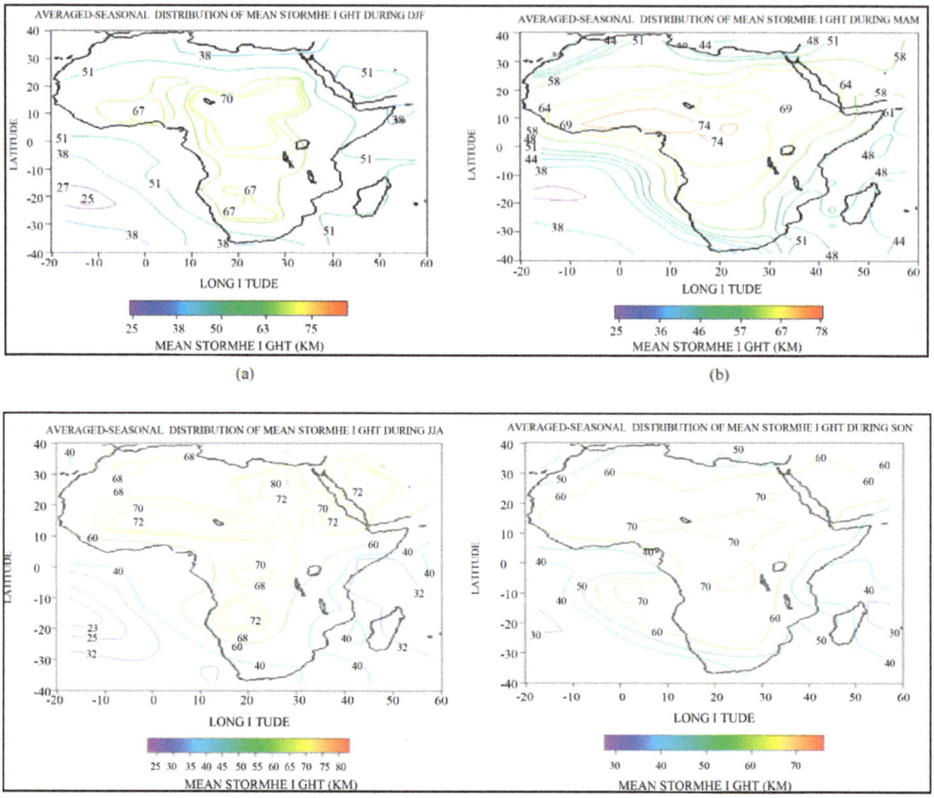

(a) (b) (c) (d)

Figure 1. Composite averaged storm height distribution during: (a) DJF; (b) MAM; (c) JJA; and (d) SON. Units are hundreds of meters with contours between 20 and 80 × 10^2 meters.

As shown in **Figure 4**, all through the year, at 700 mb, stratiform storms are majorly concentrated around the oceans and the Mediterranean sea and some parts of the land near the sea where the storm core (high storm counts) are located, in **Figure 4**(a) (DJF). In **Figure 4**(b) (MAM), we have high concentrations of storm counts in the South Atlantic Ocean extending through South Africa to Madagascar in the Indian Ocean. Similar occurrence is observed in Figures 4(c) (JJA) and 4(d) (SON) but in SON the storm histogram is very low almost disappearing in the Indian Ocean.

Figure 5 shows the seasonal distribution of convective storm height histogram at 250 mb. As compared to stratiform storm height histograms, in **Figure 5**(a) (DJF), the convective storms covers the same region but with a larger land masses than the stratiform storm, and having a core in Southern Africa; this time the storms are located on the land due to convection over the land surface but shallow storms are observed over the Ocean as seen in Madagascar for the stratiform storm above. During MAM, as shown in **Figure 5**(b), the same pattern is observed, with storm concentration migrating to West Africa where the core (region of high storm counts) extend over a large area from central to some parts of West Africa. Some small storm counts are concentrated in southern Africa region where the core was located in DJF, the previous season. The African Easterly waves are prominent during June to early October. In summer (JJA), as shown in **Figure 5**(c), the storm had spread from West Africa through central Africa to the horn of Africa having a very small core in the horn of Africa. The AEJ's instability creates ridges and troughs that form the wave features, which underlay the upper tropospheric TEJ. As the wave propagates westward above the convergence zone of the ITD, it appears to modulate the ITD features on the ground thus providing additional moisture along the west coast of Africa that wouldn't otherwise exist with the monsoon alone. Summer heating causes the development of a strong latitudinal pressure gradient and the development of an easterly jet stream (TEJ), which migrate the storms to the west coast of Africa, at an altitude of about 15 km and a latitude of 25°N. By autumn, in **Figure 5**(d), the easterly jet stream is replaced with strong westerly winds of the mid-latitude which tends to shift storm activities towards central and eastern Africa, in the upper atmosphere. Thus, movements begins in the South to West Africa, from DJF to MAM, it spreads from West Africa through central Africa to East Africa, from MAM to JJA, and by SON, there is a downward migration of storm core to the South again. Storms from deep convective systems generate latent heat.

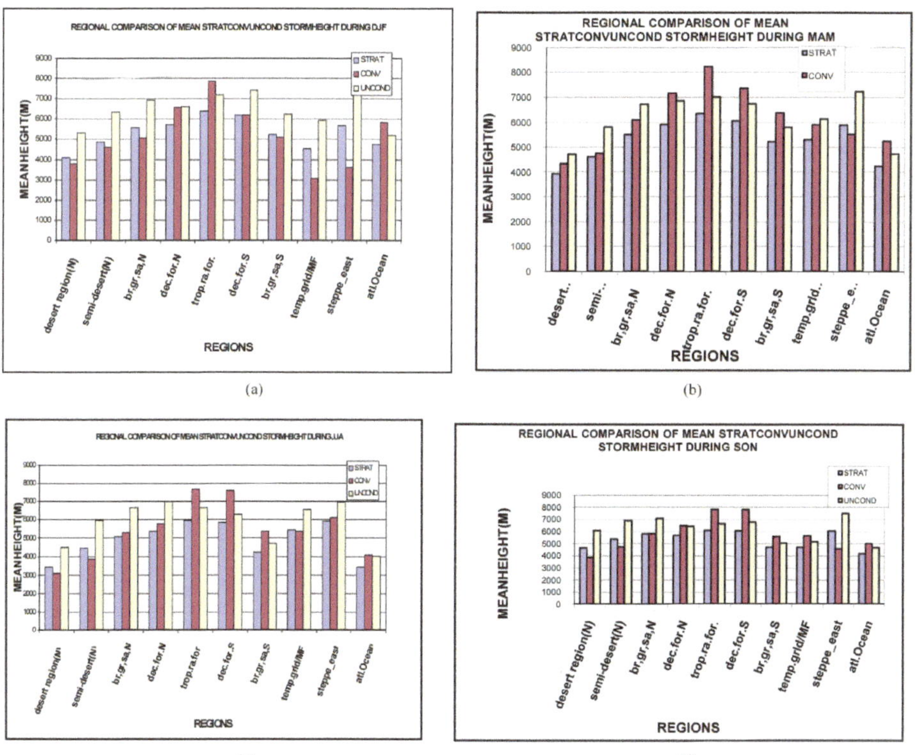

Figure 2. (a) and (b) Regional comparison of the mean seasonal stratiform, convective, and unconditioned storm height, during: (a) DJF, and (b) MAM; (c)

and (d) Regional comparison of the mean seasonal stratiform, convective, and unconditioned storm height, during: (c) JJA, and (d) SON.

As for the stratiform, at 700 mb, convective storm histograms are concentrated over the ocean rather than in the land. In **Figure 6**(a) (DJF), core is located close to the Mediterranean Sea but around 33°N inland, which is the same location as the stratiform core in DJF. In **Figure 6**(b) (MAM), the pattern is exactly the same as in stratiform, but the convective cores are stronger at South Atlantic Ocean, Lesotho and Madagascar regions. During JJA, as shown in **Figure 6**(c), storms are still concentrated in the South but are weakening with only one core at Lesotho region, by SON, in **Figure 6**(d), the core is still there but the Atlantic storm counts are decreasing already in the period.

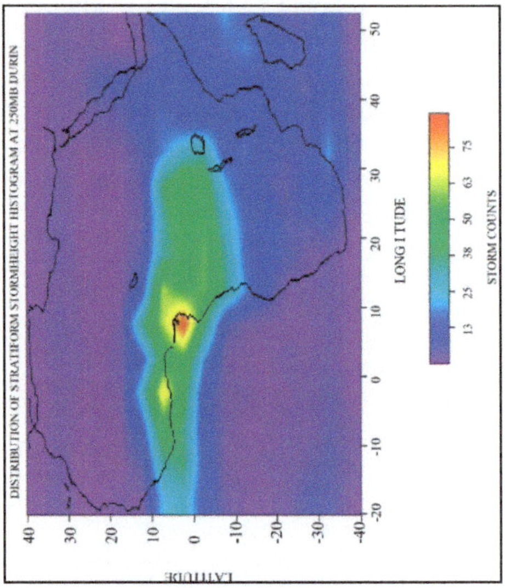

(b)

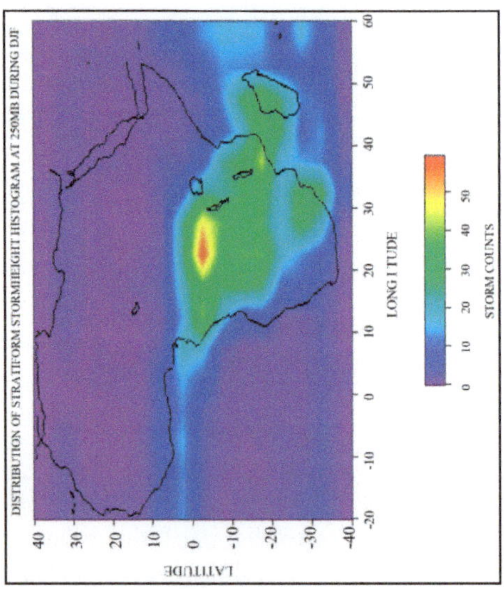

(a)

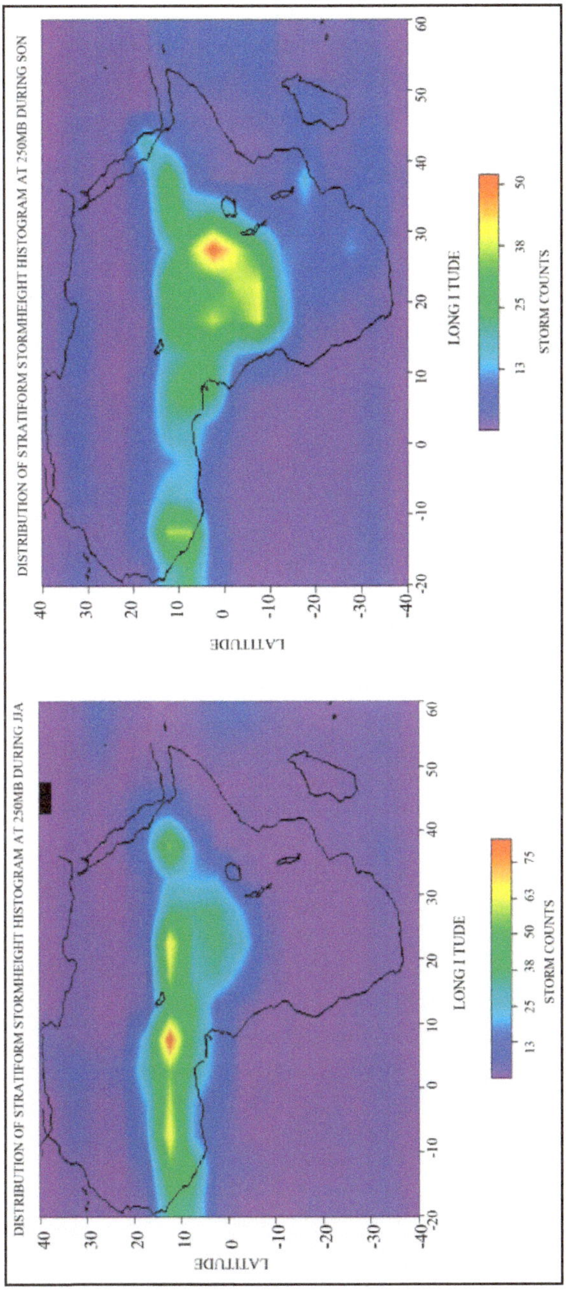

Figure 3. (a) and (b) Stratiform storm height histograms at 250 mb (11 km) during: (a) DJF, (b) MAM; (c) and (d) Stratiform storm height histograms at 250 mb (11 km) during: (c) JJA, and (d) SON.

At 250 mb, during DJF, the stratiform core is located in the Congo region while the convective core is located down South in the Kalahari region in South Africa. These two storms core are visualized when all types of storms are considered in Figures 7(a). But the region where stratiform core is located covered more land areas than when only the stratiform storm is considered, thus the presence of other storms is visualized at the location. In **Figure** 7(b) (MAM), the storms location had already moved to West Africa having two cores, one (the larger) extending from the coast into the savanna North through Nigeria/Cameroon and the other around Togo/Ghana. As expected in**Figure** 7(c) (JJA), the storm spreads from West Africa through central to the horn of Africa with four cores in the same latitude spreads across the three regions. In **Figure** 7(d) (SON), the storms have started migrating to the south with only one core in central Africa.

Just as for stratiform and convective, at 700 mb, the same pattern is observed when all types are considered, in Figures 8, with cores located close to the Mediterranean in **Figure 8**(a) (DJF), extending from the Atlantic through Lesotho (in South Africa) to the Indian Ocean in **Figure 8**(b) (MAM), and having a horizontal extension from the Atlantic through the Lesotho to the Indian Ocean in **Figure 8**(c) (JJA), and a similar pattern as well in **Figure 8**(d) (SON) with lower storm counts in the Indian ocean.

Regional Distribution of the Storm Height Histogram

Figure 9 shows the comparison of the total, convective and stratiform storm counts over the desert region. The four seasons shows a bimodal distribution especially with the total and stratiform classification. The lower modes are consistently below 2 km while the upper modes are slightly above 4 km. The amplitude of the upper mode increases from 150 to about 400 from DJF, in **Figure 9**(a), to MAM, in **Figure 9**(b), respectively and decreases from 220 to 170 from JJA, in **Figure 9**(c), to SON, in **Figure** 9(d). Analysis of mean storm height on regional basis above reveals that, in the desert, stratiform storms have tops greater than that of the convective in DJF, JJA, and SON. Convective activities are virtually very low, thus, giving way for the stratiform storm counts to dominate. The amplitude of the lower mode increases initially from DJF to MAM but decreases from JJA to SON.

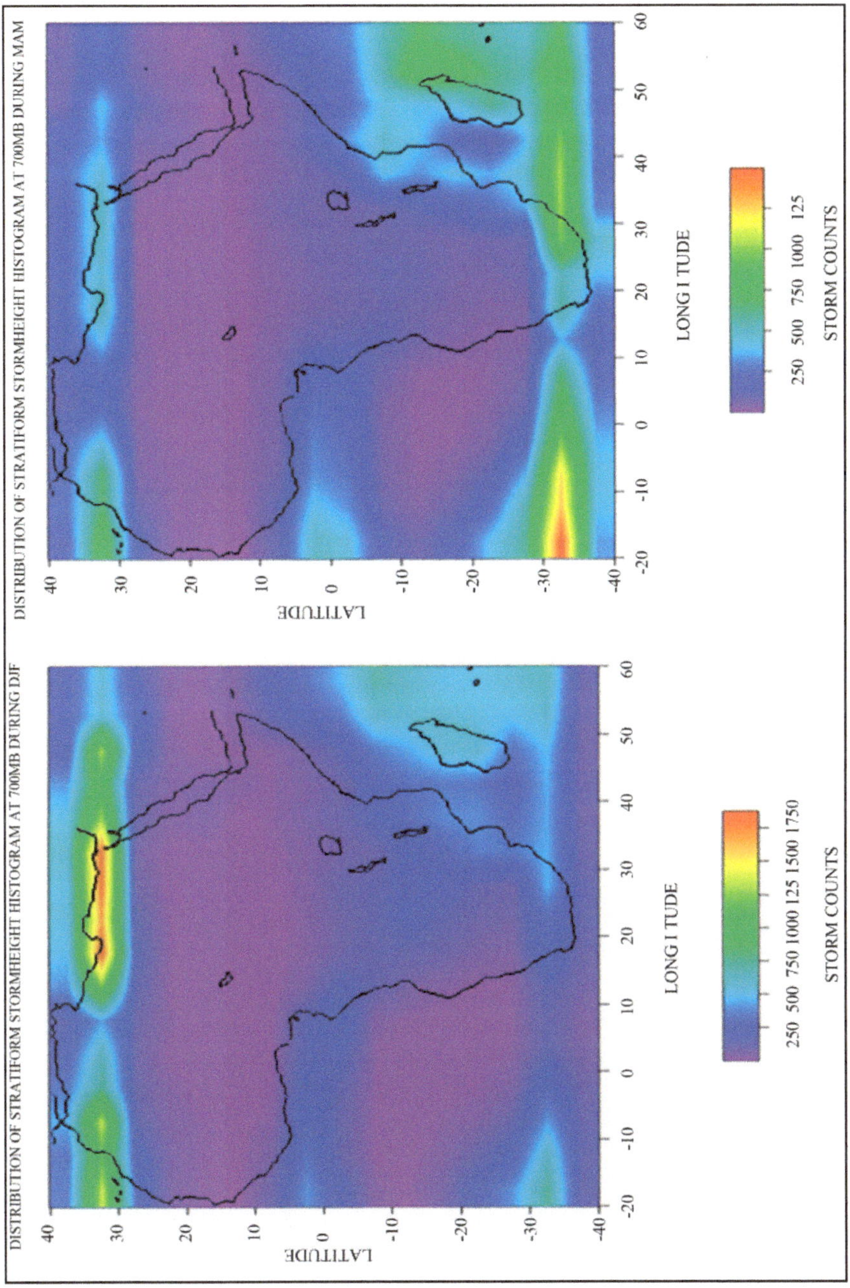
DISTRIBUTION OF STRATIFORM STORMHEIGHT HISTOGRAM AT 700MB DURING DJF
LATITUDE
LONG I TUDE
250 500 750 1000 125 1500 1750
STORM COUNTS
(a)
DISTRIBUTION OF STRATIFORM STORMHEIGHT HISTOGRAM AT 700MB DURING MAM
LATITUDE
LONG I TUDE
250 500 750 1000 125
STORM COUNTS
(b)

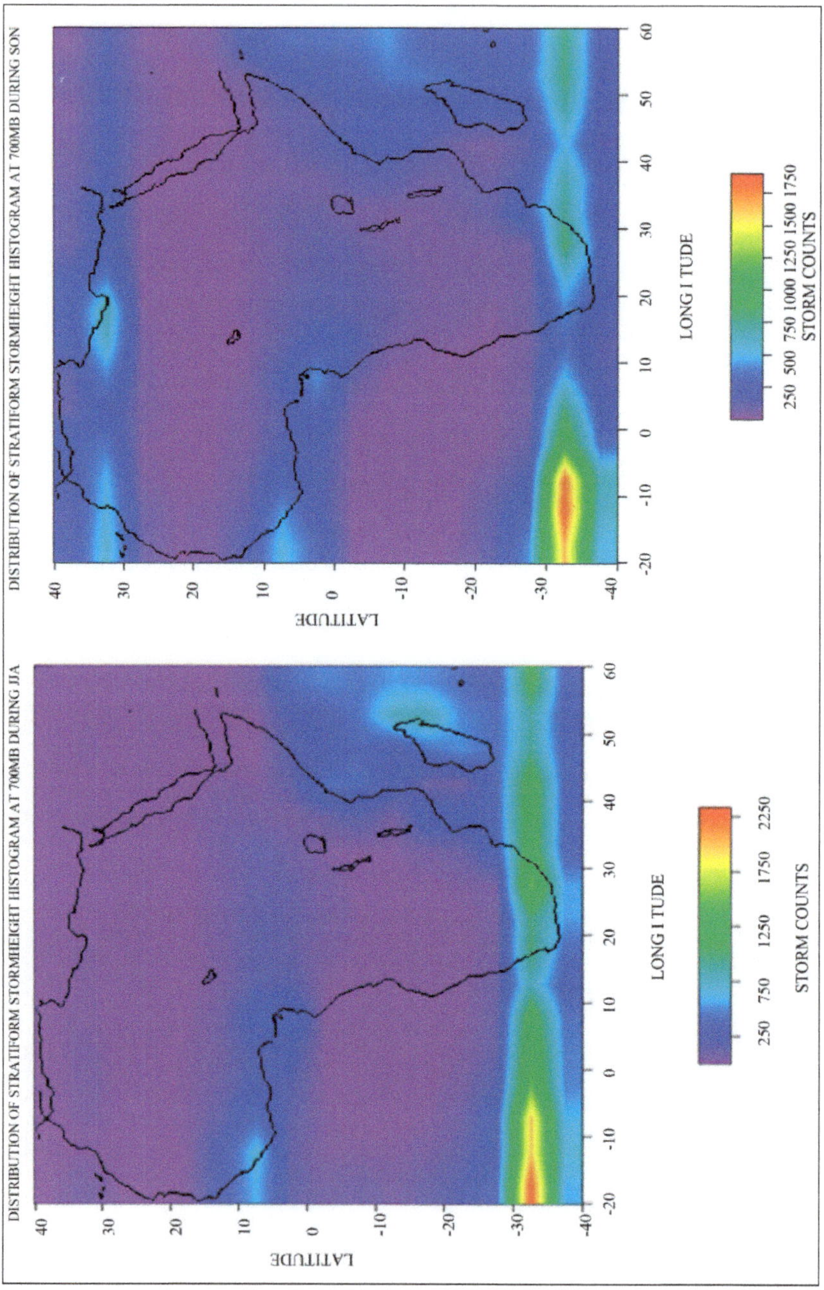

Figure 4. (a) and (b) Stratiform storm height histograms at 700 mb (3 km) during: (a) DJF; (b) MAM; (c) and (d) Stratiform storm height histograms at 700 mb (3 km) during: (c) JJA, and (d) SON.

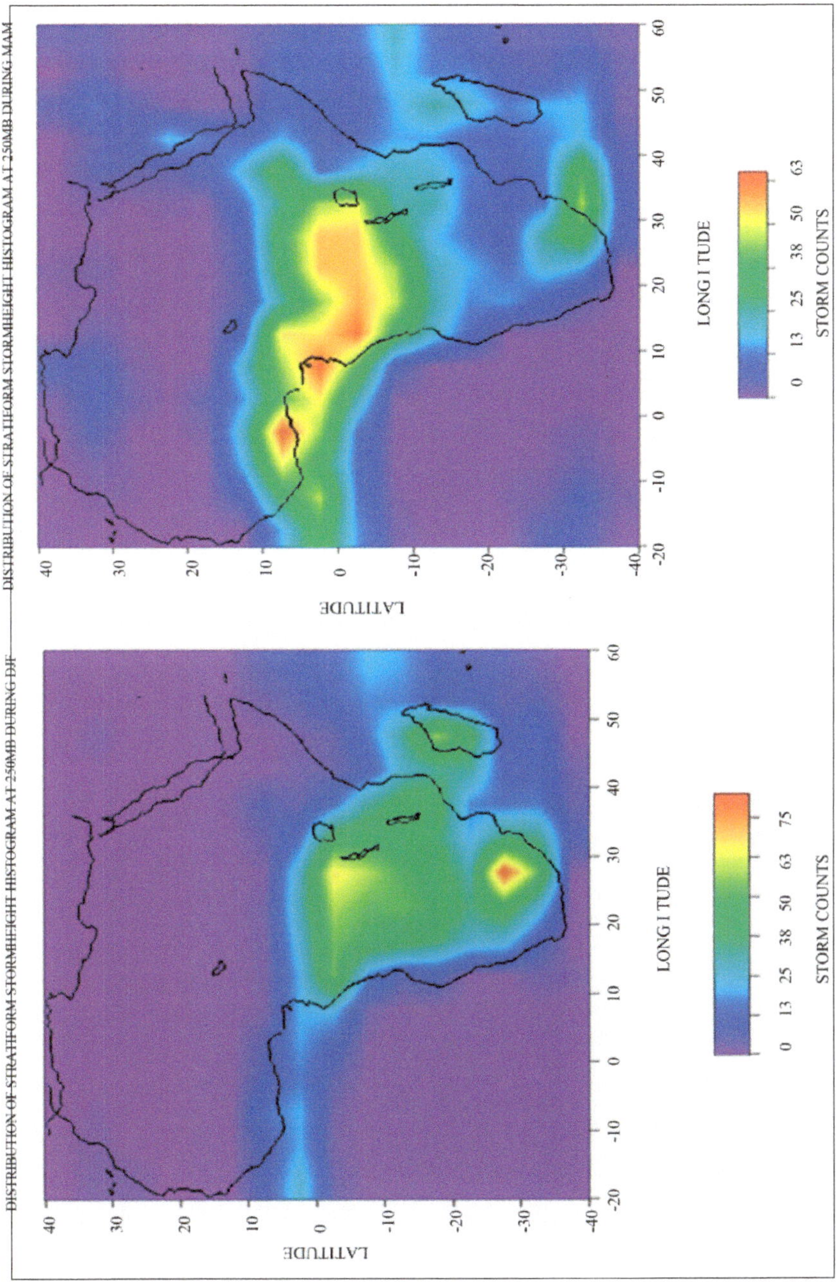
DISTRIBUTION OF STRATIFORM STORMHEIGHT HISTOGRAM AT 250MB DURING DJF
LATITUDE
LONG I TUDE
0 13 25 38 50 63 75
STORM COUNTS
(a)
DISTRIBUTION OF STRATIFORM STORMHEIGHT HISTOGRAM AT 250MB DURING MAM
LATITUDE
LONG I TUDE
0 13 25 38 50 63
STORM COUNTS
(b)

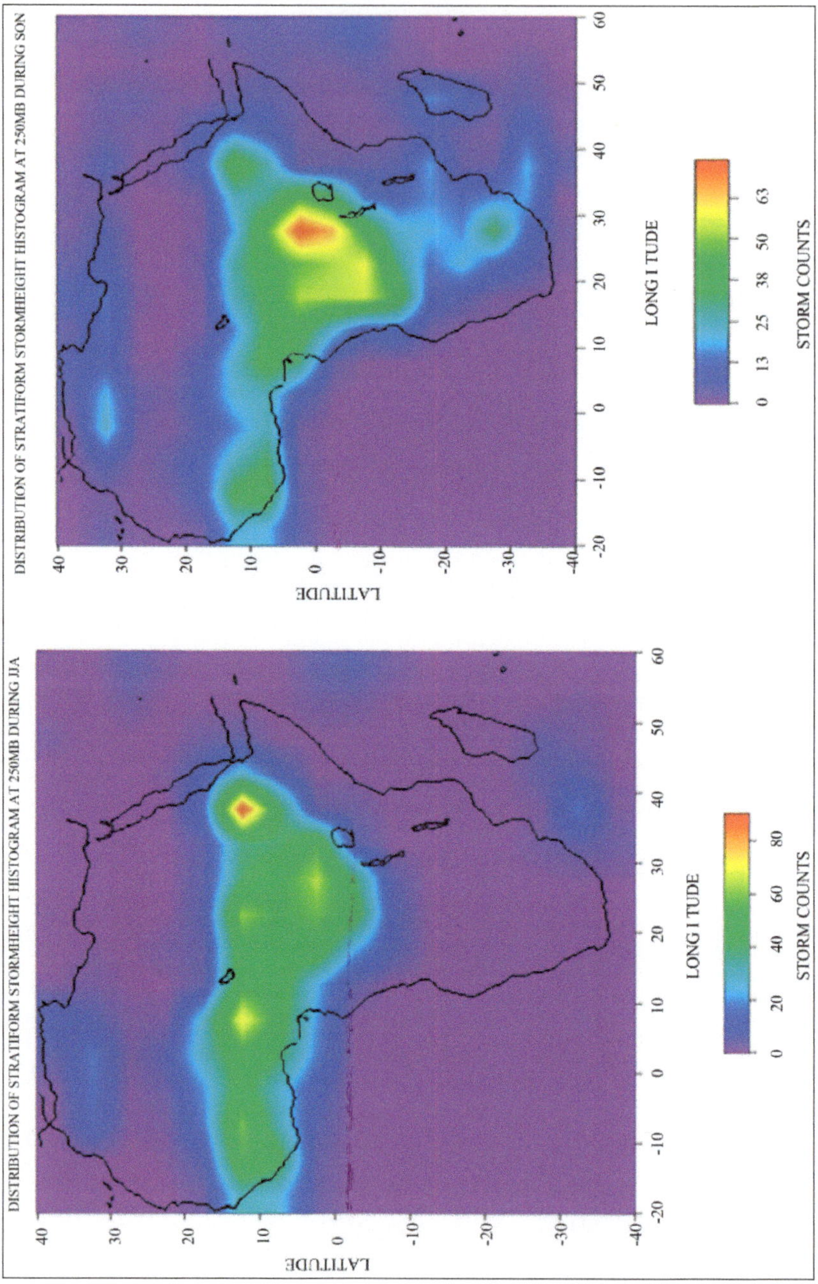

Figure 5. (a) and (b) Convective storm height histograms at 250 mb (11 km) during: (a) DJF; (b) MAM; (c) and (d) Convective storm height histograms at 250 mb (11 km) during: (c) JJA, and (d) SON.

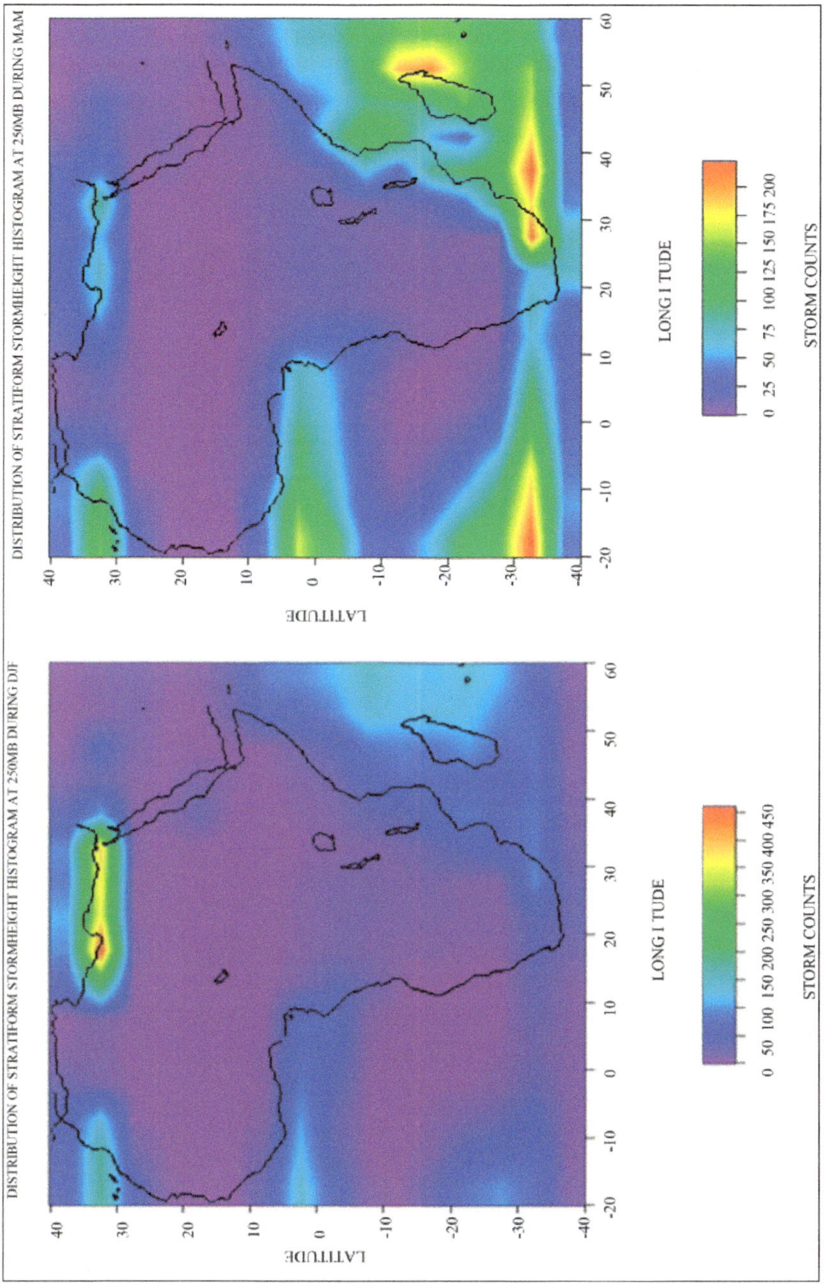
DISTRIBUTION OF STRATIFORM STORMHEIGHT HISTOGRAM AT 250MB DURING DJF
LATITUDE
LONG I TUDE
0 50 100 150 200 250 300 350 400 450
STORM COUNTS
(a)
DISTRIBUTION OF STRATIFORM STORMHEIGHT HISTOGRAM AT 250MB DURING MAM
LATITUDE
LONG I TUDE
0 25 50 75 100 125 150 175 200
STORM COUNTS
(b)

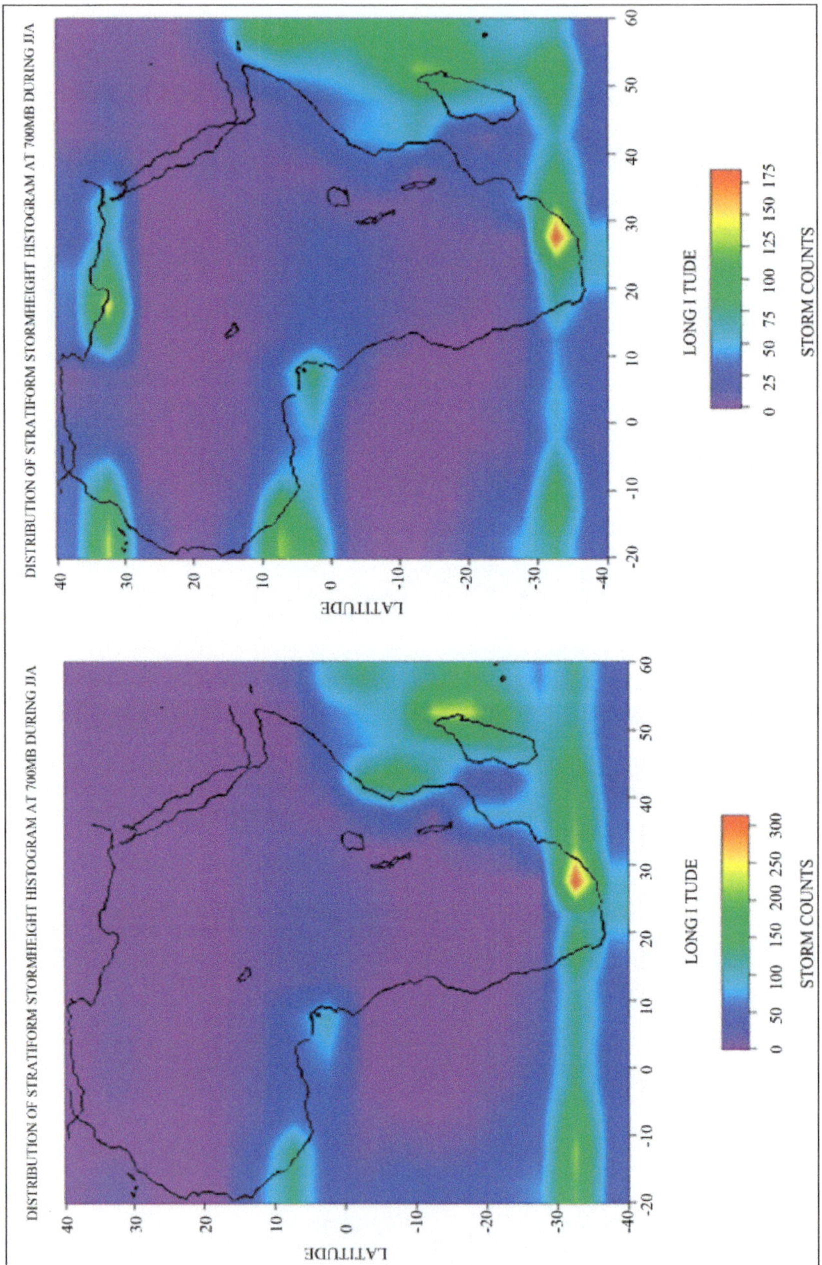

Figure 6. (a)-(d) Convective storm height histograms at 700 mb (3 km) during: (a) DJF; (b) MAM; (c) JJA; and (d) SON.

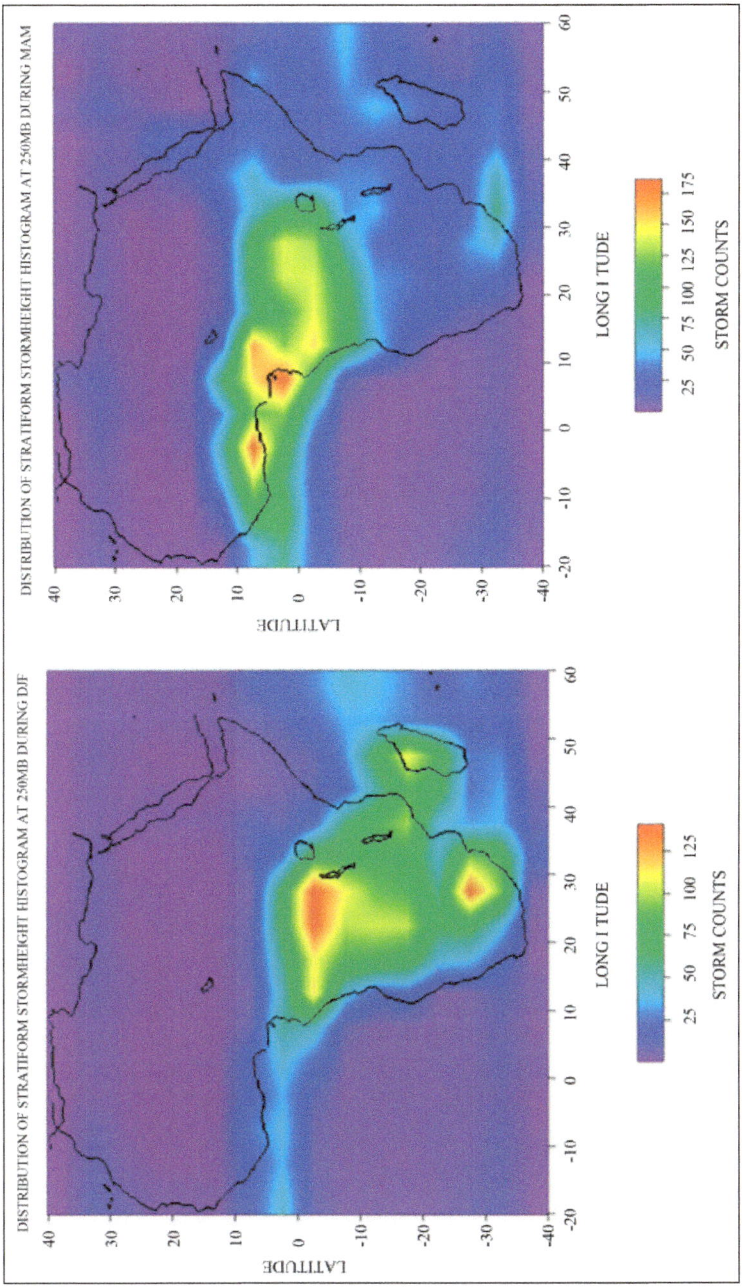
DISTRIBUTION OF STRATIFORM STORMHEIGHT HISTOGRAM AT 250MB DURING DJF
LATITUDE
LONG I TUDE
STORM COUNTS
25 50 75 100 125
(a)
DISTRIBUTION OF STRATIFORM STORMHEIGHT HISTOGRAM AT 250MB DURING MAM
LATITUDE
LONG I TUDE
STORM COUNTS
25 50 75 100 125 150 175
(b)

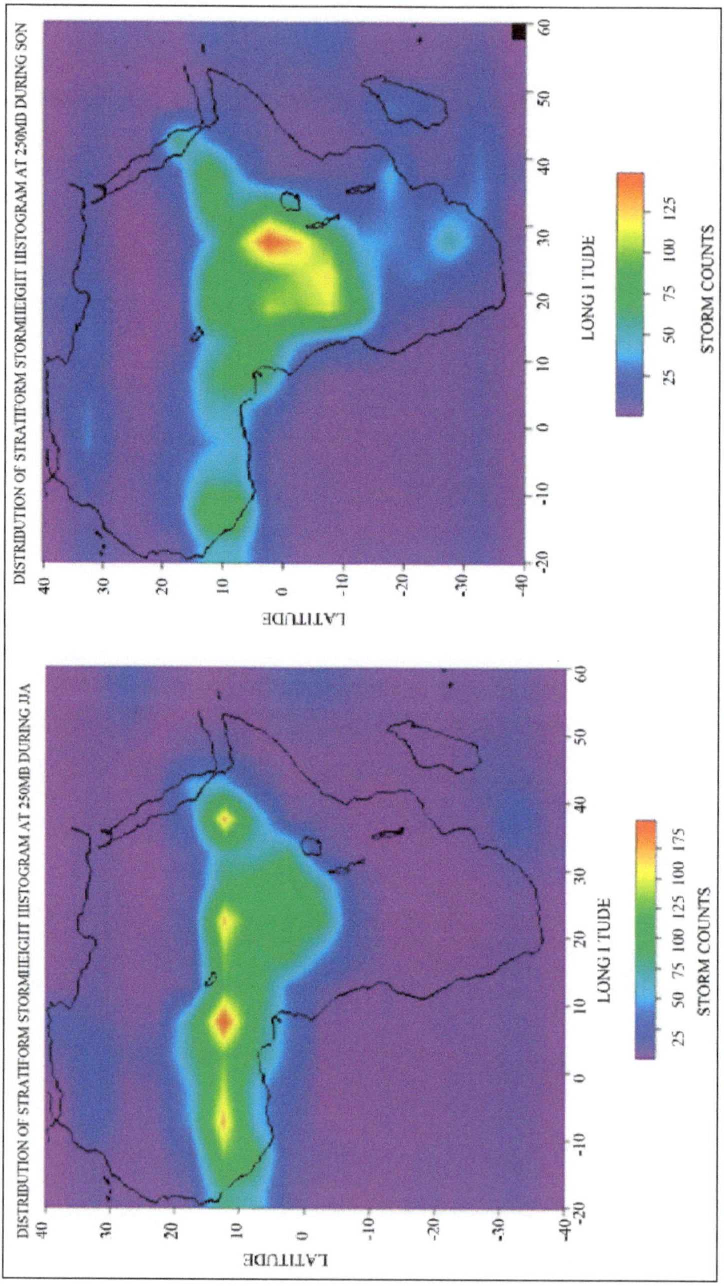

Figure 7. (a) (b) All (convective + stratiform + others) storm height histograms at 250 mb (11 km) during DJF, MAM; (c) (d) All (convective + stratiform + others) storm height histograms at 250 mb (11 km) during JJA, SON.

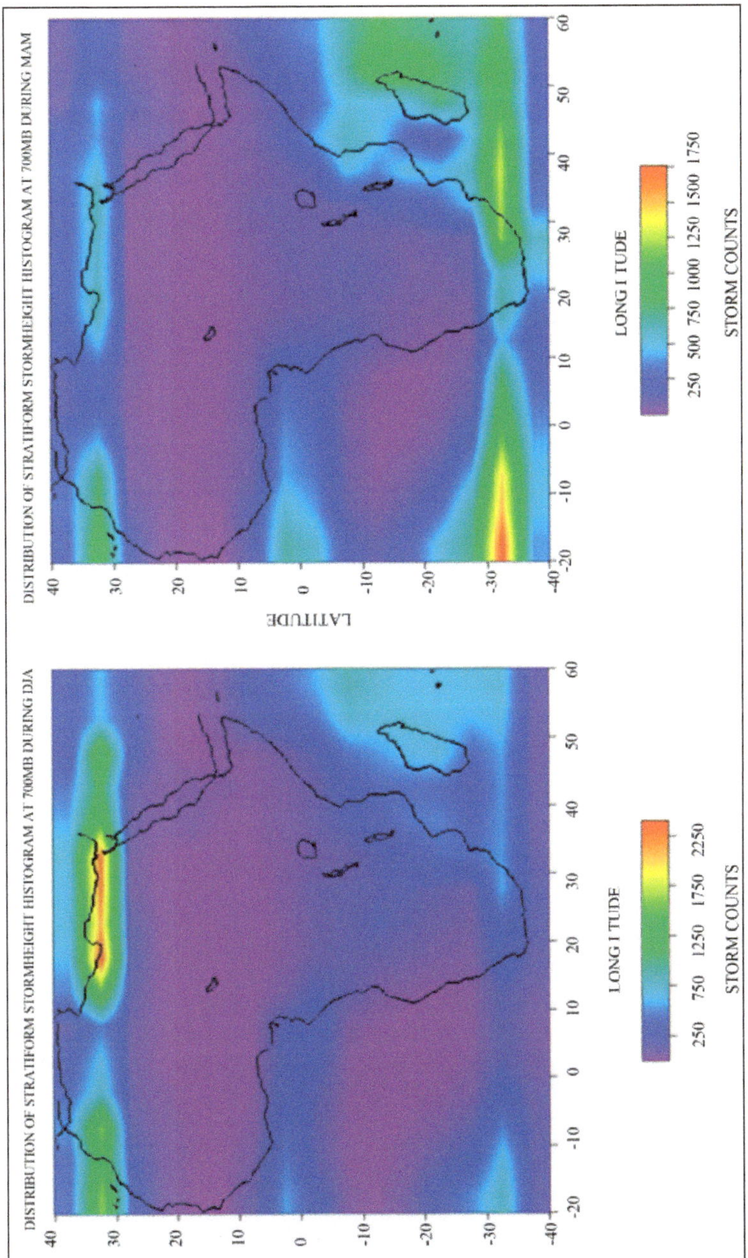
DISTRIBUTION OF STRATIFORM STORMHEIGHT HISTOGRAM AT 700MB DURING DJA
LATITUDE
LONG I TUDE
250 750 1250 1750 2250
STORM COUNTS
(a)
DISTRIBUTION OF STRATIFORM STORMHEIGHT HISTOGRAM AT 700MB DURING MAM
LATITUDE
LONG I TUDE
250 500 750 1000 1250 1500 1750
STORM COUNTS
(b)

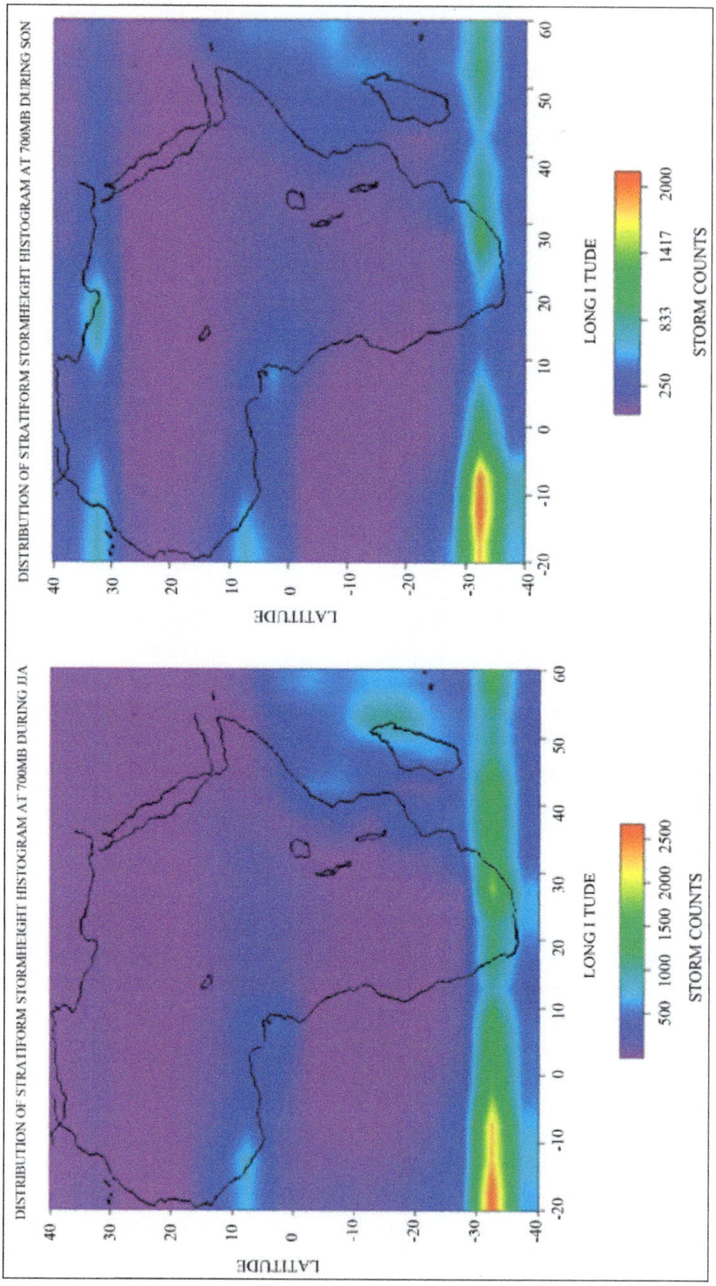

Figure 8. (a)-(d) All (convective + stratiform + others) storm height histograms at 700 mb (3 km) during DJF, MAM, JJA, SON.

Figure 10 shows the Atlantic Ocean having a bi-modal distribution all through the four seasons. The lower mode, in **Figure 10**(a) (DJF), is slightly below 2 km with peak value at around 500 while the upper mode is near 5 km with amplitude of over 1200. In **Figure 10**(b) (MAM), the lower mode of the total and stratiform is located at 2 km with amplitude of over 600 for the total and about 600 for the stratiform storms. The upper mode is located at 4 km for the total and stratiform storms while the peaks are at 700 for the total storms and below 600 for the stratiform storms. In **Figure 10**(c) (JJA), a uni-modal distribution is observed at 2 km with amplitude of over 500 with the total greater than stratiform storms. The SON season is bi-modal where the lower mode, located at 2 km, has a peak value of about 400 while the upper mode, located at 4 km, has a peak value over 500.

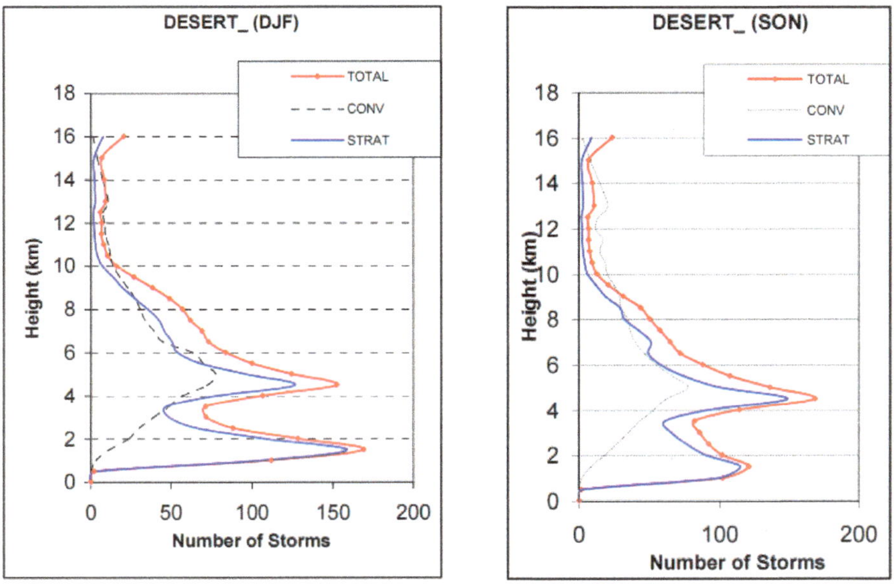

(a) (b)

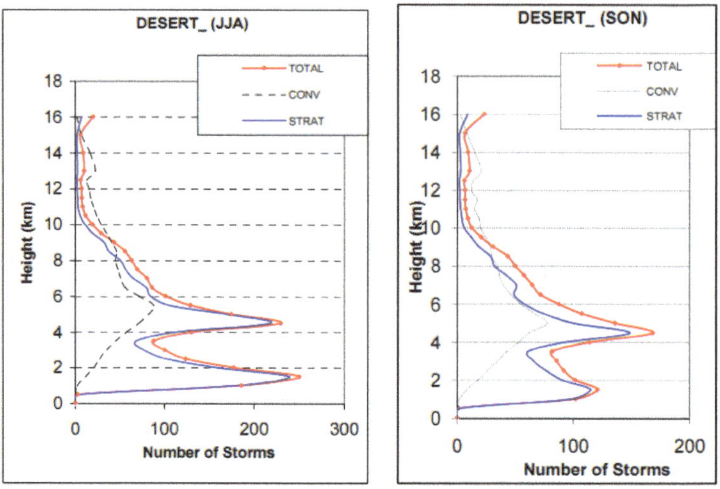

(c) (d)

Figure 9. (a) (b) Composite storm height histograms over Desert for the domain bounded by 20°N - 30°N and 10°W - 30°E of Africa, for the seasons: (a) DJF; (b) MAM; (c) (d) Composite storm height histograms over Desert for the domain bounded by 20°N - 30°N and 10°W - 30°E of Africa, for the seasons: (c) JJA; and (d) SON.

CONCLUSIONS

Mean height of stratiform storm tops ranges from 2.5 km to 5 km over the South Atlantic Ocean and 6 km to 7 km over the African continent, while the convective storm tops are as high as 8.5 km over the land, they have the lowest mode in the ocean. Relationship between stratiform and convective storms also reveals that while the stratiform storms are dominant in the North, in a particular season, the convective storms are dominant in the South. The latent heat or energy released during the formation of these storms, moves in a similar pattern as the energy in the Carnot cycle. Other types, other than the stratiform and convective storms, dominate the land in some seasons and the ocean in other seasons. In a regional perspective, the unconditioned storms have high mean storm heights over the stratiform and convective storms in all seasons as we move down south from the Northern region of the continent, but the convective storms are dominating in some cases.

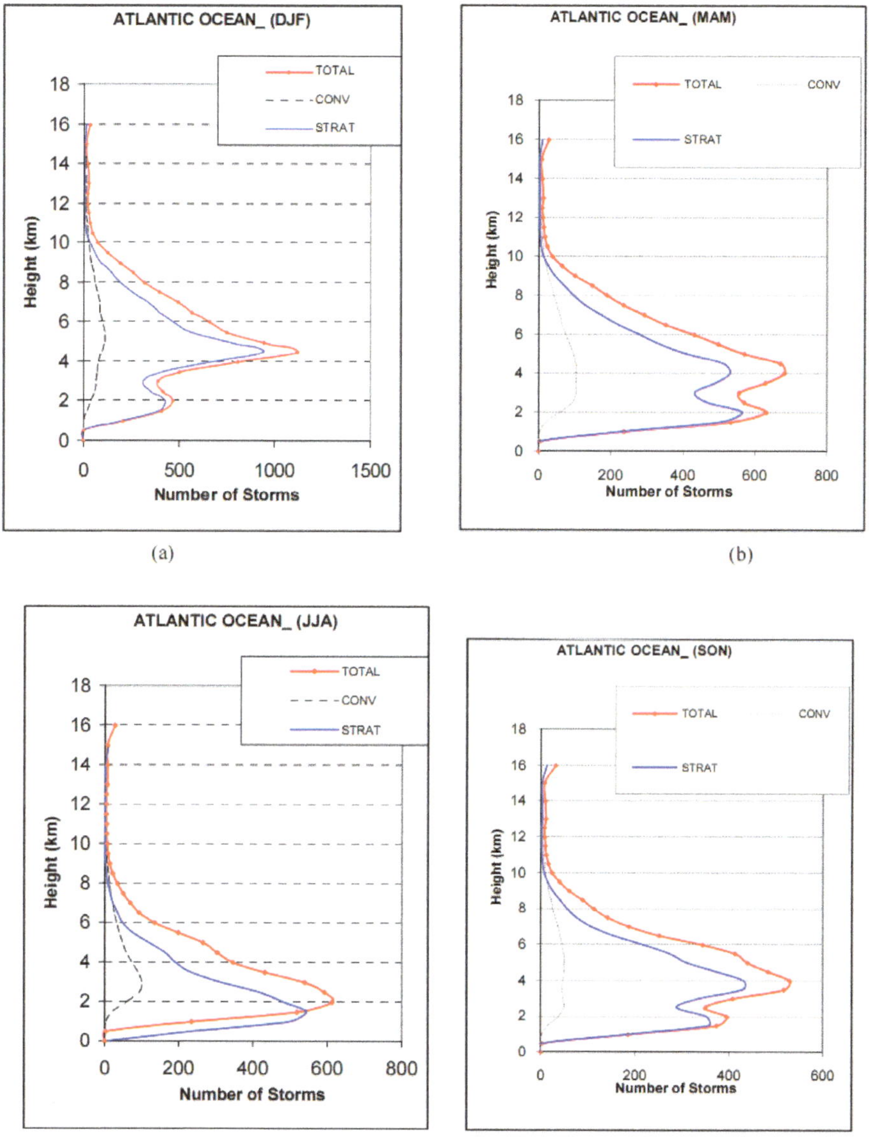

Figure 10. (a) (b) Composite storm height histograms over Atlantic Ocean for the domain bounded by 0˚S - 5˚S and 10˚W - 15˚E of Africa, for the seasons: (a) DJF; (b) MAM; Figures 10(a) and (b). Composite storm height histograms over Atlantic Ocean for the domain bounded by 0˚S - 5˚S and 10˚W - 15˚E of Africa, for the seasons: (a) DJF; (b) MAM.

The observation with the storm height histograms is that at 250 mb there are high storm counts over the land but at 700 mb the storm cores (high storm counts) are over the Ocean. On regional basis, the histograms show peak values around 5 km, which is the average height of the 0°C isotherm in the tropics. Also histograms were observed to vary with height, having a bi-modal distribution in the desert, semi-desert and brush grass savanna (north), unimodal distribution in the deciduous forest (north), tropical rainforest, deciduous forest (south), and brush grass savanna (south), again the bi-modal distribution is observed deeply in the south in temperate grassland, steppe (east), and the Atlantic Ocean. The amplitudes of these modes, indicating the number of storms, increase from DJF to MAM but decrease from JJA to SON. The values of these amplitudes are small in the south especially in the cases of unimodal histograms.

REFERENCES

1. S. Lang, W. K. Tao, J. Simpson and B. Ferrier, "Modeling of Convective-Stratiform Precipitation Processes: Sensitivity to Partitioning Methods," Journal of Applied Meteorology, Vol. 42, No. 4, 2003, pp. 505-527. http://dx.doi.org/10.1175/1520-0450(2003)042<0505:MOCSPP>2.0.CO;2
2. M. Steiner and R. A. Houze Jr., "Sensitivity of the Estimated Monthly Convective Rain Fraction to the Choice of Z-R Relation," Journal of Applied Meteorology, Vol. 36, No. 5, 1997, pp. 452-462. http://dx.doi.org/10.1175/1520-0450(1997)036<0452:SOTEMC>2.0.CO;2
3. T. Dejene, "The Regional and Diurnal Variability of the Vertical Structure of Precipitation Systems in Africa, Based on TRMM Precipitation Radar Data," 2003.
4. R. A. Houze Jr., "Cloud Clusters and Large-Scale Vertical Motions in the Tropics," Journal of the Meteorological Society of Japan, Vol. 60, No. 1, 1982, pp. 396-410.
5. R. A. Houze Jr., "Observed Structure of Mesoscale Convective Systems and Implications for Large-Scale Heating. Quart," Quarterly Journal of the Royal Meteorological Society, Vol. 15, No. 487, 1989, pp. 425-461. http://dx.doi.org/10.1002/qj.49711548702
6. B. E. Mapes and R. A. Houze Jr., "Cloud Clusters and Superclusters over the Oceanic Warm Pool," Monthly Weather Review, Vol. 121, No. 5, 1993, pp. 1398-1415.http://dx.doi.org/10.1175/1520-0493(1993)121<1398:CCASOT>2.0.CO;2
7. B. E. Mapes and R. A. Houze Jr., "Diabatic Divergence Profiles in Western Pacific Mesoscale Convective Systems," Journal of the Atmospheric Sciences, Vol. 52, No. 10, 1995, pp. 1807-1828. http://dx.doi.org/10.1175/1520-0469(1995)052<1807:DDPIWP>2.0.CO;2
8. D. L. Hartmann, H. H. Hendon and R. A. Houze Jr., "Some Implications of the Mesoscale Circulations in Tropical Cloud Clusters for Large-Scale Dynamics and Climate," Journal of the Atmospheric Sciences, Vol. 41, No. 1, 1984, pp. 113-121. http://dx.doi.org/10.1175/1520-0469(1984)041<0113:SIOTMC>2.0.CO;2
9. R. A. Houze Jr., "Stratiform Precipitation in Regions of Convection: A Meteorological Paradox?" Bulletin of the American Meteorological Society, Vol. 78, No. 10, 1997, pp. 2179-2196. http://dx.doi.org/10.1175/1520-0477(1997)078<2179:SPIROC>2.0.CO;2

10. D. A. Short and K. Nakamura, "TRMM Radar Observations of Shallow Precipitation over the Tropical Oceans," Journal of Climatee, Vol. 13, No. 23, 2000, pp. 4107- 4124.http://dx.doi.org/10.1175/1520-0442(2000)013<4107:TROOSP>2.0.CO;2

11. R. A. Houze, "Cloud Physics," Academic Press, California, 1993, 573 p.

12. G. J. Huffman and Co-Authors, "The Global Precipitation Climatology Project (GPCP) Combined Precipitation Dataset," Bulletin of the American Meteorological Society, Vol. 78, No. 1, 1997, pp. 5-20. http://dx.doi.org/10.1175/1520-0477(1997)078<0005:TGPCPG>2.0.CO;2

13. C. Kummerow and L. Giglio, "A Passive Microwave Technique for Estimating Rainfall and Vertical Structure Information from Space. Part 1: Algorithm Description," Journal of Applied Meteorology, Vol. 33, No. 1, 1994, pp. 3-18. http://dx.doi.org/10.1175/1520-0450(1994)033<0003:APMTFE>2.0.CO;2

14. F. Ferreira, P. Amayenc, S. Oury and J. Testud, Study and "Tests of Improved Rain Estimates from the TRMM Precipitation Radar," Journal of Applied Meteorology, Vol. 40, No. 11, 2001, pp. 1878-1899. http://dx.doi.org/10.1175/1520-0450(2001)040<1878:SATOIR>2.0.CO;2

15. C. Kummerow and Co-Authors, "The Status of the Tropical Rainfall Measuring Mission (TRMM) after Two Years in Orbit," Journal of Applied Meteorology, Vol. 39, No. 12, 2000, pp. 1965-1982. http://dx.doi.org/10.1175/1520-0450(2001)040<1965:TSOTTR>2.0.CO;2

16. E. J. Zipser and K. R. Lutz, "The Vertical Profile of Radar Reflectivity of Convective Cells: A Strong Indicator of Storm Intensity and Lightning Probability?" Monthly Weather Review, Vol. 122, No. 8, 1994, pp. 1751-1759. http://dx.doi.org/10.1175/1520-0493(1994)122<1751:TVPORR>2.0.CO;2

17. R. J. Donaldson, "Radar Reflectivity Profiles in Thunderstorms," Journal of Meteorology, Vol. 18, No. 3, 1961, pp. 292-305. http://dx.doi.org/10.1175/1520-0469(1961)018<0292:RRPIT>2.0.CO;2

18. Z. D. Adeyewa and K. Nakamura, "Validation of TRMM Radar Rainfall Data over Major Climatic Regions in Africa," Journal of Applied Meteorology, Vol. 42, No. 2, 2003, pp. 331-347. http://dx.doi.org/10.1175/1520-0450(2003)042<0331:VOTRRD>2.0.CO;2

19. Z. D. Adeyewa and K. Nakamura, "Preliminary Study of Rainfall and Storm Structure over Africa with TRMM Precipitation Radar Data," Meteorologische Zeitschrift, Vol. 12, No. 4, 2003, pp. 197-202. http://dx.doi.org/10.1127/0941-2948/2003/0012-0197

20. C. Prabhakara, R. Iacovazzi, J. A. Weinman and G. Dalu, "A TRMM Microwave Radiometer Rain Rates Estimation Method with Convective and Stratiform Discrimination," Journal of the Meteorological Society of Japan, Vol. 78, No. 3, 2000, pp. 241-258.

15

CHAPTER

AN IMPROVED 6S CODE FOR ATMOSPHERIC CORRECTION BASED ON WATER VAPOR CONTENT

Yin Zhang, Xiaoqin Wang, Yunzhi Chen

Keywords: Atmospheric Correction, 6S Model, Water Vapor Content, MODIS

INTRODUCTION

Since the first Earth's digital image was acquired, scientists have shown great interest in atmospheric correction [1]. There appears many atmospheric correction methods during the 40 years' development, the kind of method, take 6S [2], MODTRAN [3] and LOWTRAN [3] for instance, which is based on radiative transfer (RT) code is of high accuracy among them. One should estimate the atmospheric parameters before using these methods. The temperature, the pressure, the water vapor and ozone content and the aerosol optical depth are the necessary parameters, because the program will perform atmospheric correction by using them to simulate the transmission conditions of the solar radiation in the atmosphere. In the actual application, however, it seems very hard to obtain the atmospheric parameters synchronously with the image, especially the retrieval of the water vapor content and the aerosol optical depth (AOD), because the distribution of water vapor and aerosol varies greatly with time and space [4]. Water vapor plays an important role in atmospheric processes, which has an impact on the measured radiance because of its absorption in the solar spectrum. It is necessary to choose a standard atmospheric model or to enter the water vapor and ozone content to

Citation: Y. Zhang, X. Wang and Y. Chen, "An Improved 6S Code for Atmospheric Correction Based on Water Vapor Content," Advances in Remote Sensing, Vol. 1 No. 1, 2012, pp. 14-18, doi: 10.4236/ars.2012.11002.

define the atmospheric conditions in the atmospheric correction models in common use. Thus, the remote sensing image covers vast with a large space span, especially the low resolution images, single value of the average water vapor content cannot satisfy with the accurate description of the atmospheric conditions of the whole area.

It will be possible to derive the water vapor content from Terra MODIS by using the "water vapor" channels in addition to the existing "window" channels. Terra was launched on Dec. 12, 1999 and flies northward pass the equator at about local time 10:30 AM, which can provide the synchronous or quasi-synchronous spatial distributions of the water vapor content for atmospheric correction. Kaufman and Gao (1992) [5], Sobrino and Kharraz (2003) [6] utilized the ratios of bands radiation to estimate water vapor content. Mao and Li (2010) [7] used the RT and neural network (NN) to estimate the water vapor content.

6S is widely used today, which is based on RT theories. Zhao and Tamura (2000) [8] chose a standard atmospheric model and a standard aerosol model for atmospheric correction based on 6S code. In this study, 6S code was improved to read the water vapor content image which was retrieved from the EOS/MODIS data for atmospheric correction of the optical remote sensing data.

METHODOLOGY

The flow chart of the atmospheric correction method presented in this paper is shown in **Figure 1**. The total water vapor content was retrieved from MODIS data with 1km resolution at first. Then, 6S code was improved so that it can read the water vapor content image. At last the atmospheric parameters were obtained and atmospheric correction was implemented. The keys are to improve the 6S code and estimate the water vapor content.

Atmospheric Radiative Transfer Theory and 6S Model

In the solar spectrum, sensors on Earth remote sensing satellites measure the radiance reflected by the Atmosphere-Earth surface system. This signal mainly depends on the surface reflectance but is also perturbed by the gaseous absorption and the scattering by molecule and aerosols. The measured quantities can be expressed in terms of apparent reflectance ρ^*, defined as

$$\rho^* = \pi L / \mu_s E_s \qquad (1)$$

where L is the measured radiance, E_s is the solar flux at the top of the atmosphere and $\mu_s = \cos(\theta_s)$ where θ_s is the sun zenith angle.

Think of the lambertian uniform target with ρ_s, the apparent reflectance can be written [9]:

$$\rho^* = \rho_a + T(\theta_s)T(\theta_v)\frac{\rho_s}{1 - S\rho_s} \qquad (2)$$

where ρ_a refers to the intrinsic atmospheric reflectance, $T(\theta_s)$, [respectively $T(\theta_v)$] corresponds to the total transmission of the atmosphere on the path between the sun and the surface, (respectively between the surface and the sensor). S is the spherical albedo of the atmosphere.

In actual fact, the surface is non-uniform, we should consider the environment effects on the radiative transmission, and the apparent reflectance can be written [9]:

$$\rho^* = \rho_a + \frac{T(\theta_s)}{1 - <\rho_s(M)> S} \cdot \left(\rho_s(M) e^{-\tau/\mu v} + \langle \rho_s(M) \rangle t_d(\theta_v)\right) \quad (3)$$

where $\rho_s(M)$ corresponds to the reflectance of the target M and $< \rho_s(M)$corresponds to the reflectance of the uniform environment, $t_d(\theta_v)$ is the upward diffuse transmittance factor and $e^{-\tau/\mu v}$is the upward direct transmittance factor.

The original 6S is an improved version of 5S. It enables to simulate plane observation, to account for elevated targets, to take into account non lambertian surface conditions, and new gases (CH_4, N_2O, CO) have been integrated in the computation of the gaseous transmission. The computational accuracy for Rayleigh and aerosol scattering effects has been improved by the use of state-of-the-art approximations and implementation of successive orders of scattering (SOS) algorithm. The step size used for spectral integration has been improved to 2.5 nm. It provides some kinds of standard atmosphere models and aerosol models; in addition, people can customize a model as well.

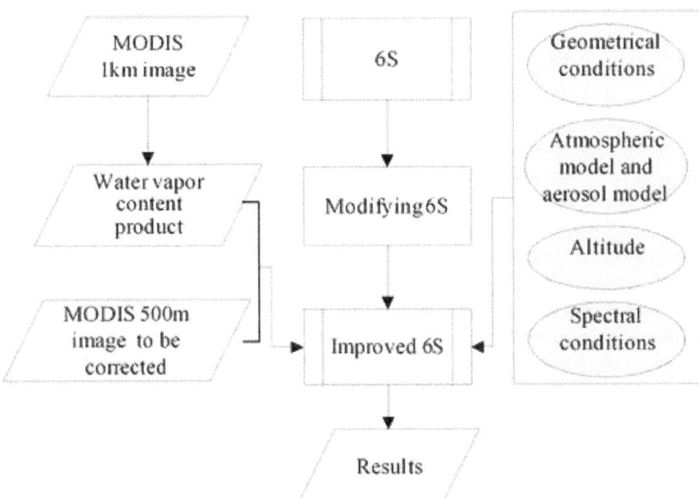

Figure 1. Flow chart of the atmospheric correction.

This study utilized the 6S code extended by Mauro Antunes who is a professor in Brazil, hereinafter to be referred as Antunes's 6S. Though Antunes's 6S model reads and writes only RAW images of 8 bits, the input mode of water vapor content is not altered. In view of this, Antunes's 6S code was improved to read the spatial distribution of water vapor content and the input image synchronously in our study. The other atmospheric parameters need not to be changed. Meanwhile, the value of the samples and lines of the image should also be given as input parameter.

Estimation of Water Vapor Content

Water vapor content is indispensable for atmospheric correction. The near-infrared at around 1 μm is sensitive to water vapor, so water vapor content can be retrieved from EOS/MODIS in the Near IR. In the paper, the algorithm developed by Sobrino (2003) [6] was chosen for water vapor content estimation. The implementation of the algorithm is very simple and includes three basic steps:

Firstly, computing the ratios between bands 17, 18, 19 and 2:

$$G_{17} = L_{17}/L_2,\ G_{18} = L_{18}/L_2,\ G_{19} = L_{19}/L_2 \quad (4)$$

where L_i is the measured radiance of the ith band.

Secondly, computing the water vapor content in each "water vapor" channel:

$$W_{17} = 26.314 - 54.434G_{17} + 28.449G_{17}^2$$
$$W_{18} = 5.012 - 23.017G_{18} + 27.884G_{18}^2$$
$$W_{19} = 9.446 - 26.887G_{19} + 19.914G_{19}^2 \quad (5)$$

where W_i refers to the water vapor content in the ith band, i = 17, 18 or 19.

Finally, calculating the total water vapor content by giving a weight to each channel:

$$W = 0.192W_{17} + 0.453W_{18} + 0.355W_{19} \quad (6)$$

where W corresponds to the total water vapor content.

ATMOSPHERIC CORRECTION

Study Site

The data we used is MODIS multi-spatial scale data with resolution 500 m and 1 km. The original image size is 560 × 480 pixels with resolution 500 m, covering the city of Fuzhou, in Fujian province, China. MODIS imaging date is 25th, May, 2010. The gray image of MODIS band 1 with resolution 500 m was shown in the left of **Figure 2**.

Water vapor content retrieval experiments were carried out from MODIS with resolution 1 km using the method introduced above. The estimated water vapor content image was shown in the right of **Figure 2**.

Model Implementation

Before running the improved 6S model, we should obtain the atmospheric parameters first. The main parameters are as follows:

- Geometrical conditions. To define geometrical conditions when using MODIS data, input values for month, day, universal time, center latitude and longitude of the image were entered. The code would automatically estimate the solar zenith angle, view zenith angle and the solar and view azimuth angle.

- Atmospheric model. There are 7 standard atmospheric models and 2 user-defined models proposed by 6S. We used the latter with assuming that the ozone content was 0.3 cm/atm.
- Aerosol model (type and concentration). We chose the continental model to define aerosol type. To define the concentration of aerosols, visibility (km) should be entered directly into 6S. The aerosol optical thickness at 550 nm should then be computed from a standard aerosol profile by using subroutine ODA550 provided by 6S.
- The altitude of the target and the sensor. We assumed that the target is at the sea level and the sensor is out of the atmosphere.
- The spectral conditions. We chose to enter the wavelength range of MODIS band1.

Using the data and parameters above, the improved 6S code was implemented for atmospheric correction of the MODIS band1 with resolution 500 m. Meanwhile, the Antunes's 6S code was also performed in the same conditions with the average water vapor content of 0.9843 g/cm^2.

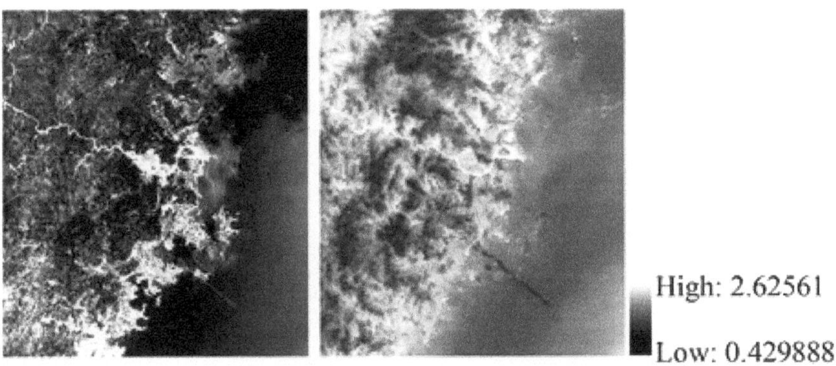

Figure 2. Image to be corrected (left) and water vapor content (right) (g/cm^2).

RESULTS AND DISCUSSION

The two corrected images were shown in **Figure 3**. The remarkable differences can be seen between the two images. There are apparent stripes in the left image corrected by Antunes's 6S model. Therefore, we can infer that the right image which was corrected by the improved 6S model is more reasonable in the spatial distribution than the left image, especially in the ocean area. Then histograms (**Figure 4**) and difference images (**Figure 5**) were used to compare the images before and after atmospheric correction. In addition, the statistics was computed and shown in **Table 1**.

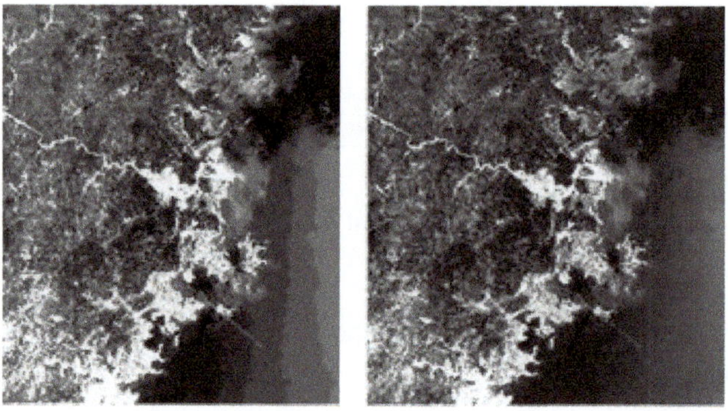

Figure 3. Image corrected by Antunes's 6S (left) and image corrected by improved 6S (right).

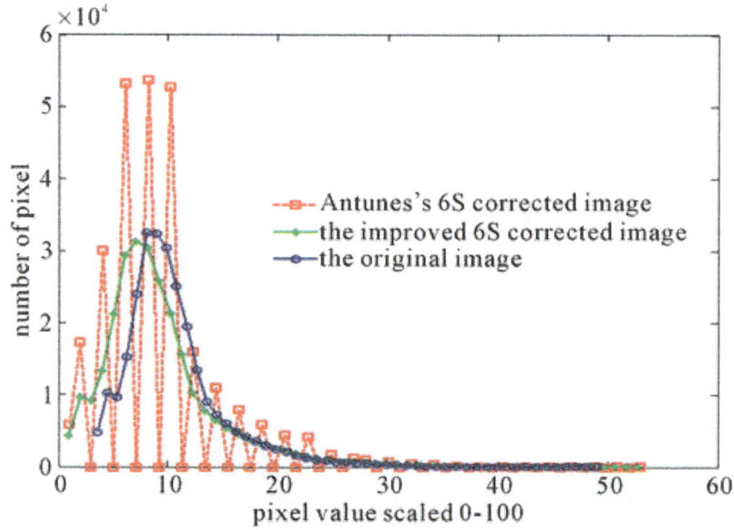

Figure 4. Histograms of the three images before and after atmospheric correction.

Figure 5. Difference image between the original and Antunes's 6S corrected image (left), the difference image between the original and improved 6S corrected image (middle) and the difference image between the two 6S corrected images (right).

Table 1. Statistics of the images

Images	Statistics			
	Min (%)	Max (%)	Mean (%)	Std (%)
original image	3.605567	48.95111	7.313808	2.27023
Antunes's 6S corrected image	1	53	5.413397	2.68078
improved 6S corrected image	0.964300	52.66310	5.384489	2.69217
difference image between the original and the Antunes's 6S	−4.048885	3.116226	1.900412	0.50168
difference image between the original and the improved 6S	−3.711987	2.641267	1.929320	0.42220
difference image between the improved 6S and Antunes's 6S	−0.696700	0.775700	−0.028908	0.31402

From **Figure 4**, it is clear that the pixel value distribution of the image corrected by Antunes's 6S model is concentrated, which means that there are a large number of pixels have the some reflectance range, while the other intervals have few pixels. The histogram of the image corrected by the improved 6S model is closer to the histogram of the original image. So the distribution of pixel values is more uniform than that corrected by Antunes's 6S model. It also can be seen that the range of pixel values becomes wide after atmospheric correction.

From the left and the middle difference images in **Figure 5**, it can be seenthat the changes between the improved 6S corrected image and the original image were homogeneous, while the changes were sharp between Antunes's 6S corrected image and the original image. What's more, the two corrected images are different due to the different water vapor content. The influences made by the water vapor content can obviously see from the right image in **Figure 5**.

From **Table 1**, we can summarize that 1) in comparison with the original image, the pixel value range of the corrected images (both the two 6S codes) extends and the mean value decreases sharply while the standard deviation grows slightly; 2) after atmospheric correction, the mean value of the two corrected images decreases by 1.900412 and 1.929320, respectively; 3) there are a lot of changes of the statistics between the two corrected images, but the mean value decrease by 0.028909 after atmospheric correction using the improved 6S code.

CONCLUSIONS

The atmospheric correction 6S code was improved by reading the water vapor content retrieved from MODIS data with 1 km resolution. After atmospheric correction using the improved 6S model, the pixel value range of the image extended. Compared with the image corrected by Antunes's 6S model, the pixel value is more successive and the histogram was closer to the histogram of the original image. Therefore, the improved 6S model is more reasonable than the Antunes's 6S model in atmospheric correction.

Though this improved 6S code was only applied to MODIS, it can be applied to other satellite data. Validation need to be done and the aerosol optical thickness need to be considered in this code. Further work is ongoing.

ACKNOWLEDGEMENTS

This work was supported by the Natural Science Foundation for Distinguished Young Scholars of Fujian Province (Grant No. 2009J06024). We also thank Mauro A. Homem Antunes for his extended 6S code.

REFERENCES

1. B. Cairns and B. E. Carlson, "Atmospheric Correction and Its Application to an Analysis of Hyperion Data," IEEE Transaction on Geoscience and Remote Sensing, Vol. 41, No. 6, 2003, pp. 1232-1245. doi:10.1109/TGRS.2003.813134
2. E. F. Vermote, D. Tanre, J. L. Deuze, M. Herman, and J.-J. Morcrette, "Second Simulation of the Satellite Signal in the Solar Spectrum, 6S: An Overview," IEEE Transactions on Geoscience and Remote Sensing, Vol. 35, No. 3, 1997, pp. 675-686.doi:10.1109/36.581987
3. B. Y. Wu, W. Li, H. B. Chen, F. Li, W. X. Zhang and D. Lu, "Practical Algorithms of Atmospheric Radiative Transfer," Meteorological Press, Beijing, 1998, pp. 21-40 (in Chinese).
4. S. L. Liang, H. L. Fang and M. Z. Chen, "Atmospheric Correction of Landsat ETM+ Land Surface Imagery-Part I: Method," IEEE Transactions on Geoscience and Remote Sensing, Vol. 39, No. 11, 2001, pp. 2490-2498. doi:10.1109/36.964986
5. Y. J. Kaufman and B. C. Gao, "Remote Sensing of Water Vapor in the Near IR from EOS/MODIS," IEEE Transactions on Geosciences and Remote Sensing, Vol. 30, No. 5, 1992, pp. 871-884. doi:10.1109/36.175321
6. J. A. Sobrino and J. E. Kharraz, "Surface Temperature and Water Vapor Retrieval from MODIS Data," International Journal of Remote Sensing, Vol. 24, No. 24, 2003, pp. 5161-5182. doi:10.1080/0143116031000102502
7. K. B. Mao, H. D. Li, D. Y. Hu, J. Wang, J. X. Huang, et al., "Estimation of Water Vapor Content in Near-Infrared Bands around 1 mm from MODIS Data by Using RMMN," Optics Express, Vol. 18, No. 9, 2010, pp. 9542- 9554. doi:10.1364/OE.18.009542
8. W. J. Zhao, M. Tamura and H. Takahashi, "Atmospheric and Spectral Corrections for Estimating Surface Albedo from Satellite Data Using 6S Code," Remote Sensing of environment, Vol. 76, No. 2, 2000, pp. 202-212. doi:10.1016/S0034-4257(00)00204-2
9. 6S User Guide Version 2, 1997.

INDEX

H

I

L

M

N

O

P

Q

R

S

T

U

V